Springer-Textbook

Prof. Dietmar Gross
received his Engineering Diploma in Applied Mechanics and his Doctor of Engineering degree at the University of Rostock. He was Research Associate at the University of Stuttgart and since 1976 he is Professor of Mechanics at the University of Darmstadt. His research interests are mainly focused on modern solid mechanics on the macro and micro scale, including advanced materials.

Prof. Werner Hauger
studied Applied Mathematics and Mechanics at the University of Karlsruhe and received his Ph.D. in Theoretical and Applied Mechanics from Northwestern University in Evanston. He worked in industry for several years, was a Professor at the Helmut-Schmidt-University in Hamburg and went to the University of Darmstadt in 1978. His research interests are, among others, theory of stability, dynamic plasticity and biomechanics.

Prof. Jörg Schröder
studied Civil Engineering, received his doctoral degree at the University of Hannover and habilitated at the University of Stuttgart. He was Professor of Mechanics at the University of Darmstadt and went to the University of Duisburg-Essen in 2001. His fields of research are theoretical and computer-oriented continuum mechanics, modeling of functional materials as well as the further development of the finite element method.

Prof. Wolfgang A. Wall
studied Civil Engineering at Innsbruck University and received his doctoral degree from the University of Stuttgart. Since 2003 he is Professor of Mechanics at the TU München and Head of the Institute for Computational Mechanics. His research interests cover broad fields in computational mechanics, including both solid and fluid mechanics. His recent focus is on multiphysics and multiscale problems as well as computational biomechanics.

Prof. Javier Bonet
studied Civil Engineering at the Universitat Politecnica de Catalunya in Barcelona and received his Doctorate from Swansea University in the UK. He is Professor of Computational Mechanics and Head of the School of Engineering at Swansea University where he has taught Strength of Materials, Structural Mechanics and Nonlinear Mechanics for over 20 years. His research interests are computational mechanics and finite element methods.

Dietmar Gross · Werner Hauger
Jörg Schröder · Wolfgang A. Wall
Javier Bonet

Engineering Mechanics 2

Mechanics of Materials

Prof. Dr. Dietmar Gross
TU Darmstadt
Division of Solid Mechanics
Hochschulstr. 1
64289 Darmstadt
Germany
gross@mechanik.tu-darmstadt.de

Prof. Dr. Werner Hauger
TU Darmstadt
Hochschulstr. 1
64289 Darmstadt
Germany
hauger@mechanik.tu-darmstadt.de

Prof. Dr. Jörg Schröder
Universität Duisburg-Essen
Institute of Mechanics
Universitätsstr. 15
45141 Essen
Germany
j.schroeder@uni-essen.de

Prof. Dr. Wolfgang A.Wall
TU München
Institute for Computational
Mechanics
Boltzmannstr. 15
85747 Garching
Germany
wall@lnm.mw.tum.de

Prof. Javier Bonet
Head of School
School of Engineering
Swansea University
Swansea, SA2 8PP
United Kingdom
j.bonet@swansea.ac.uk

ISBN 978-3-642-12885-1 e-ISBN 978-3-642-12886-8
DOI 10.1007/978-3-642-12886-8
Springer Heidelberg Dordrecht London New York

Library of Congress Control Number: 2011922991

Cover design: WMXDesign GmbH

Printed on acid-free paper

Springer is part of Springer Science+Business Media (www.springer.com)

Preface

Mechanics of Materials is the second volume of a three-volume textbook on Engineering Mechanics. Volume 1 deals with *Statics* while Volume 3 contains *Dynamics*. The original German version of this series has been the bestselling textbook on mechanics for more than two decades; its 11th edition is currently being published.

It is our intention to present to engineering students the basic concepts and principles of mechanics in the clearest and simplest form possible. A major objective of this book is to help the students to develop problem solving skills in a systematic manner.

The book has been developed from the many years of teaching experience gained by the authors while giving courses on engineering mechanics to students of mechanical, civil and electrical engineering. The contents of the book correspond to the topics normally covered in courses on basic engineering mechanics, also known in some countries as strength of materials, at universities and colleges. The theory is presented in as simple a form as the subject allows without becoming imprecise. This approach makes the text accessible to students from different disciplines and allows for their different educational backgrounds. Another aim of the book is to provide students as well as practising engineers with a solid foundation to help them bridge the gaps between undergraduate studies and advanced courses on mechanics and practical engineering problems.

A thorough understanding of the theory cannot be acquired by merely studying textbooks. The application of the seemingly simple theory to actual engineering problems can be mastered only if the student takes an active part in solving the numerous examples in this book. It is recommended that the reader tries to solve the problems independently without resorting to the given solutions. In order to focus on the fundamental aspects of how the theory is applied, we deliberately placed no emphasis on numerical solutions and numerical results.

We gratefully acknowledge the support and the cooperation of the staff of the Springer Verlag who were responsive to our wishes and helped to create the present layout of the books.

Darmstadt, Essen, Munich and Swansea,
December 2010

D. Gross
W. Hauger
J. Schröder
W.A. Wall
J. Bonet

Table of Contents

Introduction 1

1 **Tension and Compression in Bars**
1.1 Stress 7
1.2 Strain 13
1.3 Constitutive Law 14
1.4 Single Bar under Tension or Compression 18
1.5 Statically Determinate Systems of Bars 29
1.6 Statically Indeterminate Systems of Bars 33
1.7 Supplementary Examples 40
1.8 Summary 46

2 **Stress**
2.1 Stress Vector and Stress Tensor 49
2.2 Plane Stress 52
2.2.1 Coordinate Transformation 53
2.2.2 Principal Stresses 56
2.2.3 Mohr's Circle 62
2.2.4 The Thin-Walled Pressure Vessel 68
2.3 Equilibrium Conditions 70
2.4 Supplementary Examples 73
2.5 Summary 75

3 **Strain, Hooke's Law**
3.1 State of Strain 79
3.2 Hooke's Law 84
3.3 Strength Hypotheses 90
3.4 Supplementary Examples 92
3.5 Summary 95

4 **Bending of Beams**
4.1 Introduction 99
4.2 Second Moments of Area 101
4.2.1 Definitions 101
4.2.2 Parallel-Axis Theorem 108

4.2.3	Rotation of the Coordinate System, Principal Moments of Inertia	**113**
4.3	Basic Equations of Ordinary Bending Theory	**117**
4.4	Normal Stresses	**121**
4.5	Deflection Curve	**125**
4.5.1	Differential Equation of the Deflection Curve	**125**
4.5.2	Beams with one Region of Integration	**129**
4.5.3	Beams with several Regions of Integration	**138**
4.5.4	Method of Superposition	**140**
4.6	Influence of Shear	**151**
4.6.1	Shear Stresses	**151**
4.6.2	Deflection due to Shear	**161**
4.7	Unsymmetric Bending	**162**
4.8	Bending and Tension/Compression	**171**
4.9	Core of the Cross Section	**174**
4.10	Thermal Bending	**176**
4.11	Supplementary Examples	**180**
4.12	Summary	**187**
5	**Torsion**	
5.1	Introduction	**191**
5.2	Circular Shaft	**192**
5.3	Thin-Walled Tubes with Closed Cross Sections	**203**
5.4	Thin-Walled Shafts with Open Cross Sections	**212**
5.5	Supplementary Examples	**220**
5.6	Summary	**228**
6	**Energy Methods**	
6.1	Introduction	**231**
6.2	Strain Energy and Conservation of Energy	**232**
6.3	Principle of Virtual Forces and Unit Load Method	**242**
6.4	Influence Coefficients and Reciprocal Displacement Theorem	**261**
6.5	Statically Indeterminate Systems	**265**
6.6	Supplementary Examples	**279**
6.7	Summary	**286**

7 **Buckling of Bars**

7.1 Bifurcation of an Equilibrium State **289**

7.2 Critical Loads of Bars, Euler's Column **292**

7.3 Supplementary Examples **302**

7.4 Summary **305**

Index **307**

Introduction

Volume 1 (*Statics*) showed how external and internal forces acting on structures can be determined with the aid of the equilibrium conditions alone. In doing so, real physical bodies were approximated by *rigid* bodies. However, this idealisation is often not adequate to describe the behaviour of structural elements or whole structures. In many engineering problems the deformations also have to be calculated, for example in order to avoid inadmissibly large deflections. The bodies must then be considered as being *deformable.*

It is necessary to define suitable geometrical quantities to describe the deformations. These quantities are the *displacements* and the *strains.* The geometry of deformation is given by *kinematic equations*; they connect the displacements and the strains.

In addition to the deformations, the stressing of structural members is of great practical importance. In Volume 1 we calculated the internal forces (the stress resultants). The stress resultants alone, however, allow no statement regarding the load carrying ability of a structure: a slender rod or a stocky rod, respectively, made of the same material will fail under different loads. Therefore, the concept of the *state of stress* is introduced. The amount of load that a structure can withstand can be assessed by comparing the calculated stress with an allowable stress which is based on experiments and safety requirements.

The stresses and strains are connected in the *constitutive equations.* These equations describe the behaviour of the material and can be obtained only from experiments. The most important metallic or non-metallic materials exhibit a linear relationship between the stress and the strain provided that the stress is small enough. Robert Hooke (1635–1703) first formulated this fact in the language of science at that time: *ut tensio sic vis* (lat., *as the extension, so the force*). A material that obeys *Hooke's law* is called *linearly elastic*; we will simply refer to it as *elastic.*

In the present text we will restrict ourselves to the statics of elastic structures. We will always assume that the deformations and thus the strains are very small. This assumption is satisfied in many technically important problems. It has the advantage that the equilibrium conditions can be formulated using the *undeformed* geometry of the system. In addition, the kinematic relations have a simple form in this case. Only in stability problems (see Chapter 7, Buckling) the equilibrium conditions must be formulated in the deformed geometry.

The solution of problems is based on three different types of equations: a) equilibrium conditions, b) kinematic relations and c) constitutive equations. In the case of a *statically determinate* system, these equations are uncoupled. The stress resultants and the stresses can be calculated directly from the equilibrium conditions. The strains follow subsequently from Hooke's law and the deformations are obtained from the kinematic relations.

Since we now consider the deformations of structures, we are able to analyse *statically indeterminate* systems and to calculate the forces and displacements. In such systems, the equilibrium conditions, the kinematic relations and Hooke's law represent a system of coupled equations.

We will restrict our investigations only to a few technically important problems, namely, rods subjected to tension/compression or torsion and beams under bending. In order to derive the relevant equations we frequently employ certain *assumptions* concerning the deformations or the distribution of stresses. These assumptions are based on experiments and enable us to formulate the problems with sufficient accuracy.

Special attention will be given to the notion of work and to energy methods. These methods allow a convenient solution of many problems. Their derivation and application to practical problems are presented in Chapter 6.

Investigations of the behaviour of deformable bodies can be traced back to Leonardo da Vinci (1452–1519) and Galileo Galilei (1564–1642) who derived theories on the bearing capacities of rods and beams. The first systematic investigations regarding the deformation of beams are due to Jakob Bernoulli (1655–1705) and Leonhard Euler (1707–1783). Euler also developed the theory of the buckling of columns; the importance of this theory was recognized only much later. The basis for a systematic *theory of elasticity* was laid by Augustin Louis Cauchy (1789–1857); he introduced the notions of the *state of stress* and the *state of strain.* Since then, engineers, physicists and mathematicians expanded the theory of elasticity as well as analytical and numerical methods to solve engineering problems. These developments continue to this day. In addition, theories have been developed to describe the non-elastic behaviour of materials (for example, plastic behaviour). The investigation of non-elastic behaviour, however, is not within the scope of this book.

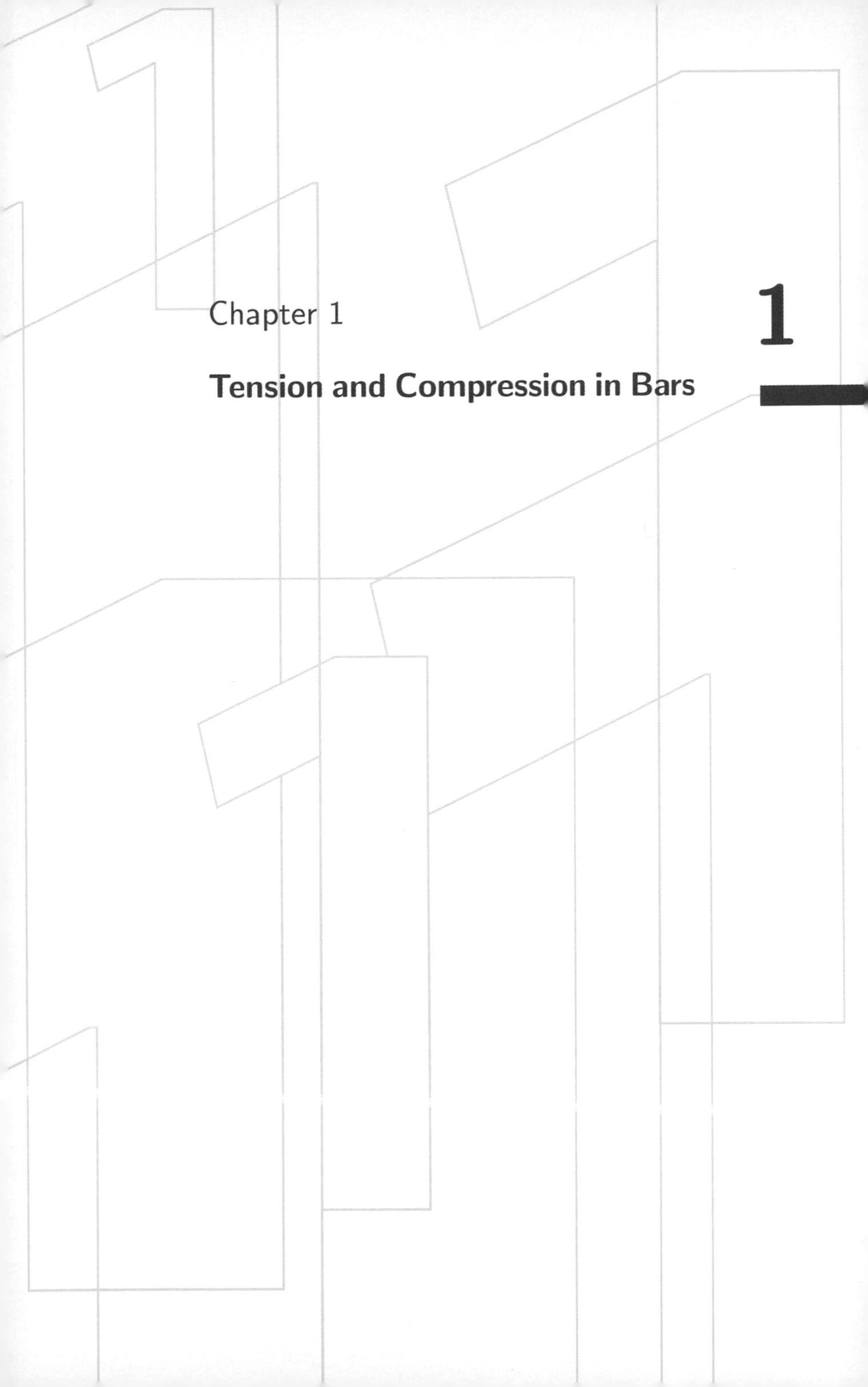

Chapter 1

Tension and Compression in Bars

1

1 Tension and Compression in Bars

1.1	Stress	**7**
1.2	Strain	**13**
1.3	Constitutive Law	**14**
1.4	Single Bar under Tension or Compression	**18**
1.5	Statically Determinate Systems of Bars	**29**
1.6	Statically Indeterminate Systems of Bars	**33**
1.7	Supplementary Examples	**40**
1.8	Summary	**46**

Objectives: In this textbook about the *Mechanics of Materials* we investigate the stressing and the deformations of elastic structures subject to applied loads. In the first chapter we will restrict ourselves to the simplest structural members, namely, bars under tension or compression.

In order to treat such problems, we need kinematic relations and a constitutive law to complement the equilibrium conditions which are known from Volume 1. The kinematic relations represent the geometry of the deformation, whereas the behaviour of the elastic material is described by the constitutive law. The students will learn how to apply these equations and how to solve statically determinate as well as statically indeterminate problems.

1.1 Stress

Let us consider a straight bar with a constant cross-sectional area A. The line connecting the centroids of the cross sections is called the *axis* of the bar. The ends of the bar are subjected to the forces F whose common line of action is the axis (Fig. 1.1a).

The *external* load causes *internal* forces. The internal forces can be visualized by an imaginary cut of the bar (compare Volume 1, Section 1.4). They are distributed over the cross section (see Fig. 1.1b) and are called *stresses*. Being area forces, they have the dimension force per area and are measured, for example, as multiples of the unit MPa (1 MPa = 1 N/mm^2). The unit "Pascal" (1 Pa = 1 N/m^2) is named after the mathematician and physicist Blaise Pascal (1623–1662); the notion of "stress" was introduced by Augustin Louis Cauchy (1789–1857). In Volume 1 (Statics) we only dealt with the resultant of the internal forces (= normal force) whereas now we have to study the internal forces (= stresses).

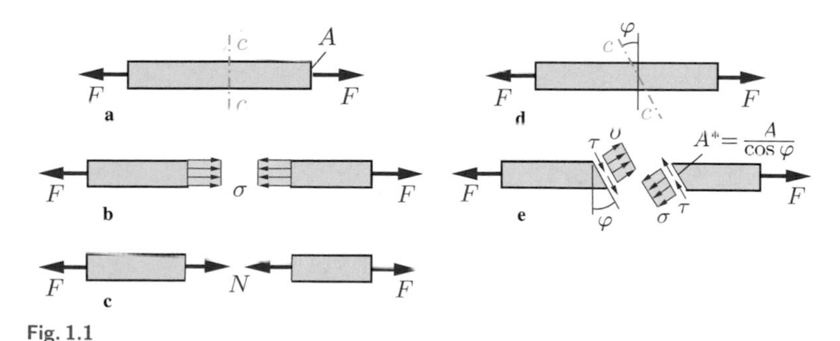

Fig. 1.1

In order to determine the stresses we first choose an imaginary cut $c - c$ perpendicular to the axis of the bar. The stresses are shown in the free-body diagram (Fig. 1.1b); they are denoted by σ. We assume that they act perpendicularly to the exposed surface A of the cross section and that they are uniformly distributed. Since they are normal to the cross section they are called *normal stresses*. Their resultant is the normal force N shown in Fig. 1.1c (compare Volume 1, Section 7.1). Therefore we have $N = \sigma A$ and the stresses σ can be calculated from the normal force N:

$$\sigma = \frac{N}{A}. \tag{1.1}$$

In the present example the normal force N is equal to the applied force F. Thus, we obtain from (1.1)

$$\sigma = \frac{F}{A}. \tag{1.2}$$

In the case of a positive normal force N (tension) the stress σ is then positive (tensile stress). Reversely, if the normal force is negative (compression) the stress is also negative (compressive stress).

Let us now imagine the bar being sectioned by a cut which is not orthogonal to the axis of the bar so that its direction is given by the angle φ (Fig. 1.1d). The internal forces now act on the exposed surface $A^* = A/\cos\varphi$. Again we assume that they are uniformly distributed. We resolve the stresses into a component σ perpendicular to the surface (the normal stress) and a component τ tangential to the surface (Fig. 1.1e). The component τ which acts *in* the direction of the surface is called *shear stress.*

Equilibrium of the forces acting on the left portion of the bar yields (see Fig. 1.1e)

$$\rightarrow: \quad \sigma A^* \cos\varphi + \tau A^* \sin\varphi - F = 0\,,$$

$$\uparrow: \quad \sigma A^* \sin\varphi - \tau A^* \cos\varphi = 0\,.$$

Note that we have to write down the equilibrium conditions for the *forces*, *not* for the *stresses.* With $A^* = A/\cos\varphi$ we obtain

$$\sigma + \tau\tan\varphi = \frac{F}{A}\,, \qquad \sigma\tan\varphi - \tau = 0\,.$$

Solving these two equations for σ and τ yields

$$\sigma = \frac{1}{1+\tan^2\varphi}\frac{F}{A}\,, \qquad \tau = \frac{\tan\varphi}{1+\tan^2\varphi}\frac{F}{A}\,.$$

It is practical to write these equations in a different form. Using the standard trigonometric relations

$$\frac{1}{1+\tan^2\varphi} = \cos^2\varphi\,, \qquad \cos^2\varphi = \frac{1}{2}(1+\cos 2\,\varphi)\,,$$
$$\sin\varphi\cos\varphi = \frac{1}{2}\sin 2\,\varphi$$

and the abbreviation $\sigma_0 = F/A$ (= normal stress in a section perpendicular to the axis) we finally get

$$\sigma = \frac{\sigma_0}{2}(1+\cos 2\,\varphi)\,, \qquad \tau = \frac{\sigma_0}{2}\sin 2\,\varphi\,. \tag{1.3}$$

Thus, the stresses depend on the direction of the cut. If σ_0 is known, the stresses σ and τ can be calculated from (1.3) for arbitrary values of the angle φ. The maximum value of σ is obtained for $\varphi = 0$, in which case $\sigma_{\max} = \sigma_0$; the maximum value of τ is found for $\varphi = \pi/4$ for which $\tau_{\max} = \sigma_0/2$.

If we section a bar near an end which is subjected to a concentrated force F (Fig. 1.2a, section $c-c$) we find that the normal stress is not distributed uniformly over the cross-sectional area. The concentrated force produces high stresses near its point of application (Fig. 1.2b). This phenomenon is known as *stress concentration*. It can be shown, however, that the stress concentration is restricted to sections in the proximity of the point of application of the concentrated force: the high stresses decay rapidly towards the average value σ_0 as we increase the distance from the end of the bar. This fact is referred to as *Saint-Venant's principle* (Adhémar Jean Claude Barré de Saint-Venant, 1797–1886).

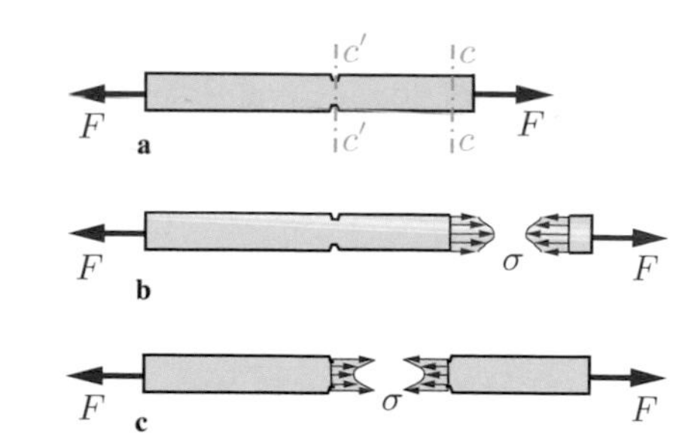

Fig. 1.2

The uniform distribution of the stress is also disturbed by holes, notches or any abrupt changes (discontinuities) of the geometry. If, for example, a bar has notches the remaining cross-sectional area (section $c' - c'$) is also subjected to a stress concentration (Fig. 1.2c). The determination of these stresses is not possible with the elementary analysis presented in this textbook.

Let us now consider a bar with only a *slight* taper (compare Example 1.1). In this case the normal stress may be calculated from (1.1) with a sufficient accuracy. Then the cross-sectional area A and the stress σ depend on the location along the axis. If volume forces act in the direction of the axis in addition to the concentrated forces, then the normal force N also depends on the location. Introducing the coordinate x in the direction of the axis we can write:

$$\sigma(x) = \frac{N(x)}{A(x)} . \tag{1.4}$$

Here it is also assumed that the stress is uniformly distributed over the cross section at a fixed value of x.

In statically determinate systems we can determine the normal force N from equilibrium conditions alone. If the cross-sectional area A is known, the stress σ can be calculated from (1.4). Statically indeterminate systems will be treated in Section 1.4.

In engineering applications structures have to be designed in such a way that a given maximum stressing is not exceeded. In the case of a bar this requirement means that the absolute value of the stress σ must not exceed a given *allowable stress* σ_{allow} : $|\sigma| \leq \sigma_{\text{allow}}$. (Note that the allowable stresses for tension and for compression are different for some materials.) The required cross section A_{req} of a bar for a given load and thus a known normal force N can then be determined from $\sigma = N/A$:

$$A_{\text{req}} = \frac{|N|}{\sigma_{\text{allow}}} . \tag{1.5}$$

This is referred to as *dimensioning* of the bar. Alternatively, the allowable load can be calculated from $|N| \leq \sigma_{\text{allow}} A$ in the case of a given cross-sectional area A.

Note that a slender bar which is subjected to compression may fail due to buckling before the stress attains an inadmissibly large value. We will investigate buckling problems in Chapter 7.

Example 1.1 A bar (length l) with a circular cross section and a slight taper (linearly varying from radius r_0 to $2\,r_0$) is subjected to the compressive forces F as shown in Fig. 1.3a. E1.1

Determine the normal stress σ in an arbitrary cross section perpendicular to the axis of the bar.

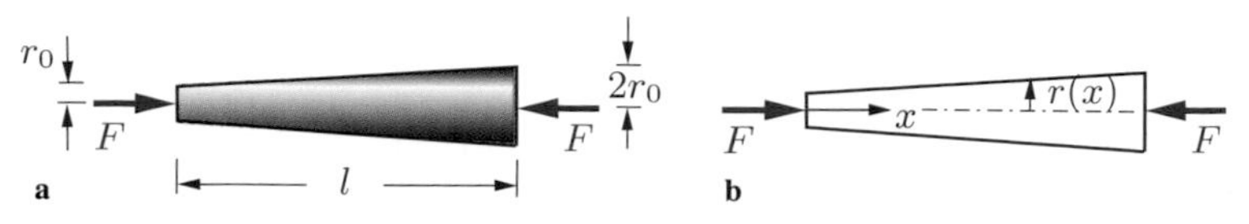

Fig. 1.3

Solution We introduce the coordinate x, see Fig. 1.3b. Then the radius of an arbitrary cross section is given by

$$r(x) = r_0 + \frac{r_0}{l}\,x = r_0\left(1 + \frac{x}{l}\right).$$

Using (1.4) with the cross section $A(x) = \pi\,r^2(x)$ and the constant normal force $N = -F$ yields

$$\underline{\underline{\sigma}} = \frac{N}{A(x)} = \underline{\underline{\frac{-F}{\pi r_0^2\left(1 + \frac{x}{l}\right)^2}}}.$$

The minus sign indicates that σ is a compressive stress. Its value at the left end ($x = 0$) is four times the value at the right end ($x = l$).

Example 1.2 A water tower (height H, density ϱ) with a cross section in the form of a circular ring carries a tank (weight W_0) as shown in Fig. 1.4a. The inner radius r_i of the ring is constant. E1.2

Determine the outer radius r in such a way that the normal stress σ_0 in the tower is constant along its height. The weight of the tower cannot be neglected.

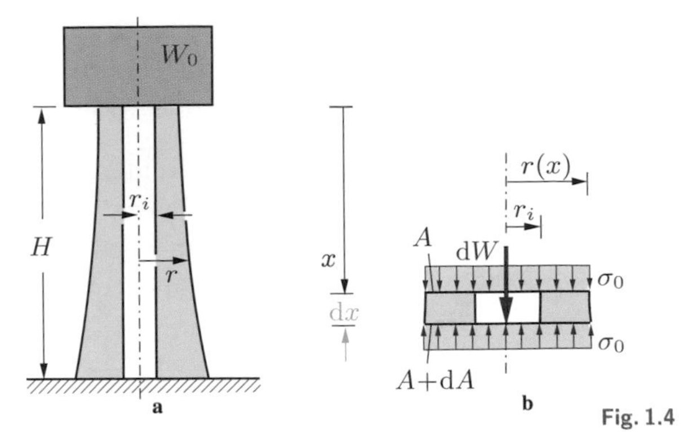

Fig. 1.4

Solution We consider the tower to be a slender bar. The relationship between stress, normal force and cross-sectional area is given by (1.4). In this example the constant compressive stress $\sigma = \sigma_0$ is given; the normal force (here counted positive as compressive force) and the area A are unknown.

The equilibrium condition furnishes a second equation. We introduce the coordinate x as shown in Fig. 1.4b and consider a slice element of length $\mathrm{d}x$. The cross-sectional area of the circular ring as a function of x is

$$A = \pi(r^2 - r_i^2) \tag{a}$$

where $r = r(x)$ is the unknown outer radius. The normal force at the location x is given by $N = \sigma_0 A$ (see 1.4). At the location $x + \mathrm{d}x$, the area and the normal force are $A + \mathrm{d}A$ and $N + \mathrm{d}N = \sigma_0(A + \mathrm{d}A)$.

The weight of the element is $\mathrm{d}W = \varrho\, g\, \mathrm{d}V$ where $\mathrm{d}V = A\,\mathrm{d}x$ is the volume of the element. Note that terms of higher order are neglected (compare Volume 1, Section 7.2.2). Equilibrium in the vertical direction yields

$$\uparrow: \quad \sigma_0(A + \mathrm{d}A) - \varrho\, g\, \mathrm{d}V - \sigma_0 A = 0 \quad \rightarrow \quad \sigma_0\, \mathrm{d}A - \varrho\, g\, A\, \mathrm{d}x = 0\,.$$

Separation of variables and integration lead to

$$\int \frac{\mathrm{d}A}{A} = \int \frac{\varrho\, g}{\sigma_0}\, \mathrm{d}x \quad \rightarrow \quad \ln \frac{A}{A_0} = \frac{\varrho\, g\, x}{\sigma_0} \quad \rightarrow \quad A = A_0\, \mathrm{e}^{\frac{\varrho\, g\, x}{\sigma_0}}\,. \tag{b}$$

The constant of integration A_0 follows from the condition that the stress at the upper end of the tower (for $x = 0$ we have $N = W_0$) also has to be equal to σ_0:

$$\frac{W_0}{A_0} = \sigma_0 \qquad \rightarrow \qquad A_0 = \frac{W_0}{\sigma_0}\,. \tag{c}$$

Equations (a) to (c) yield the outer radius:

$$\underline{\underline{r^2(x) = r_i^2 + \frac{W_0}{\pi\,\sigma_0} e^{\frac{\varrho g x}{\sigma_0}}}}\,.$$

1.2 Strain

1.2

We will now investigate the deformations of an elastic bar. Let us first consider a bar with a constant cross-sectional area which has the undeformed length l. Under the action of tensile forces (Fig. 1.5) it gets slightly longer. The elongation is denoted by Δl and is assumed to be much smaller than the original length l. As a measure of the amount of deformation, it is useful to introduce, in addition to the elongation, the ratio between the elongation and the original (undeformed) length:

$$\varepsilon = \frac{\Delta l}{l}\,. \tag{1.6}$$

The dimensionless quantity ε is called *strain*. If, for example, a bar of the length $l = 1$ m undergoes an elongation of $\Delta l = 0.5$ mm then we have $\varepsilon = 0.5 \cdot 10^{-3}$. This is a strain of 0.05%. If the bar gets longer ($\Delta l > 0$) the strain is positive; it is negative in the case of a shortening of the bar. In what follows we will consider only small deformations: $|\Delta l| \ll l$ or $|\varepsilon| \ll 1$, respectively.

The definition (1.6) for the strain is valid only if ε is constant over the entire length of the bar. If the cross-sectional area is not

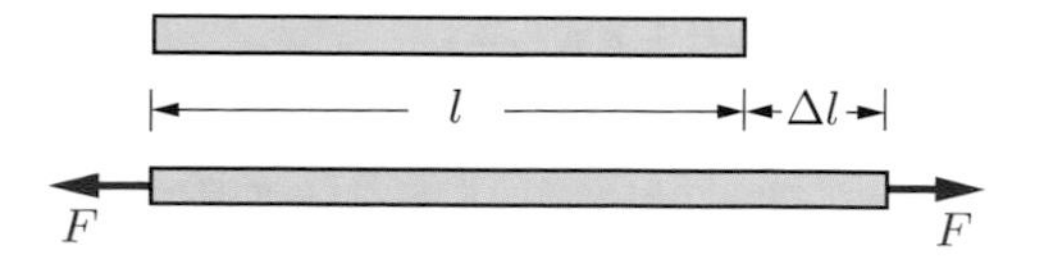

Fig. 1.5

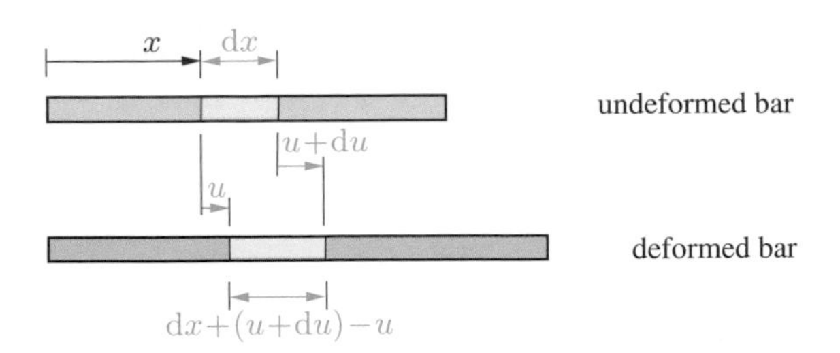

Fig. 1.6

constant or if the bar is subjected to volume forces acting along its axis, the strain may depend on the location. In this case we have to use a *local* strain which will be defined as follows. We consider an element of the bar (Fig. 1.6) instead of the whole bar. It has the length dx in the undeformed state. Its left end is located at x, the right end at $x + dx$. If the bar is elongated, the cross sections undergo displacements in the x-direction which are denoted by u. They depend on the location: $u = u(x)$. Thus, the displacements are u at the left end of the element and $u + du$ at the right end. The length of the elongated element is $dx+(u+du)-u = dx+du$. Hence, the elongation of the element is given by du. Now the local strain can be defined as the ratio between the elongation and the undeformed length of the element:

$$\varepsilon(x) = \frac{du}{dx}\,. \tag{1.7}$$

If the displacement $u(x)$ is known, the strain $\varepsilon(x)$ can be determined through differentiation. Reversely, if $\varepsilon(x)$ is known, the displacement $u(x)$ is obtained through integration.

The displacement $u(x)$ and the strain $\varepsilon(x)$ describe the geometry of the deformation. Therefore they are called *kinematic quantities*. Equation (1.7) is referred to as a kinematic relation.

1.3

1.3 Constitutive Law

Stresses are quantities derived from statics; they are a measure for the stressing in the material of a structure. On the other hand, strains are kinematic quantities; they measure the deformation of a body. However, the deformation depends on the load which acts on the body. Therefore, the stresses and the strains are not independent. The physical relation that connects these quantities is called *constitutive law*. It describes the behaviour of the material of the body under a load. It depends on the material and can be obtained only with the aid of experiments.

One of the most important experiments to find the relationship between stress and strain is the tension or compression test. Here, a small specimen of the material is placed into a testing machine and elongated or shortened. The force F applied by the machine onto the specimen can be read on the dial of the machine; it causes the normal stress $\sigma = F/A$. The change Δl of the length l of the specimen can be measured and the strain $\varepsilon = \Delta l/l$ can be calculated.

The graph of the relationship between stress and strain is shown schematically (not to scale) for a steel specimen in Fig. 1.7. This graph is referred to as *stress-strain diagram*. One can see that for small values of the strain the relationship is linear (straight line) and the stress is proportional to the strain. This behaviour is valid until the stress reaches the *proportional limit* σ_P. If the stress exceeds the proportional limit the strain begins to increase more rapidly and the slope of the curve decreases. This continues until the stress reaches the *yield stress* σ_Y. From this point of the stress-strain diagram the strain increases at a practically constant stress: the material begins to *yield*. Note that many materials do

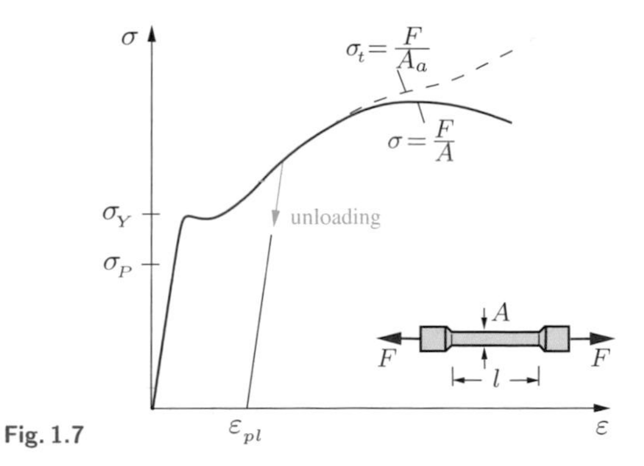

Fig. 1.7

not exhibit a pronounced yield point. At the end of the yielding the slope of the curve increases again which shows that the material can sustain an additional load. This phenomenon is called *strain hardening*.

Experiments show that an elongation of the bar leads to a reduction of the cross-sectional area A. This phenomenon is referred to as *lateral contraction*. Whereas the cross-sectional area decreases uniformly over the entire length of the bar in the case of small stresses, it begins to decrease locally at very high stresses. This phenomenon is called *necking*. Since the actual cross section A_a may then be considerably smaller than the original cross section A, the stress $\sigma = F/A$ does not describe the real stress any more. It is therefore appropriate to introduce the stress $\sigma_t = F/A_a$ which is called *true stress* or *physical stress*. It represents the true stress in the region where necking takes place. The stress $\sigma = F/A$ is referred to as *nominal* or *conventional* or *engineering stress*. Fig. 1.7 shows both stresses until fracture occurs.

Consider a specimen being first *loaded* by a force which causes the stress σ. Assume that σ is smaller than the yield stress σ_Y, i.e., $\sigma < \sigma_Y$. Subsequently, the load is again removed. Then the specimen will return to its original length: the strain returns to zero. In addition, the curves during the loading and the unloading coincide. This behaviour of the material is called *elastic*; the behaviour in the region $\sigma \leq \sigma_P$ is referred to as *linearly elastic*. Now assume that the specimen is loaded beyond the yield stress, i.e., until a stress $\sigma > \sigma_Y$ is reached. Then the curve during the unloading is a straight line which is parallel to the straight line in the linear-elastic region, see Fig. 1.7. If the load is completely removed the strain does not return to zero: a *plastic strain* ε_{pl} remains after the unloading. This material behaviour is referred to as *plastic*.

In the following we will always restrict ourselves to a linearly-elastic material behaviour. For the sake of simplicity we will refer to this behaviour shortly as elastic, i.e., in what follows "elastic" always stands for "linearly elastic". Then we have the linear relationship

Table 1.1 Material Constants

Material	E in MPa	α_T in 1/°C
Steel	$2{,}1 \cdot 10^5$	$1{,}2 \cdot 10^{-5}$
Aluminium	$0{,}7 \cdot 10^5$	$2{,}3 \cdot 10^{-5}$
Concrete	$0{,}3 \cdot 10^5$	$1{,}0 \cdot 10^{-5}$
Wood (in fibre direction)	$0{,}7 \ldots 2{,}0 \cdot 10^4$	$2{,}2 \ldots 3{,}1 \cdot 10^{-5}$
Cast iron	$1{,}0 \cdot 10^5$	$0{,}9 \cdot 10^{-5}$
Copper	$1{,}2 \cdot 10^5$	$1{,}6 \cdot 10^{-5}$
Brass	$1{,}0 \cdot 10^5$	$1{,}8 \cdot 10^{-5}$

$$\sigma = E\,\varepsilon \tag{1.8}$$

between the stress and the strain. The proportionality factor E is called *modulus of elasticity* or *Young's modulus* (Thomas Young, 1773–1829). The constitutive law (1.8) is called *Hooke's law* after Robert Hooke (1635–1703). Note that Robert Hooke could not present this law in the form (1.8) since the notion of stress was introduced only in 1822 by Augustin Louis Cauchy (1789–1857).

The relation (1.8) is valid for tension and for compression: the modulus of elasticity has the same value for tension and compression. However, the stress must be less than the proportional limit σ_P which may be different for tension or compression.

The modulus of elasticity E is a constant which depends on the material and which can be determined with the aid of a tension test. It has the dimension of force/area (which is also the dimension of stress); it is given, for example, in the unit MPa. Table 1.1 shows the values of E for several materials at room temperature. Note that these values are just a guidance since the modulus of elasticity depends on the composition of the material and on the temperature.

A tensile or a compressive force, respectively, causes the strain

$$\varepsilon = \sigma / E \tag{1.9}$$

in a bar, see (1.8). Changes of the length and thus strains are not only caused by forces but also by changes of the temperature. Experiments show that the *thermal strain* ε_T is proportional to the change ΔT of the temperature if the temperature of the bar is changed uniformly across its section and along its length:

$$\varepsilon_T = \alpha_T \Delta T\,. \tag{1.10}$$

The proportionality factor α_T is called *coefficient of thermal expansion*. It is a material constant and is given in the unit 1/°C. Table 1.1 shows several values of α_T.

If the change of the temperature is not the same along the entire length of the bar (if it depends on the location) then (1.10) represents the local strain $\varepsilon_T(x) = \alpha_T \Delta T(x)$.

If a bar is subjected to a stress σ as well as to a change ΔT of the temperature, the total strain ε is obtained through a superposition of (1.9) and (1.10):

$$\varepsilon = \frac{\sigma}{E} + \alpha_T \Delta T\,. \tag{1.11}$$

This relation can also be written in the form

$$\sigma = E(\varepsilon - \alpha_T \Delta T)\,. \tag{1.12}$$

1.4

1.4 Single Bar under Tension or Compression

There are three different types of equations that allow us to determine the stresses and the strains in a bar: the equilibrium condition, the kinematic relation and Hooke's law. Depending on the problem, the equilibrium condition may be formulated for the entire bar, a portion of the bar (see Section 1.1) or for an element of the bar. We will now derive the equilibrium condition for an

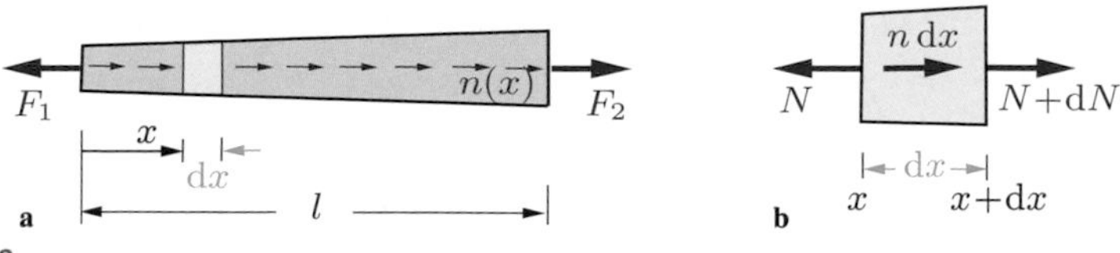

Fig. 1.8

element. For this purpose we consider a bar which is subjected to two forces F_1 and F_2 at its ends and to a line load $n = n(x)$, see Fig. 1.8a. The forces are assumed to be in equilibrium. We imagine a slice element of infinitesimal length $\mathrm{d}x$ separated from the bar as shown in Fig. 1.8b. The free-body diagram shows the normal forces N and $N + \mathrm{d}N$, respectively, at the ends of the element; the line load is replaced by its resultant $n\mathrm{d}x$ (note that n may be considered to be constant over the length $\mathrm{d}x$, compare Volume 1, Section 7.2.2). Equilibrium of the forces in the direction of the axis of the bar

$$\rightarrow: \qquad N + \mathrm{d}N + n\,\mathrm{d}x - N = 0$$

yields the *equilibrium condition*

$$\frac{\mathrm{d}N}{\mathrm{d}x} + n = 0\,. \qquad (1.13)$$

In the special case of a vanishing line load ($n \equiv 0$) the normal force in the bar is constant.

The *kinematic relation* for the bar is (see (1.7))

$$\varepsilon = \frac{\mathrm{d}u}{\mathrm{d}x}\,,$$

and *Hooke's law* is given by (1.11):

$$\varepsilon = \frac{\sigma}{E} + \alpha_T \Delta T\,.$$

If we insert the kinematic relation and $\sigma = N/A$ into Hooke's law we obtain

$$\frac{\mathrm{d}u}{\mathrm{d}x} = \frac{N}{EA} + \alpha_T \Delta T\,. \qquad (1.14)$$

This equation relates the displacements $u(x)$ of the cross sections and the normal force $N(x)$. It may be called the *constitutive law for the bar*. The quantity EA is known as *axial rigidity*. Equations (1.13) and (1.14) are the basic equations for a bar under tension or compression.

The displacement u of a cross section is found through integration of the strain:

$$\varepsilon = \frac{\mathrm{d}u}{\mathrm{d}x} \quad \rightarrow \quad \int \mathrm{d}u = \int \varepsilon\,\mathrm{d}x \quad \rightarrow \quad u(x) - u(0) = \int_0^x \varepsilon\,\mathrm{d}\bar{x}\,.$$

The elongation Δl follows as the difference of the displacements at the ends $x = l$ and $x = 0$ of the bar:

$$\Delta l = u(l) - u(0) = \int_0^l \varepsilon\,\mathrm{d}x\,. \qquad (1.15)$$

With $\varepsilon = \mathrm{d}u/\mathrm{d}x$ and (1.14) this yields

$$\Delta l = \int_0^l \left(\frac{N}{EA} + \alpha_T \Delta T\right) \mathrm{d}x\,. \qquad (1.16)$$

In the special case of a bar (length l) with constant axial rigidity ($EA = \text{const}$) which is subjected only to forces at its end ($n \equiv 0, N = F$) and to a uniform change of the temperature ($\Delta T = \text{const}$), the elongation is given by

$$\Delta l = \frac{F\,l}{EA} + \alpha_T \Delta T\,l\,. \qquad (1.17)$$

If, in addition, $\Delta T = 0$ we obtain

$$\Delta l = \frac{F\, l}{EA}\,, \tag{1.18}$$

and if $F = 0$, (1.17) reduces to

$$\Delta l = \alpha_T \Delta T\, l\,. \tag{1.19}$$

If we want to apply these equations to specific problems, we have to distinguish between statically determinate and statically indeterminate problems. In a *statically determinate* system we can always calculate the normal force $N(x)$ with the aid of the equilibrium condition. Subsequently, the strain $\varepsilon(x)$ follows from $\sigma = N/A$ and Hooke's law $\varepsilon = \sigma/E$. Finally, integration yields the displacement $u(x)$ and the elongation Δl. A change of the temperature causes only *thermal strains* (no stresses!) in a statically determinate system.

In a *statically indeterminate* problem the normal force cannot be calculated from the equilibrium condition alone. In such problems the basic equations (equilibrium condition, kinematic relation and Hooke´s law) are a system of *coupled* equations and have to be solved simultaneously. A change of the temperature in general causes additional stresses; they are called *thermal stresses*.

Finally we will reduce the basic equations to a single equation for the displacement u. If we solve (1.14) for N and insert into (1.13) we obtain

$$(EA\, u')' = -\, n + (EA\, \alpha_T \Delta T)'\,. \tag{1.20a}$$

Here, the primes denote derivatives with respect to x. Equation (1.20a) simplifies in the special case $EA = \text{const}$ and $\Delta T = \text{const}$ to

$$EA\, u'' = -\, n\,. \tag{1.20b}$$

If the functions $EA(x)$, $n(x)$ and $\Delta T(x)$ are given, the displacement $u(x)$ of an arbitrary cross section can be determined through integration of (1.20). The constants of integration are calculated from the boundary conditions. If, for example, one end of the bar is fixed then $u = 0$ at this end. If, on the other hand, one end of the bar can move and is subjected to a force F_0, then applying (1.14) and $N = F_0$ yields the boundary condition $u' = F_0/EA + \alpha_T \Delta T$. This reduces to the boundary condition $u' = 0$ in the special case of a stress-free end ($F_0 = 0$) of a bar whose temperature is not changed ($\Delta T = 0$).

Frequently, one or more of the quantities in (1.20) are given through different functions of x in different portions of the bar (e.g., if there exists a jump of the cross section). Then the bar must be divided into several regions and the integration has to be performed separately in each of theses regions. In this case the constants of integration can be calculated from boundary conditions and matching conditions (compare Volume 1, Section 7.2.4).

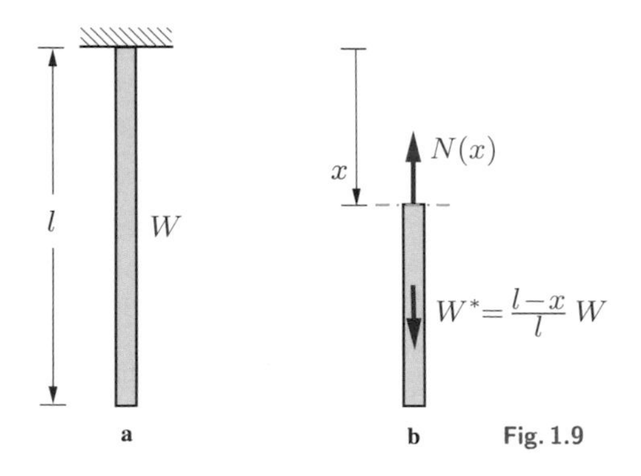

Fig. 1.9

As an illustrative example of a statically determinate system let us consider a slender bar (weight W, cross-sectional area A) that is suspended from the ceiling (Fig. 1.9a). First we determine the normal force caused by the weight of the bar. We cut the bar at an arbitrary position x (Fig. 1.9b). The normal force N is equal to the weight W^* of the portion of the bar below the imaginary cut. Thus, it is given by $N(x) = W^*(x) = W(l - x)/l$. Equation (1.4) now yields the normal stress

$$\sigma(x) = \frac{N(x)}{A} = \frac{W}{A}\left(1 - \frac{x}{l}\right) .$$

Accordingly, the normal stress in the bar varies linearly; it decreases from the value $\sigma(0) = W/A$ at the upper end to $\sigma(l) = 0$ at the free end.

The elongation Δl of the bar due to its own weight is obtained from (1.16):

$$\Delta l = \int_0^l \frac{N}{EA}\,\mathrm{d}x = \frac{W}{EA}\int_0^l \left(1 - \frac{x}{l}\right)\mathrm{d}x = \frac{1}{2}\frac{W\,l}{EA}\,.$$

It is half the elongation of a bar with negligible weight which is subjected to the force W at the free end.

We may also solve the problem by applying the differential equation (1.20b) for the displacements $u(x)$ of the cross sections of the bar. Integration with the constant line load $n = W/l$ yields

$$\begin{aligned} EA\,u'' &= -\frac{W}{l}\,, \\ EA\,u' &= -\frac{W}{l}\,x + C_1\,, \\ EA\,u &= -\frac{W}{2\,l}\,x^2 + C_1\,x + C_2\,. \end{aligned}$$

The constants of integration C_1 and C_2 can be determined from the boundary conditions. The displacement of the cross section at the upper end of the bar is equal to zero: $u(0) = 0$. Since the stress σ vanishes at the free end, we have $u'(l) = 0$. This leads to $C_2 = 0$ and $C_1 = W$. Thus, the displacement and the normal force are given by

$$u(x) = \frac{1}{2}\frac{W\,l}{EA}\left(2\,\frac{x}{l} - \frac{x^2}{l^2}\right), \quad N(x) = EA\,u'(x) = W\left(1 - \frac{x}{l}\right).$$

Since $u(0) = 0$, the elongation is equal to the displacement of the free end:

$$\Delta l = u(l) = \frac{1}{2}\frac{W\,l}{EA}\,.$$

The stress is obtained as

$$\sigma(x) = \frac{N(x)}{A} = \frac{W}{A}\left(1 - \frac{x}{l}\right).$$

As an illustrative example of a statically indeterminate system let us consider a bar which is placed stress-free between two rigid walls (Fig. 1.10a). It has the cross-sectional areas A_1 and A_2, respectively. We want to determine the support reactions if the temperature of the bar is raised uniformly by an amount ΔT in region ① .

The free-body diagram (Fig. 1.10b) shows the *two* support reactions B and C. They cannot be calculated from only *one* equilibrium condition:

$$\rightarrow: \quad B - C = 0.$$

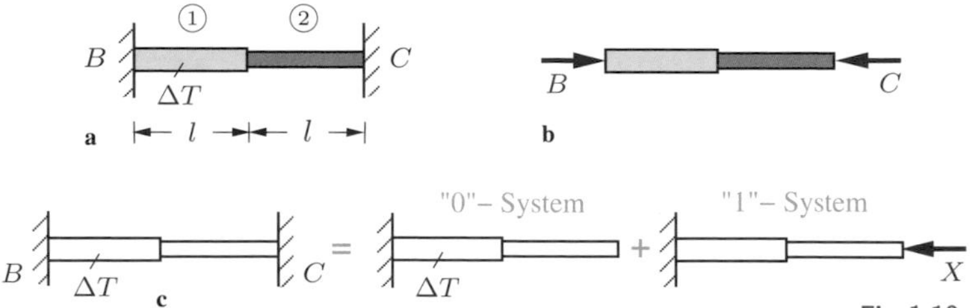

Fig. 1.10

Therefore we have to take into account the deformation of the bar. The elongations in the regions ① and ② are given by (1.16) with $N = -B = -C$:

$$\Delta l_1 = \frac{N\,l}{EA_1} + \alpha_T \Delta T\, l\,, \qquad \Delta l_2 = \frac{N\,l}{EA_2}$$

(the temperature in region ② is not changed).

The bar is placed between two *rigid* walls. Thus, its total elongation Δl has to vanish:

$$\Delta l = \Delta l_1 + \Delta l_2 = 0.$$

This equation expresses the fact that the geometry of the deformation has to be compatible with the restraints imposed by the supports. Therefore it is called *compatibility condition.*

The equilibrium condition and the compatibility condition yield the unknown support reactions:

$$\frac{N\,l}{EA_1}+\alpha_T\Delta T\,l+\frac{N\,l}{EA_2}=0\;\rightarrow\;B=C=-N=\frac{EA_1A_2\,\alpha_T\Delta T}{A_1+A_2}\,.$$

The problem may also be solved in the following way. In a first step we generate a statically determinate system. This is achieved by removing one of the supports, for example support C. The action of this support on the bar is replaced by the action of the force $C = X$ which is as yet unknown. Note that one of the supports, for example B, is needed to have a statically determinate system. The other support, C, is in excess of the necessary support. Therefore the reaction C is referred to as being a *redundant reaction.*

Now we need to consider two different problems. First, we investigate the statically determinate system subjected to the given load (here: the change of the temperature in region ①) which is referred to as "0"-system or *primary system* (Fig. 1.10c). In this system the change of the temperature causes the thermal elongation $\Delta l_1^{(0)}$ (normal force $N = 0$) in region ① ; the elongation in region ② is zero. Thus, the displacement $u_C^{(0)}$ of the right end point of the bar is given by

$$u_C^{(0)}=\Delta l_1^{(0)}=\alpha_T\Delta T\,l\,.$$

Secondly we consider the statically determinate system subjected only to force X. It is called "1"-system and is also shown in Fig. 1.10c. Here the displacement $u_C^{(1)}$ of the right end point is

$$u_C^{(1)}=\Delta l_1^{(1)}+\Delta l_2^{(1)}=-\frac{X\,l}{EA_1}-\frac{X\,l}{EA_2}\,.$$

Both the applied load (here: ΔT) as well as the force X act in the given problem (Fig. 1.10a). Therefore, the total displacement u_C at point C follows through *superposition*:

$$u_C=u_C^{(0)}+u_C^{(1)}\,.$$

Since the rigid wall in the original system prevents a displacement at C, the geometric condition

$$u_C = 0$$

has to be satisfied. This leads to

$$\alpha_T \Delta T\, l - \frac{X\, l}{EA_1} - \frac{X\, l}{EA_2} = 0 \quad \rightarrow \quad X = C = \frac{EA_1\, A_2\, \alpha_T \Delta T}{A_1 + A_2}\,.$$

Equilibrium at the free-body diagram (Fig. 1.10b) yields the second support reaction $B = C$.

E1.3

Example 1.3 A solid circular steel cylinder (cross-sectional area A_S, modulus of elasticity E_S, length l) is placed inside a copper tube (cross-sectional area A_C, modulus of elasticity E_C, length l). The assembly is compressed between a rigid plate and the rigid floor by a force F (Fig. 1.11a).

Determine the normal stresses in the cylinder and in the tube. Calculate the shortening of the assembly.

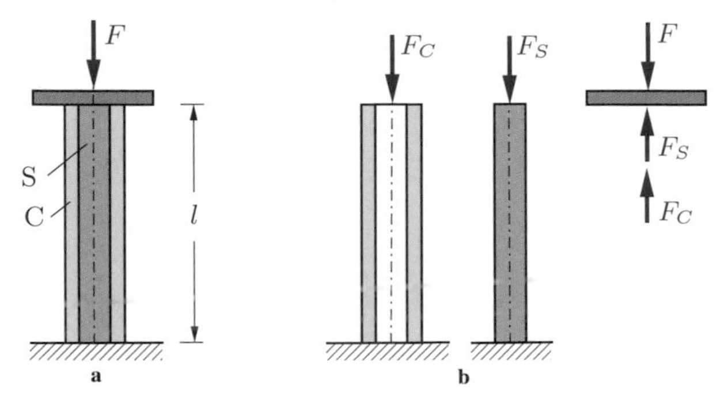

Fig. 1.11

Solution We denote the compressive forces in the steel cylinder and in the copper tube by F_S and F_C, respectively (Fig. 1.11b). Equilibrium at the free-body diagram of the plate yields

$$F_C + F_S = F\,. \tag{a}$$

Since equilibrium furnishes only one equation for the two unknown forces F_S and F_C, the problem is statically indeterminate. We obtain a second equation by taking into account the deformation of the system. The shortenings (here counted positive) of the two parts are given according to (1.18) by

$$\Delta l_C = \frac{F_C\, l}{EA_C}\,, \qquad \Delta l_S = \frac{F_S\, l}{EA_S} \tag{b}$$

where, for simplicity, we have denoted the axial rigidity $E_C A_C$ of the copper tube byEA_C and the axial rigidity $E_S A_S$ of the steel cylinder by EA_S.

The plate and the floor are assumed to be rigid. Therefore the geometry of the problem requires that the shortenings of the copper tube and of the steel cylinder coincide. This gives the compatibility condition

$$\Delta l_C = \Delta l_S\,. \tag{c}$$

Solving the Equations (a) to (c) yields the forces

$$F_C = \frac{EA_C}{EA_C + EA_S}\, F\,, \qquad F_S = \frac{EA_S}{EA_C + EA_S}\, F\,. \tag{d}$$

The compressive stresses follow according to (1.2):

$$\underline{\underline{\sigma_C = \frac{E_C}{EA_C + EA_S}\, F}}\,, \qquad \underline{\underline{\sigma_S = \frac{E_S}{EA_C + EA_S}\, F}}\,.$$

Inserting (d) into (b) leads to the shortening:

$$\underline{\underline{\Delta l_C = \Delta l_S = \frac{F\, l}{EA_C + EA_S}}}\,.$$

Example 1.4 A copper tube ② is placed over a threaded steel bolt ① of length l. The pitch of the threads is given by h. A nut fits snugly against the tube without generating stresses in the system (Fig. 1.12a). Subsequently, the nut is given n full turns and the temperature of the entire assembly is increased by the amount ΔT. The axial rigidities and the coefficients of thermal expansion of the bolt and the tube are given.

E1.4

Determine the force in the bolt.

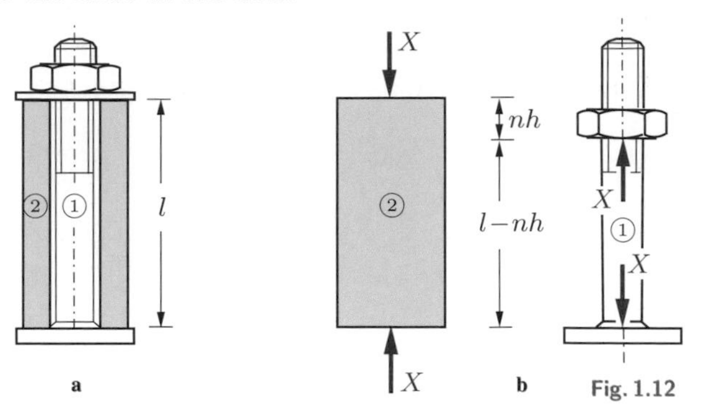

Fig. 1.12

Solution After the nut has been turned it exerts a compressive force X on the tube which causes a shortening of the tube. According to Newton's third axiom (action = reaction) a force of equal magnitude and opposite direction acts via the nut on the bolt which elongates. The free-body diagrams of bolt and tube are shown in Fig. 1.12b.

The problem is statically indeterminate since force F cannot be determined from equilibrium alone. Therefore we have to take into account the deformations. The length of the bolt after the nut has been turned, see the free-body diagram in Fig. 1.12b, is given by $l_1 = l - n\,h$. Its elongation Δl_1 follows from

$$\Delta l_1 = \frac{X(l - n\,h)}{EA_1} + \alpha_{T1}\Delta T(l - n\,h)\,.$$

Since $n\,h \ll l$, this can be reduced to

$$\Delta l_1 = \frac{X\,l}{EA_1} + \alpha_{T1}\Delta T\,l\,.$$

The change of length Δl_2 of the tube ($l_2 = l$) is obtained from

$$\Delta l_2 = -\,\frac{X\,l}{EA_2} + \alpha_{T2}\Delta T\,l\,.$$

The length of the bolt and the length of the tube have to coincide after the deformation. This yields the compatibility condition

$$l_1 + \Delta l_1 = l_2 + \Delta l_2 \qquad \rightarrow \qquad \Delta l_1 - \Delta l_2 = l_2 - l_1 = n\,h\,.$$

Solving the equations leads to the force in the bolt:

$$X\left(\frac{l}{EA_1}+\frac{l}{EA_2}\right)+(\alpha_{T1}-\alpha_{T2})\Delta T\,l=n\,h$$

$$\rightarrow \quad \underline{\underline{X=\frac{n\,h-(\alpha_{T1}-\alpha_{T2})\Delta T\,l}{\left(\frac{1}{EA_1}+\frac{1}{EA_2}\right)l}}}\,.$$

1.5

1.5 Statically Determinate Systems of Bars

In the preceding section we calculated the stresses and deformations of single slender bars. We will now extend the investigation to trusses and to structures which consist of bars and rigid bodies. In this section we will restrict ourselves to statically determinate systems where we can first calculate the forces in the bars with the aid of the equilibrium conditions. Subsequently, the stresses in the bars and the elongations are determined. Finally, the displacements of arbitrary points of the structure can be found. Since it is assumed that the elongations are small as compared with the lengths of the bars, we can apply the equilibrium conditions to the *undeformed* system.

As an illustrative example let us consider the truss in Fig. 1.13a. Both bars have the axial rigidity EA. We want to determine the displacement of pin C due to the applied force F. First we calculate the forces S_1 and S_2 in the bars. The equilibrium conditions, applied to the free-body diagram (Fig. 1.13b), yield

$$\begin{aligned}\uparrow: &\quad S_2\sin\alpha-F=0\\ \leftarrow: &\quad S_1+S_2\cos\alpha=0\end{aligned} \quad\rightarrow\quad S_1=-\frac{F}{\tan\alpha}\,,\qquad S_2=\frac{F}{\sin\alpha}\,.$$

According to (1.17) the elongations Δl_i of the bars are given by

$$\Delta l_1=\frac{S_1\,l_1}{EA}=-\frac{F\,l}{EA}\frac{1}{\tan\alpha}\,,\qquad \Delta l_2=\frac{S_2\,l_2}{EA}=\frac{F\,l}{EA}\frac{1}{\sin\alpha\cos\alpha}\,.$$

Bar 1 becomes shorter (compression) and bar 2 becomes longer (tension). The new position C' of pin C can be found as follows. We consider the bars to be disconnected at C. Then the system becomes movable: bar 1 can rotate about point A; bar 2 can rotate about point B. The free end points of the bars then move along circular paths with radii $l_1 + \Delta l_1$ and $l_2 + \Delta l_2$, respectively. Point C' is located at the point of intersection of these arcs of circles (Fig. 1.13c).

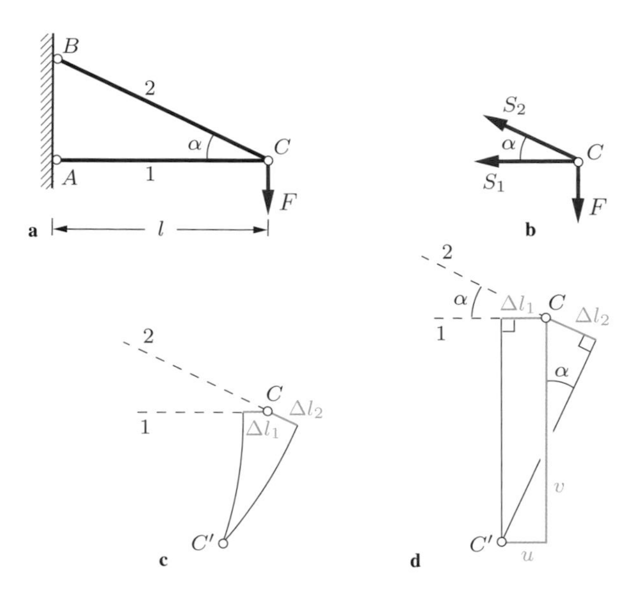

Fig. 1.13

The elongations are small as compared with the lengths of the bars. Therefore, within a good approximation the arcs of the circles can be replaced by their tangents. This leads to the *displacement diagram* as shown in Fig. 1.13d. If this diagram is drawn to scale, the displacement of pin C can directly be taken from it. We want to apply a "graphic-analytical" solution. It suffices then to draw a sketch of the diagram. Applying trigonometric relations we obtain the horizontal and the vertical components of the displacement:

$$u = |\Delta l_1| = \frac{F\,l}{EA}\frac{1}{\tan\alpha}\,, \qquad (1.21)$$

$$v = \frac{\Delta l_2}{\sin\alpha} + \frac{u}{\tan\alpha} = \frac{F\,l}{EA}\frac{1+\cos^3\alpha}{\sin^2\alpha\cos\alpha}\,.$$

To determine the displacement of a pin of a truss with the aid of a displacement diagram is usually quite cumbersome and can be recommended only if the truss has very few members. In the case of trusses with many members it is advantageous to apply an energy method (see Chapter 6).

The method described above can also be applied to structures which consist of bars and rigid bodies.

E1.5

Example 1.5 A rigid beam (weight W) is mounted on three elastic bars (axial rigidity EA) as shown in Fig. 1.14a.

Determine the angle of slope of the beam that is caused by its weight after the structure has been assembled.

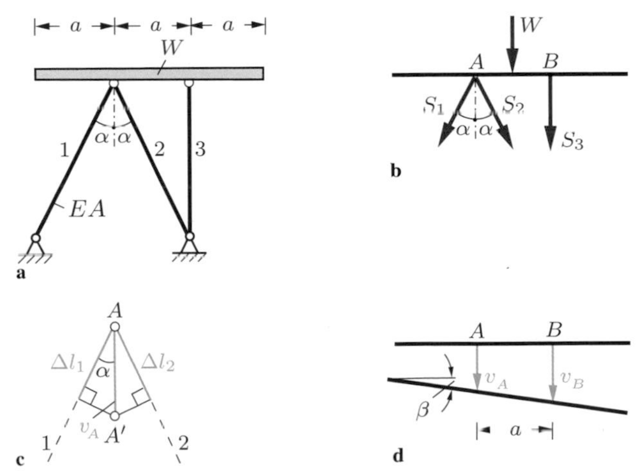

Fig. 1.14

Solution First we calculate the forces in the bars with the aid of the equilibrium conditions (Fig. 1.14b):

$$S_1 = S_2 = -\frac{W}{4\cos\alpha}\,, \qquad S_3 = -\frac{W}{2}\,.$$

With $l_1 = l_2 = l/\cos\alpha$ and $l_3 = l$ we obtain the elongations:

$$\Delta l_1 = \Delta l_2 = \frac{S_1\, l_1}{EA} = -\frac{W\, l}{4EA\cos^2\alpha}\,, \qquad \Delta l_3 = \frac{S_3\, l_3}{EA} = -\frac{W\, l}{2\,EA}\,.$$

Point B of the beam is displaced downward by $v_B = |\Delta l_3|$. To determine the vertical displacement v_A of point A we sketch a displacement diagram (Fig. 1.14c). First we plot the changes Δl_1 and Δl_2 of the lengths in the direction of the respective bar. The lines perpendicular to these directions intersect at the displaced position A' of point A. Thus, its vertical displacement is given by $v_A = |\Delta l_1|/\cos\alpha$.

Since the displacements v_A and v_B do not coincide, the beam does not stay horizontal after the structure has been assembled. The angle of slope β is obtained with the approximation $\tan\beta \approx \beta$ (small deformations) and $l = a\cot\alpha$ as (see Fig. 1.14d)

$$\underline{\underline{\beta}} = \frac{v_B - v_A}{a} = \underline{\underline{\frac{2\cos^3\alpha - 1}{4\cos^3\alpha}\,\frac{W\cot\alpha}{EA}}}\,.$$

If $\cos^3\alpha > \frac{1}{2}$ (or $\cos^3\alpha < \frac{1}{2}$), then the beam is inclined to the right (left). In the special case $\cos^3\alpha = \frac{1}{2}$, i.e. $\alpha = 37.5^\circ$, it stays horizontal.

E1.6

Example 1.6 The truss in Fig. 1.15a is subjected to a force F. Given: $E = 2\cdot 10^2$ GPa, $F = 20\,$kN.

Determine the cross-sectional area of the three members so that the stresses do not exceed the allowable stress $\sigma_{\text{allow}} = 150$ MPa and the displacement of support B is smaller than 0.5 ‰ of the length of bar 3.

Solution First we calculate the forces in the members. The equilibrium conditions for the free-body diagrams of pin C and support B (Fig. 1.15b) yield

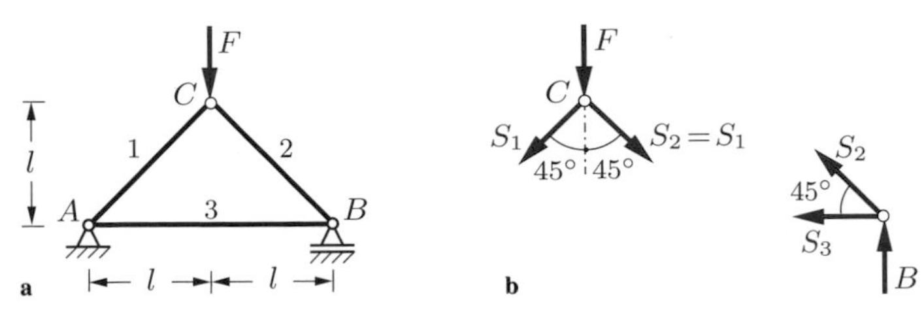

Fig. 1.15

$$S_1 = S_2 = -\frac{\sqrt{2}}{2} F\,, \qquad S_3 = \frac{F}{2}\,.$$

The stresses do not exceed the allowable stress if

$$|\sigma_1| = \frac{|S_1|}{A_1} \le \sigma_{\text{allow}}\,,\ |\sigma_2| = \frac{|S_2|}{A_2} \le \sigma_{\text{allow}}\,,\ \sigma_3 = \frac{S_3}{A_3} \le \sigma_{\text{allow}}\,.$$

This leads to the cross-sectional areas

$$\underline{\underline{A_1 = A_2}} = \frac{|S_1|}{\sigma_{\text{allow}}} = \underline{\underline{94.3\ \text{mm}^2}}\,,\ A_3 = \frac{S_3}{\sigma_{\text{allow}}} = 66.7\ \text{mm}^2\,. \quad \text{(a)}$$

In addition, the displacement of support B has to be smaller than 0.5 ‰ of the length of bar 3. This displacement is equal to the elongation $\Delta l_3 = S_3\, l_3 / EA_3$ of bar 3 (support A is fixed). From $\Delta l_3 < 0.5 \cdot 10^{-3}\, l_3$ we obtain

$$\frac{\Delta l_3}{l_3} = \frac{S_3}{EA_3} < 0.5{\cdot}10^{-3} \;\to\; \underline{\underline{A_3}} > \frac{2\,S_3}{E} 10^3 = \frac{F}{E} 10^3 = \underline{\underline{100\text{mm}^2}}\,.$$

Comparison with (a) yields the required area $A_3 = 100\,\text{mm}^2$.

1.6 Statically Indeterminate Systems of Bars

1.6

We will now investigate statically indeterminate systems for which the forces in the bars cannot be determined with the aid of the equilibrium conditions alone since the number of the unknown quantities exceeds the number of the equilibrium conditions. In such systems the basic equations (equilibrium conditions, kinematic equations (compatibility) and Hooke's law) are coupled equations.

Let us consider the symmetrical truss shown in Fig. 1.16a. It is stress-free before the load is applied. The axial rigidities EA_1, EA_2, $EA_3 = EA_1$ are given; the forces in the members are unknown. The system is statically indeterminate to the first degree (the decomposition of a force into three directions cannot be done uniquely in a coplanar problem, see Volume 1, Section 2.2). The two equilibrium conditions applied to the free-body diagram of pin K (Fig. 1.16b) yield

$$\begin{aligned} \rightarrow: &\quad -S_1 \sin\alpha + S_3 \sin\alpha = 0 \qquad\qquad \rightarrow \quad S_1 = S_3\,, \\ \uparrow: &\quad S_1 \cos\alpha + S_2 + S_3 \cos\alpha - F = 0 \rightarrow \quad S_1 = S_3 = \frac{F - S_2}{2\cos\alpha}\,. \end{aligned} \tag{a}$$

The elongations of the bars are given by

$$\Delta l_1 = \Delta l_3 = \frac{S_1\, l_1}{EA_1}\,, \qquad \Delta l_2 = \frac{S_2\, l}{EA_2}\,. \tag{b}$$

To derive the compatibility condition we sketch a displacement diagram (Fig. 1.16c) from which we find

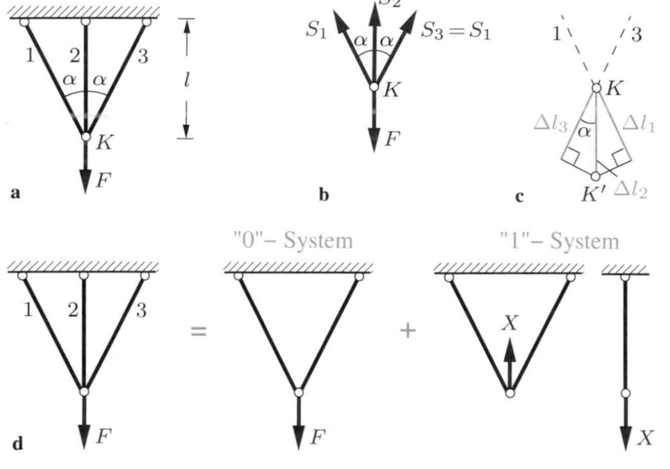

Fig. 1.16

$$\Delta l_1 = \Delta l_2 \cos\alpha \,. \tag{c}$$

With (a), (b) and $l_1 = l/\cos\alpha$ we obtain from (c)

$$\frac{(F - S_2)\, l}{2\, EA_1 \cos^2\alpha} = \frac{S_2\, l}{EA_2} \cos\alpha$$

which leads to

$$S_2 = \frac{F}{1 + 2\,\dfrac{EA_1}{EA_2}\cos^3\alpha} \,.$$

The remaining two forces in the bars follow from (a):

$$S_1 = S_3 = \frac{\dfrac{EA_1}{EA_2}\cos^2\alpha}{1 + 2\,\dfrac{EA_1}{EA_2}\cos^3\alpha}\, F \,.$$

Note that now the vertical displacement v of pin K can also be written down:

$$v = \Delta l_2 = \frac{S_2\, l}{EA_2} = \frac{\dfrac{F\, l}{EA_2}}{1 + 2\,\dfrac{EA_1}{EA_2}\cos^3\alpha} \,.$$

The problem may also be solved using the method of superposition. In a first step we remove bar 2 to obtain a statically determinate system, the “0“-system. It consists of the two bars 1 and 3 and it is subjected to the given force F (Fig. 1.16d). The forces $S_1^{(0)}$ and $S_3^{(0)}$ in these bars follow from the equilibrium conditions as

$$S_1^{(0)} = S_3^{(0)} = \frac{F}{2\cos\alpha} \,.$$

The corresponding elongations are obtained with $l_1 = l/\cos\alpha$:

$$\Delta l_1^{(0)} = \Delta l_3^{(0)} = \frac{S_1^{(0)}\, l_1}{EA_1} = \frac{F\, l}{2\, EA_1 \cos^2\alpha} \,. \tag{d}$$

In a second step we consider the statically determinate system under the action of an unknown force X ("1"-system, see also Fig. 1.16d). Note that this force acts in the opposite direction on bar 2 (actio = reactio). Now we get

$$\begin{aligned} S_1^{(1)} &= S_3^{(1)} = -\frac{X}{2\cos\alpha}\,, \qquad S_2^{(1)} = X\,, \\ \Delta l_1^{(1)} &= \Delta l_3^{(1)} = -\frac{Xl}{2\,EA_1\cos^2\alpha}\,, \qquad \Delta l_2^{(1)} = \frac{Xl}{EA_2}\,. \end{aligned} \tag{e}$$

The total elongation of the bars is obtained through superposition of the systems "0" and "1":

$$\Delta l_1 = \Delta l_3 = \Delta l_1^{(0)} + \Delta l_1^{(1)}\,, \qquad \Delta l_2 = \Delta l_2^{(1)}\,. \tag{f}$$

The compatibility condition (c) is again taken from the displacement diagram (Fig. 1.16c). It leads with (d) - (f) to the unknown force $X = S_2^{(1)} = S_2$:

$$\frac{F\,l}{2\,EA_1\cos^2\alpha} - \frac{X\,l}{2\,EA_1\cos^2\alpha} = \frac{X\,l}{EA_2}\cos\alpha$$

$$\rightarrow X = S_2 = \frac{F}{1 + 2\,\dfrac{EA_1}{EA_2}\cos^3\alpha}\,.$$

The forces S_1 and S_3 follow from superposition:

$$S_1 = S_3 = S_1^{(0)} + S_1^{(1)} = \frac{\dfrac{EA_1}{EA_2}\cos^2\alpha}{1 + 2\,\dfrac{EA_1}{EA_2}\cos^3\alpha}\,F\,.$$

A system of bars is statically indeterminate of degree n if the number of the unknowns exceeds the number of the equilibrium conditions by n. In order to determine the forces in the bars of such a system, n compatibility conditions are needed in addition to the equilibrium conditions. Solving this system of equations yields the unknown forces in the bars.

A statically indeterminate system of degree n can also be solved with the method of superposition. Then n bars are removed in order to obtain a statically determinate system. The action of the bars which are removed is replaced by the action of the static redundants $S_i = X_i$. Next $n+1$ different auxiliary systems are considered. The given load acts in the "0"-system, whereas the "i"-system ($i = 1, 2, ..., n$) is subjected only to the force X_i. In each of the statically determinate auxiliary problems the forces in the bars and thus the elongations can be calculated. Applying the n compatibility conditions yields a system of equations for the n unknown forces X_i. The forces in the other bars can subsequently be determined through superposition.

E1.7

Example 1.7 A rigid beam (weight negligible) is suspended from three vertical bars (axial rigidity EA) as shown in Fig. 1.17a.

Determine the forces in the originally stress-free bars if

a) the beam is subjected to a force F ($\Delta T = 0$),

b) the temperature of bar 1 is changed by ΔT ($F = 0$).

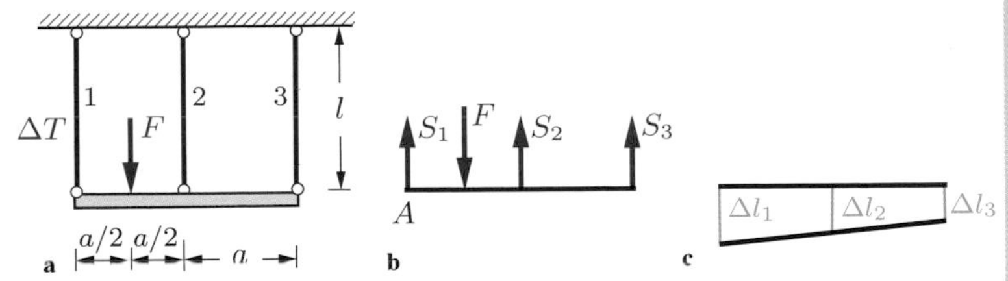

Fig. 1.17

Solution The system is statically indeterminate to the first degree: there are only two equilibrium conditions for the three unknown forces S_i (Fig. 1.17b). a) If the structure is subjected to force F the equilibrium conditions are

$$\uparrow: \quad S_1 + S_2 + S_3 - F = 0\,,$$

$$\overset{\curvearrowleft}{A}: \quad -\frac{a}{2}\,F + a\,S_2 + 2\,a\,S_3 = 0\,. \qquad \text{(a)}$$

The elongations of the bars are given by ($\Delta T = 0$)

$$\Delta l_1 = \frac{S_1\,l}{EA}\,, \qquad \Delta l_2 = \frac{S_2\,l}{EA}\,, \qquad \Delta l_3 = \frac{S_3\,l}{EA}\,. \qquad \text{(b)}$$

We sketch a displacement diagram (Fig. 1.17c) and find the compatibility condition

$$\Delta l_2 = \frac{\Delta l_1 + \Delta l_3}{2} \,. \tag{c}$$

Now we have six equations for the three forces S_j and the three elongations Δl_j . Solving for the forces yields

$$\underline{\underline{S_1 = \frac{7}{12} F}}\,, \qquad \underline{\underline{S_2 = \frac{1}{3} F}}\,, \qquad \underline{\underline{S_3 = \frac{1}{12} F}}\,.$$

b) If bar 1 is heated ($F = 0$), the equilibrium conditions are

$$\begin{aligned} \uparrow: \quad & S_1 + S_2 + S_3 = 0\,, \\ \overset{\frown}{A}: \quad & aS_2 + 2aS_3 = 0\,, \end{aligned} \tag{a'}$$

and the elongations are given by

$$\Delta l_1 = \frac{S_1\, l}{EA} + \alpha_T \Delta T l\,, \qquad \Delta l_2 = \frac{S_2\, l}{EA}\,, \qquad \Delta l_3 = \frac{S_3\, l}{EA}\,. \tag{b'}$$

The compatibility condition (c) is still valid. Solving (a′), (b′) and (c) yields

$$\underline{\underline{S_1 = S_3 = -\frac{1}{6} EA\, \alpha_T \Delta T}}\,, \qquad \underline{\underline{S_2 = \frac{1}{3} EA\, \alpha_T \Delta T}}\,.$$

E1.8

Example 1.8 To assemble the truss in Fig. 1.18a, the free end of bar 3 (length $l - \delta$, $\delta \ll l$) has to be connected with pin C.

a) Determine the necessary force F acting at pin C (Fig. 1.18b).

b) Calculate the forces in the bars after the truss has been assembled and force F has been removed.

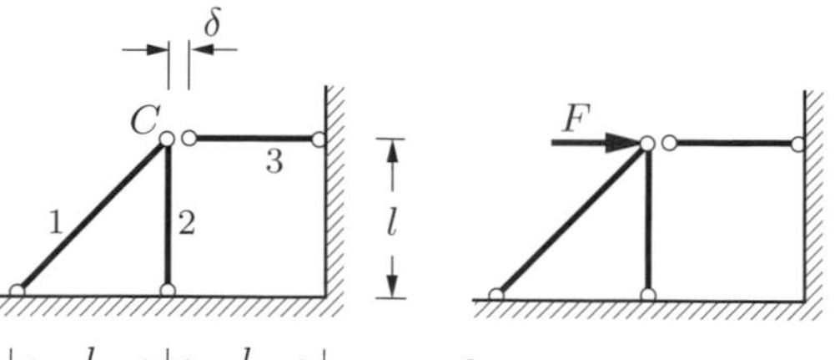

Fig. 1.18

Solution a) The force F causes a displacement of pin C. The horizontal component v of this displacement has to be equal to δ to allow assembly. The required force follows with $\alpha = 45°$ from (1.21):

$$v = \frac{F\,l}{EA}\frac{1+\sqrt{2}/4}{\sqrt{2}/4} = \delta \qquad \rightarrow \qquad \underline{\underline{F = \frac{EA\,\delta}{(2\sqrt{2}+1)\,l}}}\,.$$

b) The force F is removed after the truss has been assembled. Then pin C undergoes another displacement. Since now a force S_3 in bar 3 is generated, pin C does not return to its original position: it is displaced to position C^* (Fig. 1.18c). The distance between points C and C^* is given by

$$v^* = \frac{S_3\,l}{EA}\frac{1+\sqrt{2}/4}{\sqrt{2}/4}\,.$$

The compatibility condition

$$v^* + \Delta l_3 = \delta$$

can be taken from Fig. 1.18c. With the elongation

$$\Delta l_3 = \frac{S_3(l-\delta)}{EA} \approx \frac{S_3 l}{EA}$$

of bar 3 we reach

$$\frac{S_3\,l}{EA}\frac{1+\sqrt{2}/4}{\sqrt{2}/4} + \frac{S_3\,l}{EA} = \delta \qquad \rightarrow \qquad \underline{\underline{S_3 = \frac{EA\,\delta}{2(\sqrt{2}+1)l}}}\,.$$

The other two forces follow from the equilibrium condition at pin C:

$$\underline{\underline{S_1 = \sqrt{2}\,S_3}}\,, \qquad \underline{\underline{S_2 = -\,S_3}}\,.$$

1.7

1.7 Supplementary Examples

Detailed solutions to the following examples are given in (**A**) D. Gross et al. *Formeln und Aufgaben zur Technischen Mechanik 2*, Springer, Berlin 2010, or (**B**) W. Hauger et al. *Aufgaben zur Technischen Mechanik 1-3*, Springer, Berlin 2008.

E1.9

Example 1.9 A slender bar (density ρ, modulus of elasticity E) is suspended from its upper end as shown in Fig. 1.19. It has a rectangular cross section with a constant depth and a linearly varying width. The cross section at the upper end is A_0.

Determine the stress $\sigma(x)$ due to the force F and the weight of the bar. Calculate the minimum stress $\sigma_{\min}$ and its location.

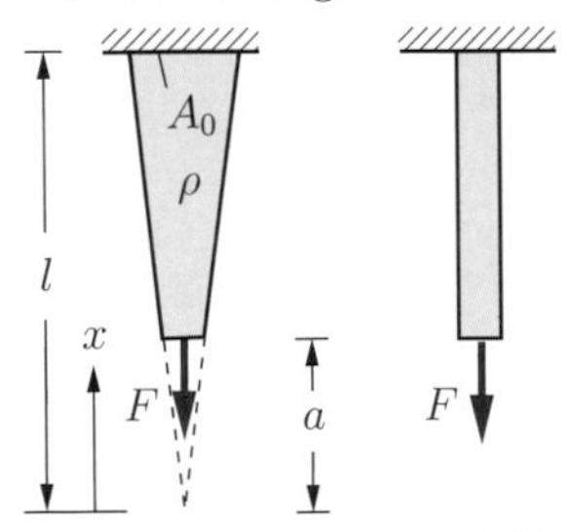

Fig. 1.19

Results: see (**A**)

$$\sigma(x) = \frac{Fl + \rho g \frac{A_0}{2}(x^2 - a^2)}{A_0 x}, \quad \sigma_{\min} = \rho g x^*, \; x^* = \sqrt{\frac{2Fl}{\rho g A_0} - a^2}\,.$$

E1.10

Example 1.10 Determine the elongation Δl of the tapered circular shaft (modulus of elasticity E) shown in Fig. 1.20 if it is subjected to a tensile force F.

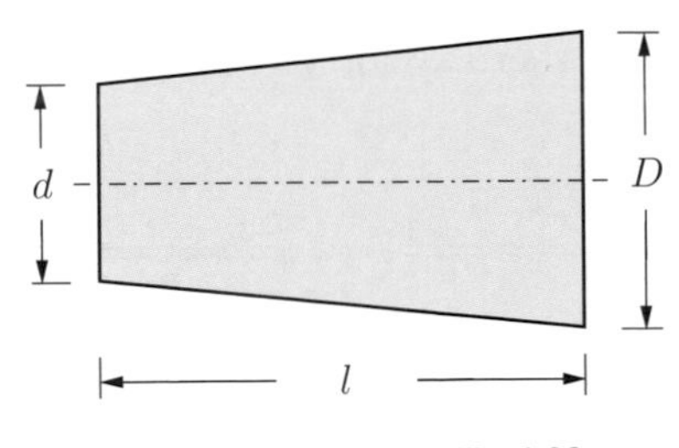

Fig. 1.20

Result: see (**A**) $\Delta l = \dfrac{4Fl}{\pi E D d}\,.$

E1.11

Example 1.11 A slender bar (weight W_0, modulus of elasticity E, coefficient of thermal expansion α_T) is suspended from its upper end. It just touches the ground as shown in Fig. 1.21 without generating a contact force.

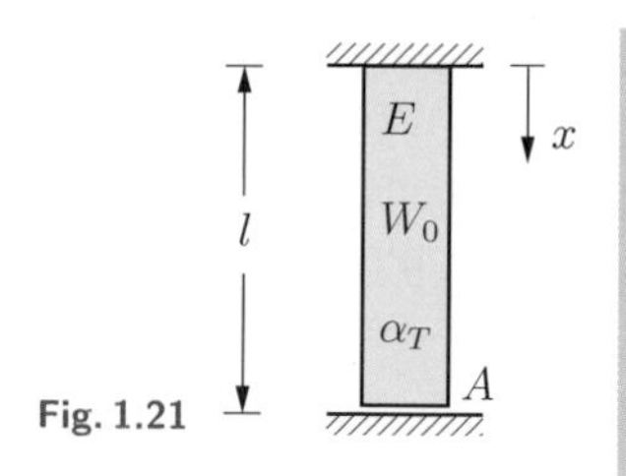

Fig. 1.21

Calculate the stress $\sigma(x)$ if the temperature of the bar is uniformly increased by ΔT. Determine ΔT so that there is compression in the whole bar.

Results: see (**A**)

$$\sigma(x) = \frac{W_0}{A}\left(1 - \frac{x}{l}\right) - E\alpha_T\Delta T, \qquad \Delta T > \frac{W_0}{EA\alpha_T}.$$

E1.12

Example 1.12 The bar (cross sectional area A) shown in Fig. 1.22 is composed of steel and aluminium. It is placed stress-free between two rigid walls. Given: $E_{st}/E_{al} = 3, \alpha_{st}/\alpha_{al} = 1/2$.

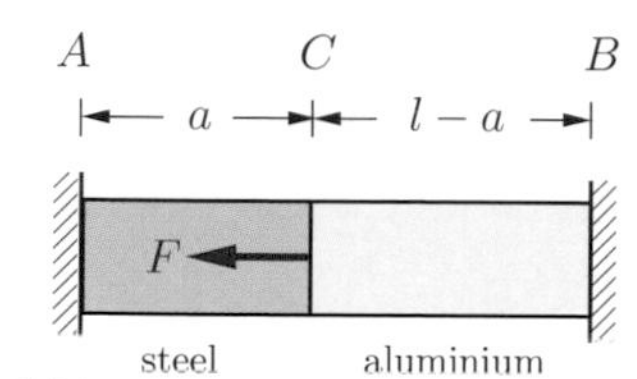

Fig. 1.22

a) Calculate the support reactions if the bar is subjected to a force F at point C.
b) Calculate the normal force in the whole bar if it is subjected only to a change of temperature ΔT ($F = 0$).

Results: see (**A**)

$$\text{a) } N_A = -F\frac{3(l-a)}{3l-2a}, \quad N_B = F\frac{a}{3l-2a},$$

$$\text{b) } N = -\frac{2l-a}{3l-2a}E_{st}\alpha_{st}A\Delta T\,.$$

E1.13

Example 1.13 The column in Fig. 1.23 consists of reinforced concrete. It is subjected to a tensile force F. Given: $E_{st}/E_c = 6, A_{st}/A_c = 1/9$.

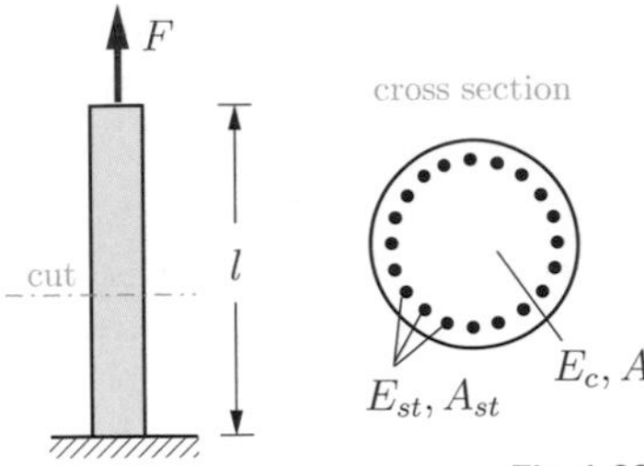

Fig. 1.23

Determine the stresses in the steel and in the concrete and the elongation Δl of the column if

a) the bonding between steel and concrete is perfect,
b) the bonding is damaged so that only the steel carries the load.

Results: see (**A**)

$$\text{a)}\ \sigma_{st} = 4\frac{F}{A}, \quad \sigma_c = \frac{2}{3}\frac{F}{A}, \quad \Delta l = \frac{2}{5}\frac{Fl}{EA_{st}},$$

$$\text{b)}\ \sigma_{st} = 10\frac{F}{A}, \quad \Delta l = \frac{Fl}{EA_{st}}.$$

E1.14

Example 1.14 A slender bar (density ρ, modulus of elasticity E, length l) is suspended from its upper end as shown in Fig. 1.24. It has a rectangular cross section with a constant depth a. The width b varies linearly from $2b_0$ at he fixed end to b_0 at the free end.

Determine the stresses $\sigma(x)$ and $\sigma(l)$ and the elongation Δl of the bar due to its own weight.

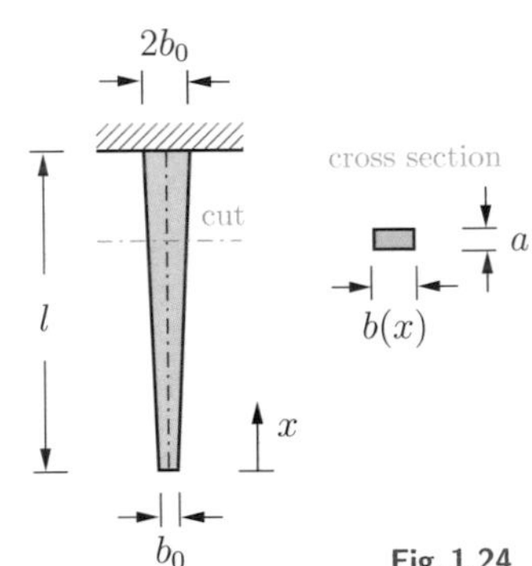

Fig. 1.24

Results: see (**B**)

$$\sigma(x) = \frac{1}{2}\rho g\frac{(2l+x)x}{l+x}, \quad \sigma(l) = \frac{3\rho g l}{4}, \quad \Delta l = \frac{\rho g l^2}{4E}(3 - 2\ln 2).$$

E1.15

Example 1.15 A rigid chair (weight negligible) is supported by three bars (axial rigidity EA) as shown in Fig. 1.25. It is subjected to a force F at point B.

a) Calculate the forces S_i in the bars and the elongations Δl_i of the bars.

b) Determine the displacement of point C.

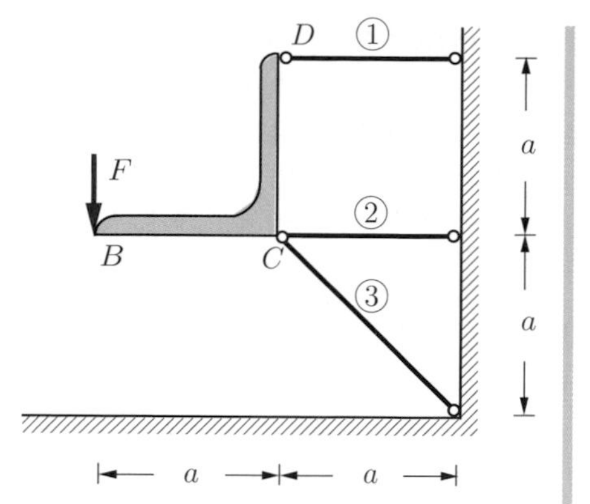

Fig. 1.25

Results: see (**A**)

$$\text{a)}\ S_1 = F, \quad S_2 = 0, \quad S_3 = -\sqrt{2}F,$$

$$\Delta l_1 = \frac{Fa}{EA}, \quad \Delta l_2 = 0, \quad \Delta l_3 = -2\frac{Fa}{EA},$$

$$\text{b)}\ u_C = 0, \quad v_C = 2\sqrt{2}\frac{Fa}{EA}\,.$$

E1.16

Example 1.16 Two bars (axial rigidity EA) are pin-connected and supported at C (Fig. 1.26).

a) Calculate the support reaction at C due to the force F.

b) Determine the displacement of the support.

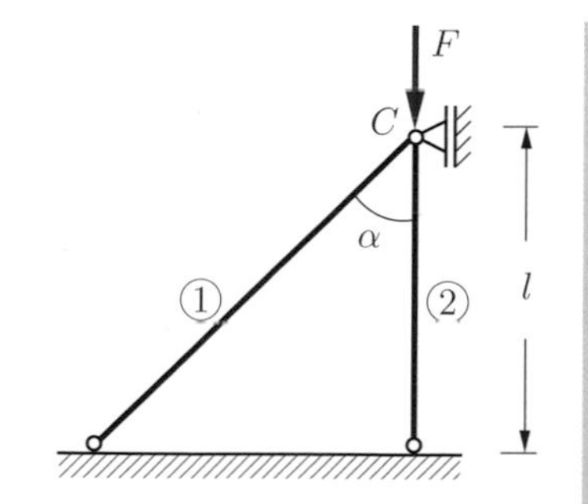

Fig. 1.26

Results: see (**A**) a) $C = \dfrac{\sin\alpha\cos^2\alpha}{1+\cos^3\alpha}F$, b) $v_C = \dfrac{1}{1+\cos^3\alpha}\dfrac{Fl}{EA}\,.$

E1.17

Example 1.17 Consider a thin circular ring (modulus of elasticity E, coefficient of thermal expansion α_T, internal radius $r-\delta, \delta \ll r$) with a rectangular cross section (width b, thickness $t \ll r$). The ring is heated in order to increase its radius which makes it possible to place it over a rigid wheel with radius r.

Determine the necessary change of temperature ΔT. Calculate the normal stress σ in the ring and the pressure p onto the wheel after the temperature has regained its original value.

Results: see (**B**) $\Delta T = \frac{\delta}{\alpha_T r}, \quad \sigma = E\frac{\delta}{r}, \quad p = \sigma\frac{t}{r}.$

E1.18

Example 1.18 The two rods (axial rigidity EA) shown in Fig. 1.27 are pin-connected at K. The system is subjected to a vertical force F.

Calculate the displacement of pin K.

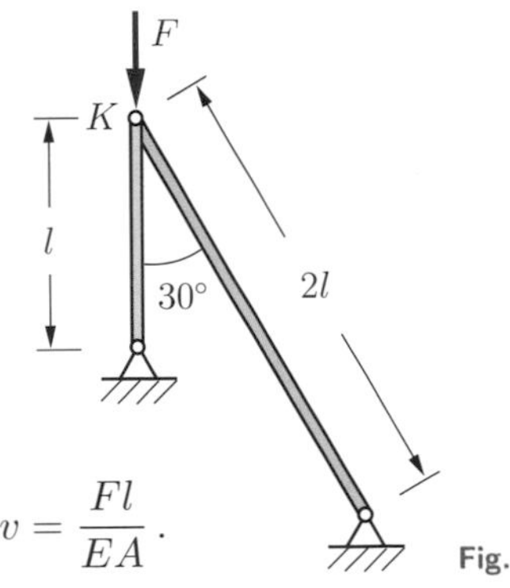

Results: see (**B**) $u = \sqrt{3}\frac{Fl}{EA}, \quad v = \frac{Fl}{EA}.$

Fig. 1.27

E1.19

Example 1.19 The structure shown in Fig. 1.28 consists of a rigid beam BC and two elastic bars (axial rigidity EA). It is subjected to a force F.

Calculate the displacement of pin C.

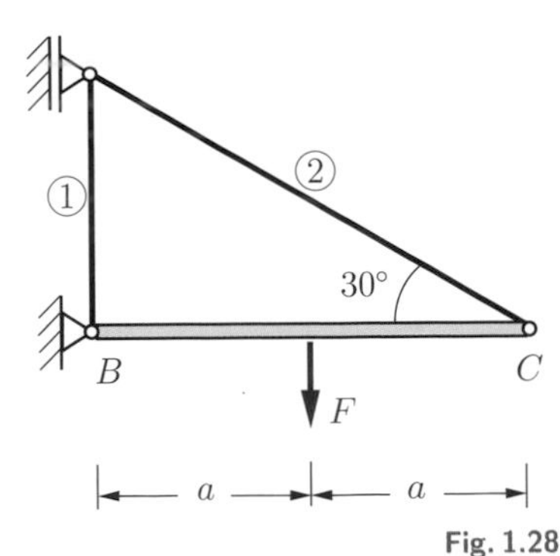

Fig. 1.28

Results: see (**B**) $u = 0, \quad v = 3\sqrt{3}\frac{Fa}{EA}.$

E1.20

Example 1.20 Fig. 1.29 shows a freight elevator. The cable (length l, axial rigidity $(EA)_1$) of the winch passes over a smooth pin K. A crate (weight W) is suspended at the end of the cable (see Example 2.13 in Volume 1). The axial rigidity $(EA)_2$ of the two bars 1 and 2 is given.

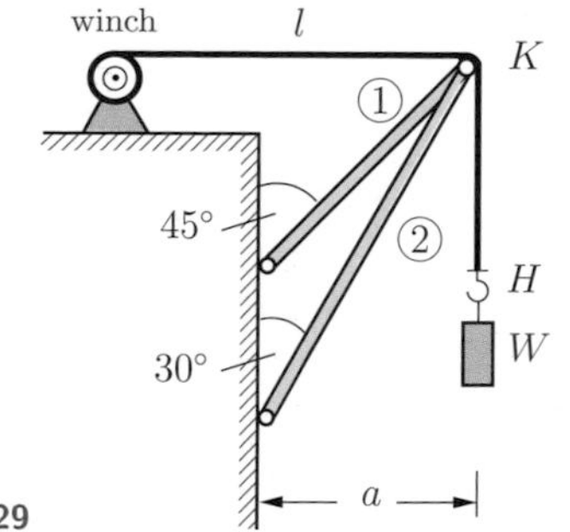

Fig. 1.29

Determine the displacements of pin K and of the end of the cable (point H) due to the weight of the crate.

Results: see (**B**)

$$u = 6.69\frac{Wa}{(EA)_2}, \quad v = 3.86\frac{Wa}{(EA)_2}, \quad f = 2.83\frac{Wa}{(EA)_2} + \frac{Wl}{(EA)_1}.$$

E1.21

Example 1.21 To assemble the truss (axial rigidity EA of the three bars) in Fig. 1.30 the end point P of bar 2 has to be connected with pin K. Assume $\delta \ll h$.

Determine the forces in the bars after the truss has been assembled.

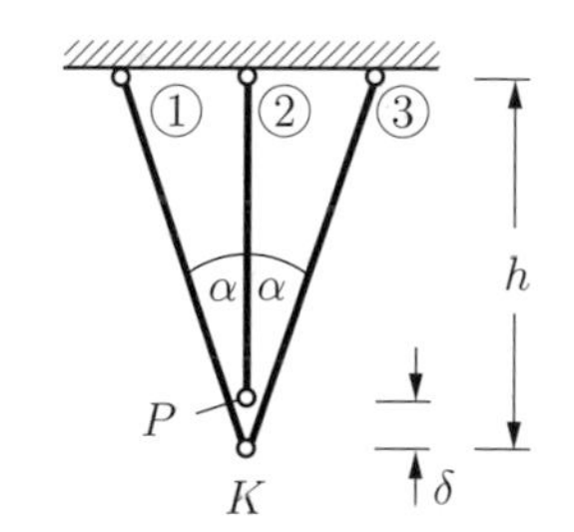

Fig. 1.30

Results: see (**B**)

$$S_1 = S_3 = -\frac{EA\delta\cos^2\alpha}{h\,(1+2\cos^3\alpha)}, \quad S_2 = \frac{2EA\delta\cos^3\alpha}{h\,(1+2\cos^3\alpha)}.$$

1.8

1.8 Summary

- Normal stress in a section perpendicular to the axis of a bar:

$$\sigma = N/A \,,$$

N normal force, A cross-sectional area.

- Strain:

$$\varepsilon = \mathrm{d}u/\mathrm{d}x \,, \quad |\varepsilon| \ll 1 \,,$$

u displacement of a cross section.
Special case of uniform strain: $\varepsilon = \Delta l/l$.

- Hooke's law:

$$\sigma = E\,\varepsilon \,,$$

E modulus of elasticity.

- Elongation:

$$\Delta l = \int_0^l \left(\frac{N}{EA} + \alpha_T \Delta T\right) \mathrm{d}x \,,$$

EA axial rigidity, α_T coefficient of thermal expansion,
ΔT change of temperature.
Special cases:

$$N = F, \quad \Delta T = 0, \quad EA = \text{const} \quad \rightarrow \quad \Delta l = \frac{Fl}{EA} \,,$$

$$N = 0, \quad \Delta T = \text{const} \quad \rightarrow \quad \Delta l = \alpha_T \Delta T \, l \,.$$

- Statically determinate system of bars: normal forces, stresses, strains, elongations and displacements can be calculated consecutively from the equilibrium conditions, Hooke's law and kinematic equations. A change of the temperature does not cause stresses.
- Statically indeterminate system: the equations (equilibrium conditions, kinematic equations and Hooke's law) are coupled equations. A change of the temperature in general causes thermal stresses.

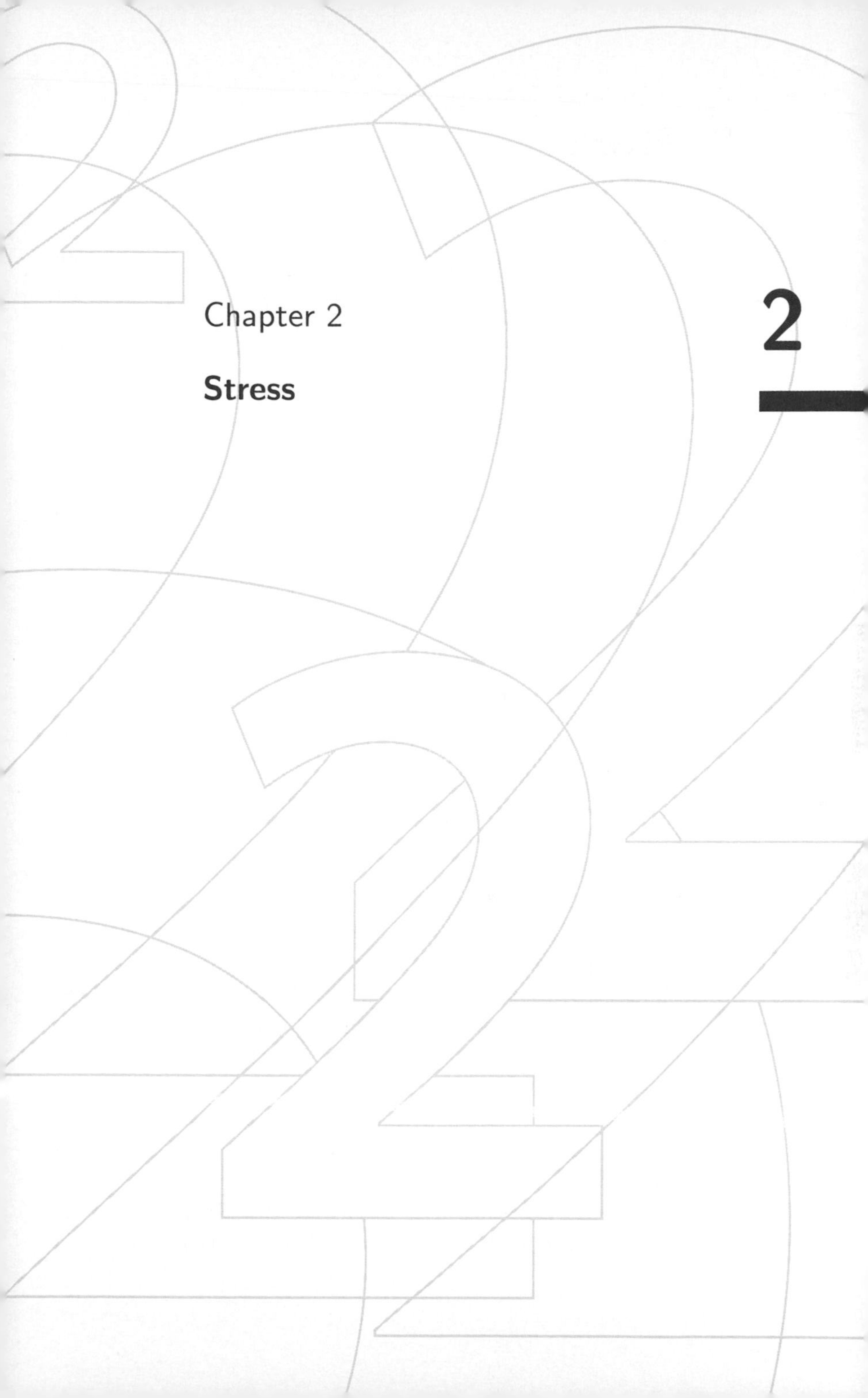

Chapter 2

Stress

2

2 Stress

2.1	Stress Vector and Stress Tensor	49
2.2	Plane Stress	52
2.2.1	Coordinate Transformation	53
2.2.2	Principal Stresses	56
2.2.3	Mohr's Circle	62
2.2.4	The Thin-Walled Pressure Vessel	68
2.3	Equilibrium Conditions	70
2.4	Supplementary Examples	73
2.5	Summary	75

Objectives: In Chapter 1 the notion of stress in a bar has been introduced. We will now generalize the concept of stress to make it applicable to arbitrary structures. For this purpose the *stress tensor* is introduced. Subsequently we will discuss in detail the *plane stress state* that appears in thin sheets or plates under in-plane loading. This state is fully determined by stress components in two sections perpendicular to each other. We will see that the normal stress and the shear stress take on extreme values for specific directions of the section.

The students will learn how to analyse the plane stress state and how to determine the stresses in different sections.

2.1 Stress Vector and Stress Tensor

So far, stresses have been calculated only in bars. To be able to determine stresses also in other structures we must generalize the concept of stress. For this purpose let us consider a body which is loaded arbitrarily, e.g. by single forces $\boldsymbol{F}_i$ and area forces $\boldsymbol{p}$ (Fig. 2.1a). The external load generates internal forces. In an imaginary section $s-s$ through the body the internal area forces (stresses) are distributed over the entire area A. In contrast to the bar where these stresses are constant over the cross section (see Section 1.1) they now generally vary throughout the section.

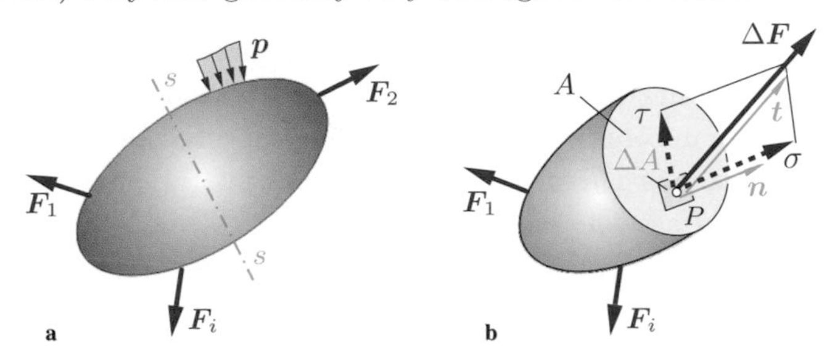

Fig. 2.1 a b

Since the stress is no longer the same everywhere in the section, it must be defined at an arbitrary point P of the cross section (Fig. 2.1b). The area element ΔA containing P is subjected to the resultant internal force $\Delta \boldsymbol{F}$ (note: according to the law of action and reaction the same force acts in the opposite cross section with opposite direction). The average stress in the area element is defined as the ratio $\Delta \boldsymbol{F}/\Delta A$ (force per area). We assume that the ratio $\Delta \boldsymbol{F}/\Delta A$ in the limit $\Delta A \to 0$ tends to a finite value:

$$\boldsymbol{t} = \lim_{\Delta A \to 0} \frac{\Delta \boldsymbol{F}}{\Delta A} = \frac{\mathrm{d}\boldsymbol{F}}{\mathrm{d}A} . \qquad (2.1)$$

This limit value is called *stress vector* $\boldsymbol{t}$.

The stress vector can be decomposed into a component normal to the cross section at point P and a component tangential to the cross section. We call the normal component *normal stress* σ and the tangential component *shear stress* τ.

In general, the stress vector $\boldsymbol{t}$ depends on the location of point P in the section area A. The stress distribution in the section is known when the stress vector $\boldsymbol{t}$ is known for all points of A. However, the *stress state* at a point P of the section is not yet sufficiently determined by $\boldsymbol{t}$ for the following reason. If we choose sections through P having *different directions*, different forces will act in the sections because of the different orientation of the area elements. Therefore, the stresses also depend on the orientation of the section which is characterized by the normal vector $\boldsymbol{n}$ (cf. stresses (1.3) in a bar for different directions of the section).

It can be shown that the stress state at point P is uniquely determined by three stress vectors for three sections through P, perpendicular to each other. It is useful to choose the directions of a Cartesian coordinate system for the respective orientations. The three sections can most easily be visualized if we imagine them to be the surfaces of a volume element with edge lengths $\mathrm{d}x$, $\mathrm{d}y$ and $\mathrm{d}z$ at point P (Fig. 2.2a). A stress vector acts on each of its six surfaces. It can be decomposed into its components perpendicular to the section (= normal stress) and tangential to the section (= shear stress). The shear stress subsequently can be further decomposed into its components according to the coordinate directions. To characterize the components double subscripts are used: the first subscript indicates the orientation of the section by the direction of its normal vector whereas the second subscript indicates the direction of the stress component. For example, τ_{yx} is a shear stress acting in a section whose normal points in y-direction; the stress itself points in x-direction (Fig. 2.2a).

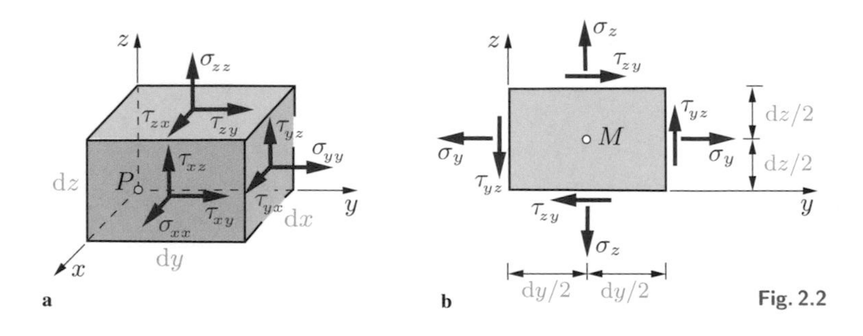

Fig. 2.2

The notation can be simplified for the normal stresses. In this case the directions of the normal to the section and of the stress component coincide. Thus, both subscripts are always equal and one of them can be omitted without losing information:

$$\sigma_{xx} = \sigma_x\,, \qquad \sigma_{yy} = \sigma_y\,, \qquad \sigma_{zz} = \sigma_z\,.$$

From now on we will adopt this shorter notation.

Using the introduced notation, the stress vector, for example in the section with the normal vector pointing in y-direction, can be written as

$$\boldsymbol{t} = \tau_{yx}\,\boldsymbol{e}_x + \sigma_y\,\boldsymbol{e}_y + \tau_{yz}\,\boldsymbol{e}_z\,. \tag{2.2}$$

The *sign convention* for the stresses is the same as for the stress resultants (cf. Volume 1, Section 7.1):

Positive stresses at a *positive (negative)* face point in *positive (negative)* directions of the coordinates.

Accordingly, positive (negative) normal stresses cause tension (compression) in the volume element. Figure 2.2a shows positive stresses acting on the positive faces.

By means of the decomposition of the three stress vectors into their components we have obtained three normal stresses $(\sigma_x, \sigma_y, \sigma_z)$ and six shear stresses $(\tau_{xy}, \tau_{xz}, \tau_{yx}, \tau_{yz}, \tau_{zx}, \tau_{zy})$. However, not all shear stresses are independent of each other. This can be shown by formulating the equilibrium condition for the moments about an axis parallel to the x-axis through the center of the volume element (cf. Fig. 2.2b). Since equilibrium statements are valid for forces, the stresses must be multiplied by the associated area elements:

$$\overset{\curvearrowleft}{C}: \quad 2\frac{\mathrm{d}y}{2}(\tau_{yz}\,\mathrm{d}x\,\mathrm{d}z) - 2\frac{\mathrm{d}z}{2}(\tau_{zy}\,\mathrm{d}x\,\mathrm{d}y) = 0 \quad \rightarrow \quad \tau_{yz} = \tau_{zy}\,.$$

Two further relations are obtained from the moment equilibrium about the other axes:

$$\tau_{xy} = \tau_{yx}\,, \qquad \tau_{xz} = \tau_{zx}\,, \qquad \tau_{yz} = \tau_{zy}\,. \tag{2.3}$$

In words:

The shear stresses with the same subscripts in two orthogonal sections (e.g. τ_{xy} and τ_{yx}) are equal.

They are sometimes called *complementary shear stresses*. Since they have the same algebraic sign they are directed either towards or away from the common edge of the cubic volume element (cf. Fig. 2.2). As a result of (2.3) there exist only six independent stress components.

The components of the three stress vectors can be arranged in a matrix:

$$\boldsymbol{\sigma} = \begin{bmatrix} \sigma_x & \tau_{xy} & \tau_{xz} \\ \tau_{yx} & \sigma_y & \tau_{yz} \\ \tau_{zx} & \tau_{zy} & \sigma_z \end{bmatrix} = \begin{bmatrix} \sigma_x & \tau_{xy} & \tau_{xz} \\ \tau_{xy} & \sigma_y & \tau_{yz} \\ \tau_{xz} & \tau_{yz} & \sigma_z \end{bmatrix}\,. \tag{2.4}$$

The main diagonal contains the normal stresses; the remaining elements are the shear stresses. The matrix (2.4) is *symmetric* because of (2.3).

The quantity $\boldsymbol{\sigma}$ is called *stress tensor* (the concept *tensor* will be explained in Section 2.2.1). The elements of (2.4) are the components of the stress tensor. The *stress state* at a material point is uniquely defined by the stress vectors for three sections, orthogonal to each other, and consequently by the stress tensor (2.4).

2.2

2.2 Plane Stress

We will now examine the state of stress in a *disk*. This plane structural element has a thickness t much smaller than its in-plane dimensions and it is loaded solely *in* its plane by in-plane forces (Fig. 2.3). The upper and the lower face of the disk are load-free. Since no external forces in the z-direction exist, we can assume with sufficient accuracy that also no stresses will appear in this direction:

$$\tau_{xz} = \tau_{yz} = \sigma_z = 0\,.$$

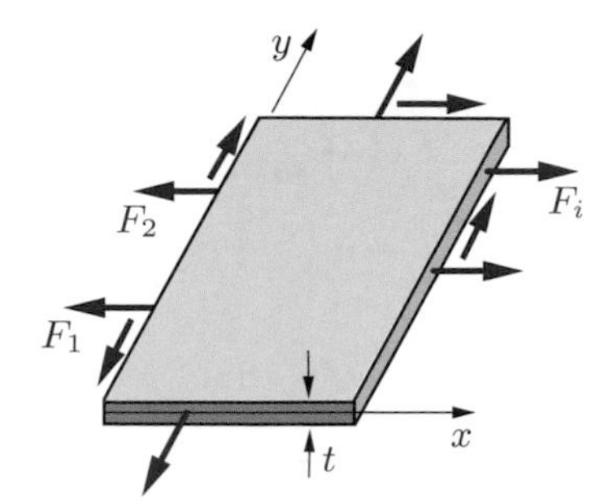

Fig. 2.3

Because of the small thickness we furthermore can assume that the stresses σ_x, σ_y and $\tau_{xy} = \tau_{yx}$ are constant across the thickness of the disk. Such a stress distribution is called a *state of plane stress*. In this case, the third row and the third column of the matrix (2.4) vanish and we get

$$\boldsymbol{\sigma} = \begin{bmatrix} \sigma_x & \tau_{xy} \\ \tau_{xy} & \sigma_y \end{bmatrix}.$$

In general, the stresses depend on the location, i.e. on the coordinates x and y. In the special case when the stresses are independent of the location, the stress state is called *homogeneous*.

2.2.1 Coordinate Transformation

Up to now only stresses in sections parallel to the coordinate axes have been considered. Now we will show how from these stresses, the stresses in an arbitrary section perpendicular to the disk can be determined. For this purpose we consider an infinitesimal wedge-shaped element of thickness t cut out from the disk (Fig. 2.4). The directions of the sections are characterized by the x, y-coordinate system and the angle φ. We introduce a ξ, η-system which is rotated with respect to the x, y-system by the angle φ and whose ξ-axis is normal to the inclined section. Here φ is counted positive *counterclockwise*.

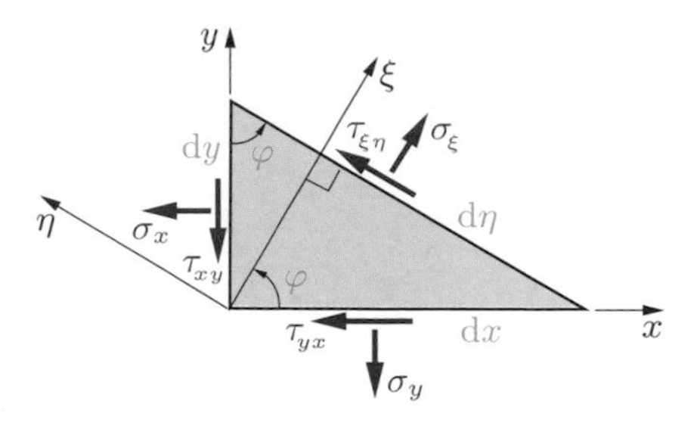

Fig. 2.4

According to the coordinate directions, the stresses in the inclined section are denoted as σ_ξ and $\tau_{\xi\eta}$. The corresponding cross section is given by $\mathrm{d}A = \mathrm{d}\eta\, t$. The other two cross sections perpendicular to the y- and x-axis, respectively, are $\mathrm{d}A \sin\varphi$ and $\mathrm{d}A\cos\varphi$. The equilibrium conditions for the forces in ξ- and in η-direction are

$$
\begin{aligned}
\nearrow: \quad \sigma_\xi \,\mathrm{d}A \; &- (\sigma_x \,\mathrm{d}A \cos\varphi)\cos\varphi - (\tau_{xy}\,\mathrm{d}A\cos\varphi)\sin\varphi \\
&- (\sigma_y\,\mathrm{d}A\sin\varphi)\sin\varphi - (\tau_{yx}\,\mathrm{d}A\sin\varphi)\cos\varphi = 0\,,
\end{aligned}
$$

$$
\begin{aligned}
\nwarrow: \quad \tau_{\xi\eta}\,\mathrm{d}A &+ (\sigma_x\,\mathrm{d}A\cos\varphi)\sin\varphi - (\tau_{xy}\,\mathrm{d}A\cos\varphi)\cos\varphi \\
&- (\sigma_y\,\mathrm{d}A\sin\varphi)\cos\varphi + (\tau_{yx}\,\mathrm{d}A\sin\varphi)\sin\varphi = 0\,.
\end{aligned}
$$

Taking into account $\tau_{yx} = \tau_{xy}$, we get

$$
\begin{aligned}
\sigma_\xi &= \sigma_x\cos^2\varphi + \sigma_y\sin^2\varphi + 2\,\tau_{xy}\sin\varphi\cos\varphi\,, \\
\tau_{\xi\eta} &= -\,(\sigma_x - \sigma_y)\sin\varphi\cos\varphi + \tau_{xy}(\cos^2\varphi - \sin^2\varphi)\,.
\end{aligned}
\tag{2.5a}
$$

Additionally, we will now determine the normal stress σ_η which acts in a section with the normal pointing in η-direction. The cutting angle of this section is given by $\varphi + \pi/2$. Therefore, σ_η is obtained by replacing in the first equation of (2.5a) the normal stress σ_ξ by σ_η and the angle φ by $\varphi+\pi/2$. Recalling that $\cos(\varphi+\pi/2) = -\sin\varphi$ and $\sin(\varphi+\pi/2) = \cos\varphi$, we obtain

$$
\sigma_\eta = \sigma_x\sin^2\varphi + \sigma_y\cos^2\varphi - 2\,\tau_{xy}\cos\varphi\sin\varphi\,. \tag{2.5b}
$$

Usually, the Equations (2.5a, b) are written in a different form. Using the standard trigonometric relations

$$\cos^2\varphi = \frac{1}{2}(1+\cos 2\varphi)\,, \qquad 2\sin\varphi\cos\varphi = \sin 2\varphi\,,$$
$$\sin^2\varphi = \frac{1}{2}(1-\cos 2\varphi)\,, \qquad \cos^2\varphi - \sin^2\varphi = \cos 2\varphi$$

we get

$$\begin{aligned}
\sigma_\xi &= \frac{1}{2}(\sigma_x+\sigma_y) + \frac{1}{2}(\sigma_x-\sigma_y)\cos 2\varphi + \tau_{xy}\sin 2\varphi\,,\\
\sigma_\eta &= \frac{1}{2}(\sigma_x+\sigma_y) - \frac{1}{2}(\sigma_x-\sigma_y)\cos 2\varphi - \tau_{xy}\sin 2\varphi\,,\\
\tau_{\xi\eta} &= \qquad\qquad\quad - \frac{1}{2}(\sigma_x-\sigma_y)\sin 2\varphi + \tau_{xy}\cos 2\varphi\,.
\end{aligned} \tag{2.6}$$

The stresses σ_x, σ_y and τ_{xy} are the components of the stress tensor in the x, y-system. From these stresses, using (2.6), the components σ_ξ, σ_η and $\tau_{\xi\eta}$ in the ξ, η-system can be determined. Equations (2.6) are called *transformation relations* for the components of the stress tensor. Fig. 2.5 shows the stresses in the x, y-system and in the ξ, η-system at the corresponding elements. Note that the stresses in either of the coordinate systems represent one and the same state of stress at a given point of the disk.

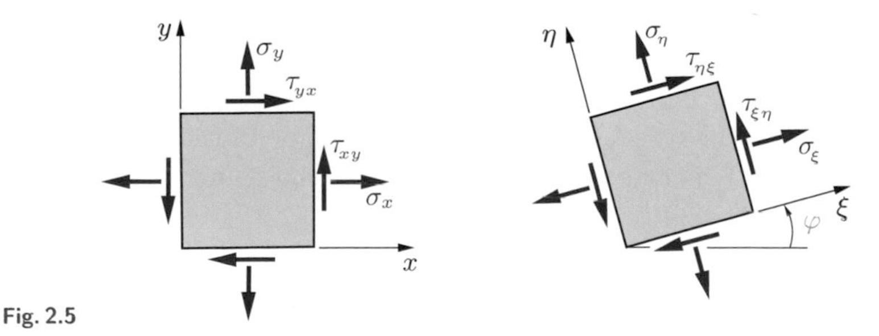

Fig. 2.5

A quantity whose components have *two* coordinate subscripts and which are transformed by a certain rule from one coordinate system to a rotated coordinate system is called a *second rank ten-*

sor. For the stress tensor the rule for the transition from the x, y-system to the ξ, η-system is given by the transformation relations (2.6). We will become familiar with other 2nd rank tensors in Sections 3.1 and 4.2. It should be mentioned that the components of vectors also fulfill specific transformation relations. Because vector components have only *one* subscript, vectors are sometimes called 1st rank tensors.

When adding the first two equations in (2.6) we obtain

$$\sigma_\xi + \sigma_\eta = \sigma_x + \sigma_y \,. \tag{2.7}$$

Thus, the sum of the normal stresses has the same value in each coordinate system. For this reason, the sum $\sigma_x + \sigma_y$ is called an *invariant* of the stress tensor. It can also be verified by simple algebraic manipulation that the determinant $\sigma_x \sigma_y - \tau_{xy}^2$ of the matrix of the stress tensor is a further invariant, that is $\sigma_x \sigma_y - \tau_{xy}^2 = \sigma_\xi \sigma_\eta - \tau_{\xi\eta}^2$.

We finally consider the special case of equal normal stresses ($\sigma_x = \sigma_y$) and vanishing shear stresses ($\tau_{xy} = 0$) in the x, y-system. Equation (2.6) then yields

$$\sigma_\xi = \sigma_\eta = \sigma_x = \sigma_y, \qquad \tau_{\xi\eta} = 0 \,.$$

Accordingly, the normal stresses for *all* directions of the sections are the same (i.e. they are independent of φ) whereas the shear stresses always vanish. Such a state of stress is called *hydrostatic* because it corresponds to the pressure in a fluid at rest where the normal stress is the same in all directions.

It should be noted that a disk also can be sectioned in such a way that the normal does not lie in the plane of the disk (slanted section). This case is not discussed here; the reader is referred to the literature.

2.2.2 Principal Stresses

According to (2.6) the stresses σ_ξ, σ_η and $\tau_{\xi\eta}$ depend on the direction of the section, i.e. on the angle φ. We now determine the angle for which these stresses have maximum and minimum values and we calculate these extreme values.

The normal stresses reach extreme values when $\mathrm{d}\sigma_\xi/\mathrm{d}\varphi = 0$ and when $\mathrm{d}\sigma_\eta/\mathrm{d}\varphi = 0$, respectively. Both conditions lead to

$$-(\sigma_x - \sigma_y)\sin 2\varphi + 2\,\tau_{xy}\cos 2\varphi = 0.$$

Hence, the angle $\varphi = \varphi^*$ that leads to a maximum or a minimum is given by

$$\tan 2\varphi^* = \frac{2\,\tau_{xy}}{\sigma_x - \sigma_y}. \tag{2.8}$$

The tangent function is π-periodic, that is, it satisfies $\tan 2\varphi^* = \tan 2(\varphi^* + \pi/2)$. Therefore, there exist two directions of the sections, φ^* and $\varphi^* + \pi/2$, perpendicular to each other, for which (2.8) is fulfilled. These directions of the sections are called *principal directions.*

The normal stresses which correspond to the principal directions are determined by introducing the condition (2.8) for φ^* into Equation (2.6) for σ_ξ or σ_η, respectively. Here, the following trigonometric relations are used:

$$\begin{aligned}\cos 2\varphi^* &= \frac{1}{\sqrt{1+\tan^2 2\varphi^*}} = \frac{\sigma_x - \sigma_y}{\sqrt{(\sigma_x-\sigma_y)^2 + 4\,\tau_{xy}^2}},\\ \sin 2\varphi^* &= \frac{\tan 2\varphi^*}{\sqrt{1+\tan^2 2\varphi^*}} = \frac{2\,\tau_{xy}}{\sqrt{(\sigma_x-\sigma_y)^2 + 4\,\tau_{xy}^2}}.\end{aligned} \tag{2.9}$$

Using the notations σ_1 and σ_2 for the extreme values of the stresses we obtain

$$\sigma_{1,2} = \frac{1}{2}(\sigma_x + \sigma_y) \pm \frac{\frac{1}{2}(\sigma_x - \sigma_y)^2}{\sqrt{(\sigma_x-\sigma_y)^2 + 4\,\tau_{xy}^2}} \pm \frac{2\,\tau_{xy}^2}{\sqrt{(\sigma_x-\sigma_y)^2 + 4\,\tau_{xy}^2}}$$

or

$$\sigma_{1,2} = \frac{\sigma_x + \sigma_y}{2} \pm \sqrt{\left(\frac{\sigma_x - \sigma_y}{2}\right)^2 + \tau_{xy}^2}\,, \tag{2.10}$$

respectively. The two normal stresses σ_1 and σ_2 are called *principal stresses*. Typically, they are numbered such that $\sigma_1 > \sigma_2$ (positive sign of the square root for σ_1).

Equation (2.8) provides two values for the angles φ^* and $\varphi^* + \pi/2$. These two angles can be assigned to the stresses σ_1 and σ_2, for example, by introducing one of them into the first equation of (2.6). Doing so, the associated normal stress, either σ_1 or σ_2, is obtained.

If the angles φ^* or $\varphi^* + \pi/2$, respectively, are introduced into the third equation of (2.6), we find $\tau_{\xi\eta} = 0$. Thus, the shear stresses vanish in sections where the normal stresses take on their extreme (principal) values σ_1 and σ_2. Inversely, when the shear stress in a section is zero, the normal stress in this section is a principal stress.

A coordinate system with its axes pointing in the principal directions is called *principal coordinate system*. We denote the axes by 1 and 2: the 1-axis points in the direction of σ_1 (first principal direction), the 2-axis in σ_2-direction (second principal direction). In Figs. 2.6a and b the stresses at an element in the x, y-system and in the principal coordinate system are displayed.

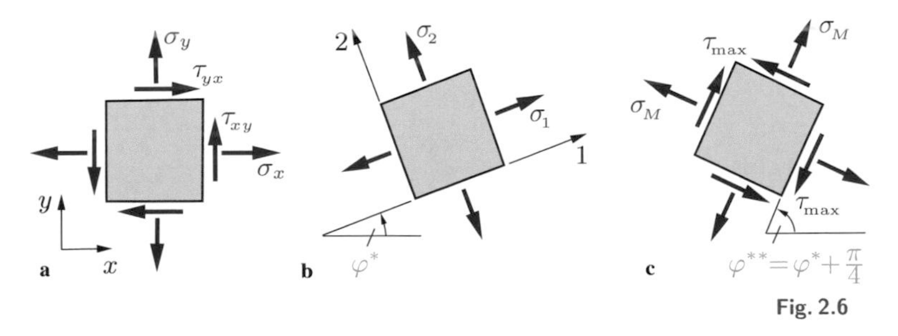

Fig. 2.6

We will now determine the extreme values of the shear stresses and the associated directions of the sections. From the condition

$$\frac{\mathrm{d}\tau_{\xi\eta}}{\mathrm{d}\varphi} = 0 \qquad \rightarrow \qquad -(\sigma_x - \sigma_y)\cos 2\varphi - 2\,\tau_{xy}\sin 2\varphi = 0$$

the angle $\varphi = \varphi^{**}$ for an extreme value is obtained:

$$\tan 2\varphi^{**} = -\frac{\sigma_x - \sigma_y}{2\,\tau_{xy}}\,. \tag{2.11}$$

Again this equation defines the two perpendicular angles φ^{**} and $\varphi^{**}+\pi/2$ where the shear stress reaches maximum or minimum values. By comparing (2.11) with (2.8) it can be seen that $\tan 2\varphi^{**} = -1/\tan 2\varphi^{*}$. Recalling of the trigonometric identity $\tan(\alpha+\pi/2) = -1/\tan\alpha$ this implies that the directions $2\varphi^{**}$ and $2\varphi^{*}$ are perpendicular to each other. As a consequence, the direction φ^{**} of the extreme shear stress is rotated by 45° with respect to the direction φ^{*} of the extreme normal stress.

The extreme shear stresses are obtained by introducing (2.11) into (2.6) and using (2.9) to give

$$\tau_{\max} = \pm\sqrt{\left(\frac{\sigma_x - \sigma_y}{2}\right)^2 + \tau_{xy}^2}\,. \tag{2.12a}$$

Since they differ only in the sign (i.e. in the sense of direction) both stresses are commonly called *maximum shear stresses*. Using the principal stresses (2.10) the maximum shear stress $\tau_{\max}$ can also be written as

$$\tau_{\max} = \pm\frac{1}{2}(\sigma_1 - \sigma_2)\,. \tag{2.12b}$$

The sense of direction of the maximum shear stress can be found by choosing the rotation angle of the ξ,η-system to be φ^{**}. Introducing φ^{**} into the third equation of (2.6) the shear stress $\tau_{\xi\eta} = \tau_{\max}$ is obtained including its correct sign.

Introducing φ^{**} into the first or second equation of (2.6) leads to a normal stress in the sections where the shear stress is maximum. We denote this stress as σ_M; it is given by

$$\sigma_M = \frac{1}{2}(\sigma_x + \sigma_y) = \frac{1}{2}(\sigma_1 + \sigma_2)\,. \tag{2.13}$$

Therefore, the normal stresses generally do *not* vanish in the sections with extreme shear stresses. Fig. 2.6c shows the stresses in the respective sections.

E2.1

Example 2.1 The homogeneous state of plane stress in a metal sheet is given by $\sigma_x = -\,64$ MPa, $\sigma_y = 32$ MPa and $\tau_{xy} = -20$ MPa. Fig. 2.7a shows the stresses and their directions as they act in the sheet.

Determine

a) the stresses in a section which is inclined at an angle of 60° to the x-axis,

b) the principal stresses and principal directions,

c) the maximum shear stress and the associated directions of the sections.

Display the stresses at an element for each case.

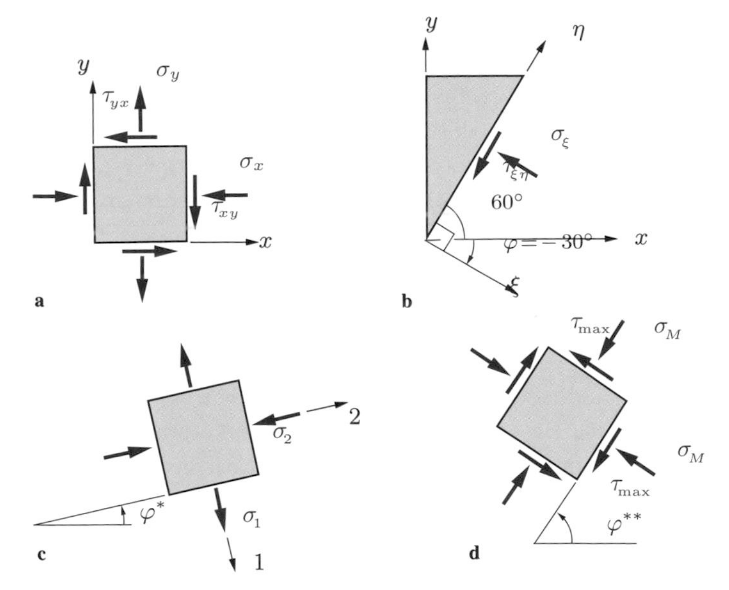

Fig. 2.7

Solution a) We cut the sheet in the prescribed direction. To characterize the section, a ξ, η-system is introduced whose ξ-axis is normal to the section (Fig. 2.7b, compare Fig. 2.5). Since it emanates from the x, y-system by a *clockwise* rotation of 30°, the rotation angle is negative: $\varphi = -30°$. Thus, from (2.6) we obtain the stresses

$$\underline{\underline{\sigma_\xi}} = \tfrac{1}{2}(-64+32) + \tfrac{1}{2}(-64-32)\cos(-60°) - 20\sin(-60°) = \underline{\underline{-22.7 \text{ MPa}}}\,,$$

$$\underline{\underline{\tau_{\xi\eta}}} = -\tfrac{1}{2}(-64-32)\sin(-60°) - 20\cos(-60°) = \underline{\underline{-51.6 \text{ MPa}}}\,.$$

Both stresses are negative. They are directed as shown in Fig. 2.7b.
b) The principal stresses are calculated by applying (2.10):

$$\sigma_{1,2} = \frac{-64+32}{2} \pm \sqrt{\left(\frac{-64-32}{2}\right)^2 + (-20)^2}$$

$$\rightarrow \quad \underline{\underline{\sigma_1 = 36 \text{ MPa}}}\,, \qquad \underline{\underline{\sigma_2 = -68 \text{ MPa}}}\,. \tag{a}$$

One of the associated principal directions follows from (2.8):

$$\tan 2\varphi^* = \frac{2(-20)}{-64-32} = 0.417 \quad \rightarrow \quad \underline{\underline{\varphi^* = 11.3°}}\,.$$

To decide which principal stress is associated with this principal direction, we introduce the angle φ^* into the first equation of (2.6) and obtain

$$\sigma_\xi(\varphi^*) = \tfrac{1}{2}\,(-64+32) + \tfrac{1}{2}\,(-64-32)\cos(22.6°) - 20\sin(22.6°) = -68 \text{ MPa} = \sigma_2\,.$$

Accordingly, the principal stress σ_2 is associated with the angle φ^*. The principal stress σ_1 acts in a section perpendicular to it (Fig. 2.7c).
c) The maximum shear stresses are determined with (a) from (2.12b):

$$\underline{\underline{\tau_{\max}}} = \pm\frac{1}{2}(36+68) = \underline{\underline{\pm 52 \text{ MPa}}}\,.$$

The associated directions of the sections are rotated by 45° with respect to the principal directions. Hence, we get

$$\underline{\underline{\varphi^{**} = 56.3^\circ\,.}}$$

The direction of $\tau_{\max}$ follows after inserting φ^{**} into (2.6) from the positive sign of $\tau_{\xi\eta}(\varphi^{**})$. The associated normal stresses are given according to (2.13) by

$$\sigma_M = \frac{1}{2}(-64 + 32) = -16\ \text{MPa}\,.$$

In Fig. 2.7d the stresses are displayed with their true directions.

2.2.3 Mohr's Circle

Using the transformation relations (2.6), the stresses σ_ξ, σ_η and $\tau_{\xi\eta}$ for a ξ,η-system can be calculated from the stresses σ_x, σ_y and τ_{xy}. These relations also allow a simple and useful geometric representation. For this purpose, in a first step, the relations (2.6) for σ_ξ and $\tau_{\xi\eta}$ are rewritten:

$$\begin{aligned} \sigma_\xi - \frac{1}{2}(\sigma_x + \sigma_y) &= \frac{1}{2}(\sigma_x - \sigma_y)\cos 2\varphi + \tau_{xy}\sin 2\varphi\,, \\ \tau_{\xi\eta} &= -\frac{1}{2}(\sigma_x - \sigma_y)\sin 2\varphi + \tau_{xy}\cos 2\varphi\,. \end{aligned} \tag{2.14}$$

By squaring and adding, the angle φ can be eliminated:

$$\left[\sigma_\xi - \frac{1}{2}(\sigma_x + \sigma_y)\right]^2 + \tau_{\xi\eta}^2 = \left(\frac{\sigma_x - \sigma_y}{2}\right)^2 + \tau_{xy}^2\,. \tag{2.15}$$

If we use in (2.14) the corresponding equation for σ_η instead of the equation for σ_ξ, we find that in (2.15) σ_ξ will be replaced by σ_η. In what follows we therefore omit the subscripts ξ and η.

For given stresses σ_x, σ_y and τ_{xy} the right-hand side of (2.15) is a fixed value which we abbreviate with r^2:

$$r^2 = \left(\frac{\sigma_x - \sigma_y}{2}\right)^2 + \tau_{xy}^2\,. \tag{2.16}$$

With $\sigma_M = \frac{1}{2}(\sigma_x + \sigma_y)$ and (2.16) Equation (2.15) then takes the form

$$(\sigma - \sigma_M)^2 + \tau^2 = r^2 \,. \tag{2.17}$$

This is the equation of a circle in the σ, τ-plane: the points (σ, τ) lie on the *stress circle* , also called *Mohr's circle* (Otto Mohr, 1835–1918). It is centered at $(\sigma_M, 0)$ and has the radius r (Fig. 2.8a).

Equation (2.16) can be rewritten as

$$r^2 = \frac{1}{4}\left[(\sigma_x + \sigma_y)^2 - 4(\sigma_x\sigma_y - \tau_{xy}^2)\right] \,.$$

Note that this equation for the radius coincides with the absolute value of the maximum shear stress given by (2.12a). That is, the radius of Mohr's circle graphically indicates the maximum shear stress at a point. Moreover, since the expressions in the round brackets are invariant (cf. Section 2.2.1), r is also an invariant.

The stress circle in the σ, τ-plane can be constructed directly if the stresses σ_x, σ_y and τ_{xy} are known, thereby avoiding the need to calculate σ_M and r. For this purpose, the stresses σ_x and σ_y, including their signs, are marked on the σ-axis. At these points the shear stress τ_{xy} is plotted according to the following rule: with the correct sign at σ_x and with the reversed sign at σ_y. This determines two points of the circle, P and P' (Fig. 2.8a). The intersection of their connecting line with the σ-axis yields the center of the circle. The circle now can be drawn using this point as its center and extending to pass through P and P'.

The stress state at a point of a disk is fully described by Mohr's circle; each section is represented by a point on the circle. For example, point P corresponds to the section where the stresses σ_x and τ_{xy} act while point P' represents the section perpendicular to the former one. The stresses in arbitrary sections as well as the extreme stresses and associated directions can be determined from the stress circle. In particular, the principal stresses σ_1, σ_2 and the maximum shear stress $\tau_{\max}$ can be directly identified (Fig. 2.8b).

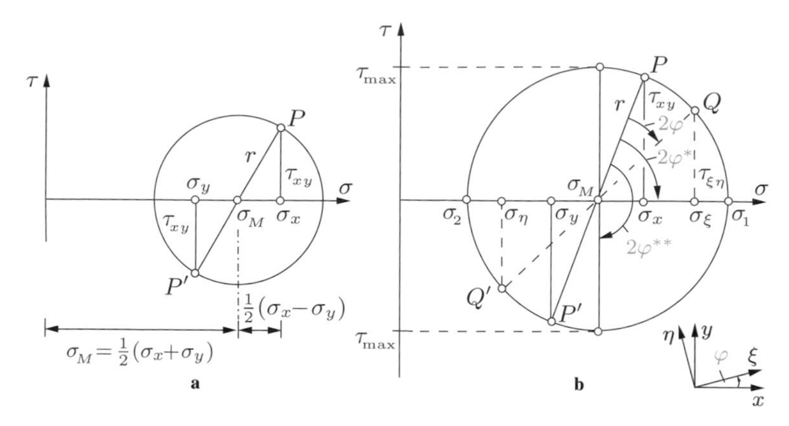

Fig. 2.8

We will now show that the stresses σ_ξ, σ_η and $\tau_{\xi\eta}$ in a ξ,η-system which is rotated with respect to the x,y-system by an angle φ (positive *counterclockwise*) are identified on Mohr's circle as follows: point Q, corresponding to a section with the stresses σ_ξ and $\tau_{\xi\eta}$ is found by plotting the *doubled* angle – i.e. 2φ – in the *reversed* sense of rotation (Fig. 2.8b); point Q' corresponding to a section perpendicular to the first one lies opposite to Q. The principal directions and the directions of maximum shear stress finally are given by the angles φ^* and φ^{**}.

To proof these statements we first find from Fig. 2.8a,b:

$$\tan 2\,\varphi^* = \frac{2\,\tau_{xy}}{\sigma_x - \sigma_y}\,,$$

$$\frac{1}{2}\,(\sigma_x - \sigma_y) = r\,\cos 2\varphi^*\,, \quad \tau_{xy} = r\,\sin 2\varphi^*\,.$$

Introducing these equations into the transformation relations (2.6) for σ_ξ and σ_η yields

$$\begin{aligned}\sigma_\xi &= \frac{1}{2}(\sigma_x + \sigma_y) + r\cos 2\varphi^* \cos 2\varphi + r\sin 2\varphi^* \sin 2\varphi \\ &= \frac{1}{2}(\sigma_x + \sigma_y) + r\cos\left(2\varphi^* - 2\varphi\right),\end{aligned}$$

$$\tau_{\xi\eta} = -\,r\cos 2\varphi^* \sin 2\varphi + r\sin 2\varphi^* \cos 2\varphi = r\sin\left(2\varphi^* - 2\varphi\right).$$

The same result follows from geometric relations in Fig. 2.8b, i.e. Mohr's circle is nothing other than the geometric representation of the transformation relations.

If Mohr's circle is used for the solution of specific problems, three quantities must be known (e.g. σ_x, τ_{xy}, σ_1) in order to draw the circle. In graphical solutions an appropriate scale for the stresses must be chosen.

In the following we finally consider three special cases. *Uniaxial tension* (Fig. 2.9a) is characterized by $\sigma_x = \sigma_0 > 0$, $\sigma_y = 0$, $\tau_{xy} = 0$. Since the shear stress is zero in the respective sections, the stresses $\sigma_1 = \sigma_x = \sigma_0$ and $\sigma_2 = \sigma_y = 0$ are the principal stresses. Mohr's circle lies just to the right of the τ-axis so that this vertical axis is its tangent. The maximum shear stress $\tau_{\max} = \sigma_0/2$ acts in sections rotated 45° with respect to the x-axis (see also Section 1.1).

The stress state characterized by $\sigma_x = 0$, $\sigma_y = 0$ and $\tau_{xy} = \tau_0$ is called *pure shear*. On account of $\sigma_M = 0$, the center of Mohr's

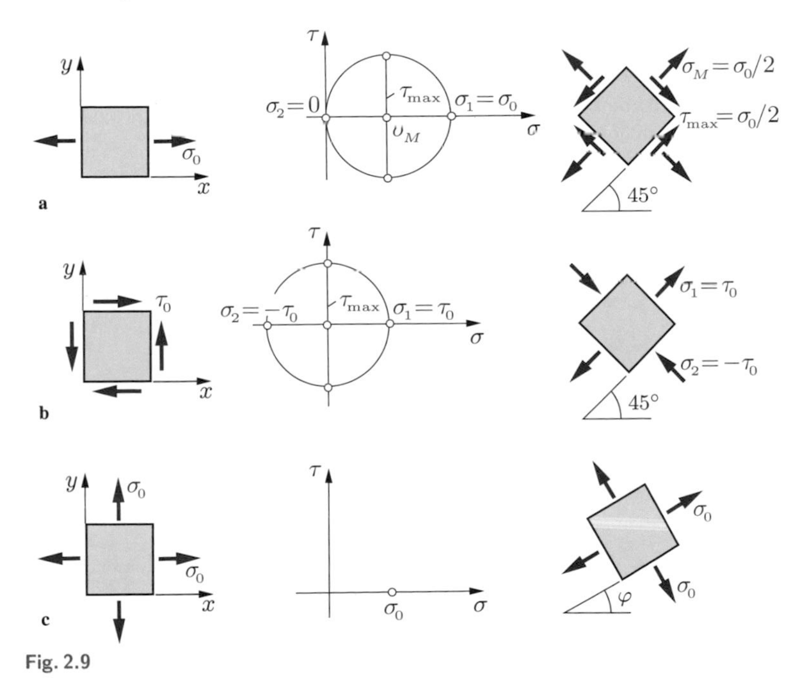

Fig. 2.9

circle in this case coincides with the origin of the coordinate system (Fig. 2.9b). The principal stresses are $\sigma_1 = \tau_0$ and $\sigma_2 = -\tau_0$; they act in sections at 45° with respect to the x-axis.

In the case of a *hydrostatic stress state* the stresses are $\sigma_x = \sigma_y = \sigma_0$ and $\tau_{xy} = 0$. Mohr's circle then is reduced to a single point on the σ-axis (Fig. 2.9c). The normal stresses for all section directions have the same value $\sigma_\xi = \sigma_\eta = \sigma_0$ and no shear stresses appear (cf. Section 2.2.1).

E2.2

Example 2.2 A plane stress state is given by $\sigma_x = 50$ MPa, $\sigma_y = -20$ MPa and $\tau_{xy} = 30$ MPa.

Using Mohr's circle, determine

a) the principal stresses and principal directions,

b) the normal and shear stress acting in a section whose normal forms the angle $\varphi = 30°$ with the x-axis.

Display the results in sketches of the sections.

Solution a) After having chosen a scale, Mohr's circle can be constructed from the given stresses (in Fig. 2.10a the given stresses are marked by green circles). From the circle, the principal stresses and directions can be directly identified:

$$\underline{\underline{\sigma_1 = 61 \text{ MPa}}}, \qquad \underline{\underline{\sigma_2 = -31 \text{ MPa}}}, \qquad \underline{\underline{\varphi^* = 20°}}.$$

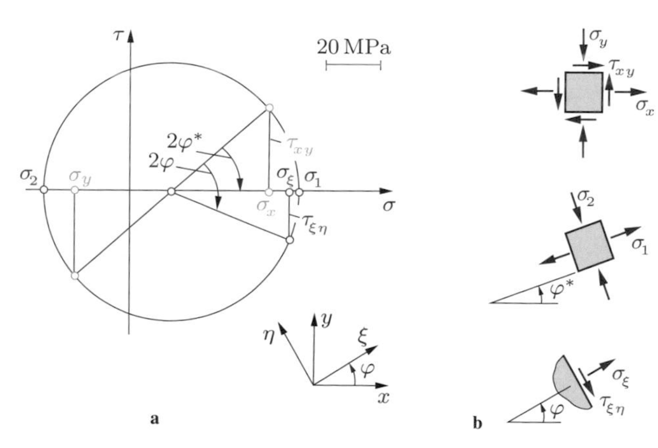

Fig. 2.10

b) To determine the stresses in the inclined section we introduce a ξ, η-coordinate system whose ξ-axis coincides with the normal of the section. The unknown stresses σ_ξ and $\tau_{\xi\eta}$ are obtained by plotting in Mohr's circle the angle 2φ in the reversed direction to φ. Doing so we obtain:

$$\underline{\underline{\sigma_\xi = 58.5 \text{ MPa}}}, \qquad \underline{\underline{\tau_{\xi\eta} = -15.5 \text{ MPa}}}.$$

The stresses with their true directions and the associated sections are displayed in Fig. 2.10b.

E2.3

Example 2.3 The two principal stresses $\sigma_1 = 40$ MPa and $\sigma_2 = -20$ MPa of a plane stress state are known.

Determine the orientation of a x, y-coordinate system with respect to the principal axes for which $\sigma_x = 0$ and $\tau_{xy} > 0$. Calculate the stresses σ_y and τ_{xy}.

Solution Using the given principal stresses σ_1 and σ_2, the properly scaled Mohr's circle can be drawn (Fig. 2.11a). From the circle the orientation of the unknown x, y-system can be obtained: the counterclockwise angle 2φ (from point σ_1 to point P) in Mohr's circle corresponds to the clockwise angle φ between the 1-axis and the x-axis. The angle and the stresses are found as

$$2\varphi = 110° \quad \rightarrow \quad \underline{\underline{\varphi = 55°}}, \quad \underline{\underline{\sigma_y = 20 \text{ MPa}}}, \quad \underline{\underline{\tau_{xy} = 28 \text{ MPa}}}.$$

The stresses and the coordinate systems are shown in Fig. 2.11b.

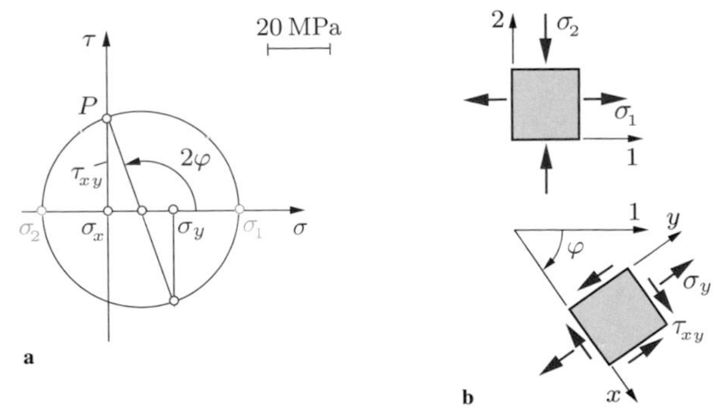

Fig. 2.11

2.2.4 The Thin-Walled Pressure Vessel

As an important application of plane stress we first consider a *thin-walled* cylindrical vessel with radius r and wall thickness $t \ll r$ (Fig. 2.12a). The vessel is subjected to an internal gage pressure p that causes stresses in its wall which need to be determined (Fig. 2.12b).

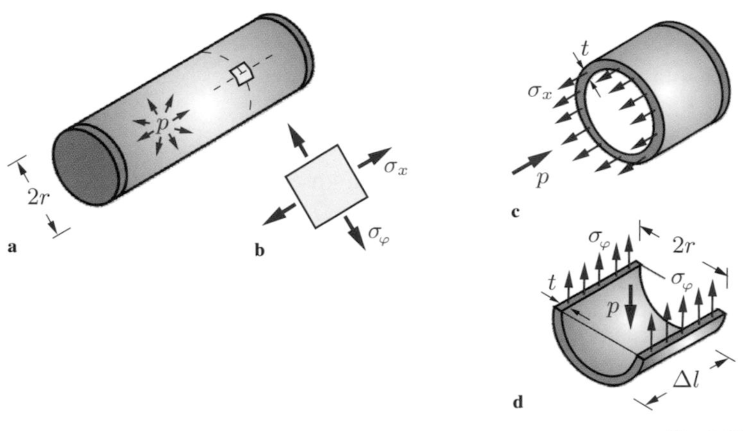

Fig. 2.12

At a sufficient distance from the end caps of the vessel, the stress state is independent of the location (homogeneous stress state). Given that $t \ll r$, the stresses in radial directions can be neglected. Thus, within a good approximation a plane stress state acts locally in the wall of the vessel (note: although the element in Fig. 2.12b is curved, it is replaced by a plane element in the tangent plane). The stress state can be described by the stresses in two sections perpendicular to each other.

First, the vessel is cut perpendicularly to its longitudinal axis (Fig. 2.12c). Since the gas or fluid pressure is independent of the location, the pressure on the section area πr^2 (of the gas or fluid) has the constant value p. Assuming that the *longitudinal stress* σ_x is constant across the wall thickness because of $t \ll r$, the equilibrium condition yields (Fig. 2.12c)

$$\sigma_x 2\,\pi\, r\, t - p\,\pi\, r^2 = 0 \qquad \rightarrow \qquad \sigma_x = \frac{1}{2}\, p\, \frac{r}{t}\,. \tag{2.18}$$

As illustrated in Fig. 2.12d we now separate a half-circular part of length Δl from the vessel. The horizontal sections of the wall are subjected to the *circumferential stress* σ_φ, also called *hoop stress*, which again is constant across the thickness. These stresses will counteract the force $p\, 2\, r\Delta l$, exerted from the gas onto the half-circular part of the vessel. Equilibrium in the vertical direction yields

$$2\,\sigma_\varphi\, t\,\Delta\, l - p\, 2\, r\Delta l = 0 \qquad \rightarrow \qquad \sigma_\varphi = p\, \frac{r}{t}\,. \tag{2.19}$$

We notice that the hoop stress is twice the longitudinal stress. This is why a cylindrical vessel under internal pressure usually fails by cracking in the longitudinal direction. A simple example is an overcooked hot dog which splits in the longitudinal direction first.

The two equations (2.18) and (2.19) for σ_x and σ_φ sometimes are called *vessel formulas*. Because of $t \ll r$ it can be seen that σ_x, $\sigma_\varphi \gg p$. Therefore, the initially made assumption that the stresses σ_r in radial direction may be neglected is justified ($|\, \sigma_r \,| \leq p$). Generally, a vessel may be called *thin-walled* when it fulfills the condition $r > 5\, t$.

The vessel formulas are also applicable to a vessel subjected to external pressure. In this case only the sign of p has to be changed, i.e. the wall is then under a compressive stress state.

Since no shear stresses are present in both sections (symmetry), the stresses σ_x and σ_φ are principal stresses: $\sigma_1 = \sigma_\varphi = p\; r/t$, $\sigma_2 = \sigma_x = p\; r/(2t)$. According to (2.12b) the maximum shear stress is given by

$$\tau_{\max} = \frac{1}{2}(\sigma_1 - \sigma_2) = \frac{1}{4}\, p\, \frac{r}{t}\,;$$

it acts in sections inclined under 45°. It should be noted that in the vicinity of the end caps more complex stress states are present which cannot be determined with an elementary theory.

Now we consider a thin-walled spherical vessel of radius r, subjected to a gage pressure p (Fig. 2.13a). Here, the stresses σ_t and σ_φ act in the wall (Fig. 2.13b). When we cut the vessel into half (Fig. 2.13c), we obtain σ_t from the equilibrium condition:

$$\sigma_t\, 2\,\pi\, r\, t - p\,\pi\, r^2 = 0 \quad \rightarrow \quad \sigma_t = \frac{1}{2}\, p\, \frac{r}{t}\,.$$

A cut, perpendicular to the first one, similarly leads to

$$\sigma_\varphi\, 2\,\pi\, r\, t - p\,\pi\, r^2 = 0 \quad \rightarrow \quad \sigma_\varphi = \frac{1}{2}\, p\, \frac{r}{t}\,.$$

Thus,

$$\sigma_t = \sigma_\varphi = \frac{1}{2}\, p\, \frac{r}{t}\,. \tag{2.20}$$

Therefore, the stress in the wall of a thin-walled spherical vessel has the value $p\,r/(2\,t)$ in any arbitrary direction. As in the foregoing case, this formula is also valid for an external pressure in which case p is negative.

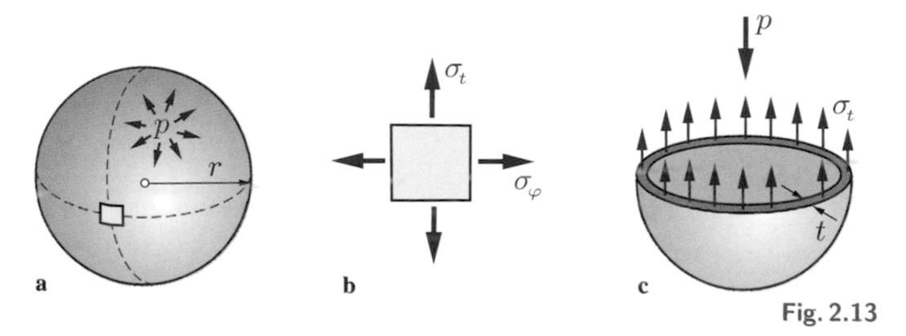

Fig. 2.13

2.3 Equilibrium Conditions

According to Section 2.1 the stress state at a material point of a body is determined by the stress tensor; its components are shown in Fig. 2.2a. In general, these components vary from point to point and these variations are not independent of each other: they are connected via the *equilibrium conditions*.

To derive the equilibrium conditions we first consider in Fig. 2.14 the stresses acting on an infinitesimal element under plane stress which is cut out from a disk of thickness t. Since the stresses in general depend on x and y, they are not the same at the opposite sections: they differ by infinitesimal increments. For example, the left face is subjected to the normal stress σ_x whereas the stress $\sigma_x + \dfrac{\partial \sigma_x}{\partial x}\,\mathrm{d}x$ (first terms of the Taylor-expansion, see e.g. Section 3.1) acts on the right face. The symbol $\partial/\partial x$ denotes the partial derivative with respect to x. Furthermore, the element may be loaded by the volume force $\boldsymbol{f}$ with the components f_x and f_y.

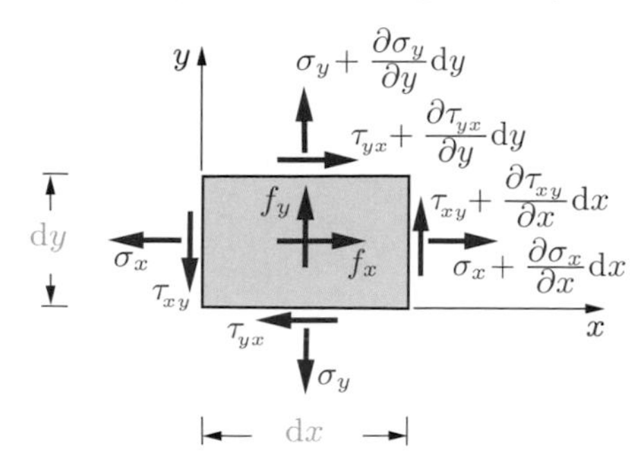

Fig. 2.14

The equilibrium condition in x-direction yields

$$-\sigma_x\,\mathrm{d}y\,t - \tau_{yx}\,\mathrm{d}x\,t + \left(\sigma_x + \frac{\partial \sigma_x}{\partial x}\,\mathrm{d}x\right)\mathrm{d}y\,t$$

$$+\left(\tau_{yx} + \frac{\partial \tau_{yx}}{\partial y}\,\mathrm{d}y\right)\mathrm{d}x\,t + f_x\,\mathrm{d}x\;\mathrm{d}y\,t = 0$$

i.e., after division by $\mathrm{d}x\mathrm{d}y$

$$\frac{\partial \sigma_x}{\partial x} + \frac{\partial \tau_{yx}}{\partial y} + f_x = 0\,. \tag{2.21a}$$

Similarly, from the equilibrium condition in y-direction we obtain

$$\frac{\partial \tau_{xy}}{\partial x} + \frac{\partial \sigma_y}{\partial y} + f_y = 0\,. \tag{2.21b}$$

Equations (2.21a, b) are called *equilibrium conditions.* In the considered case of plane stress, they consist of *two* coupled partial differential equations for the *three* components σ_x, σ_y and $\tau_{xy} = \tau_{yx}$ of the stress tensor. The stress state cannot be uniquely determined from these equations: the problem is statically indeterminate.

For a *spatial* (three dimensional) stress state the corresponding equilibrium conditions are obtained as

$$\begin{aligned} &\frac{\partial \sigma_x}{\partial x} + \frac{\partial \tau_{yx}}{\partial y} + \frac{\partial \tau_{zx}}{\partial z} + f_x = 0\,, \\ &\frac{\partial \tau_{xy}}{\partial x} + \frac{\partial \sigma_y}{\partial y} + \frac{\partial \tau_{zy}}{\partial z} + f_y = 0\,, \\ &\frac{\partial \tau_{xz}}{\partial x} + \frac{\partial \tau_{yz}}{\partial y} + \frac{\partial \sigma_z}{\partial z} + f_z = 0\,. \end{aligned} \tag{2.22}$$

These are three coupled partial differential equations for the six components of the stress tensor.

The components of the stress tensor are constant in a *homogeneous* stress state . In this case all partial derivatives in (2.21a, b) and (2.22), respectively, vanish. The equilibrium conditions are then only fulfilled if $f_x = f_y = f_z = 0$. Thus, a homogeneous stress state under the action of volume forces is not possible.

It should be mentioned that from the equilibrium of moments, applied to the element, the symmetry of the stress tensor follows even when the stress increments are taken into account (cf. Section 2.1).

2.4 Supplementary Examples

Detailed solutions to the following examples are given in (**A**) D. Gross et al. *Formeln und Aufgaben zur Technischen Mechanik 2*, Springer, Berlin 2010 or (**B**) W. Hauger et al. *Aufgaben zur Technischen Mechanik 1-3*, Springer, Berlin 2008.

E2.4

Example 2.4 The stresses $\sigma_x = 20$ MPa, $\sigma_y = 30$ MPa and $\tau_{xy} = 10$ MPa in a metal sheet are known (Fig. 2.15).

Determine the principal stresses and their directions.

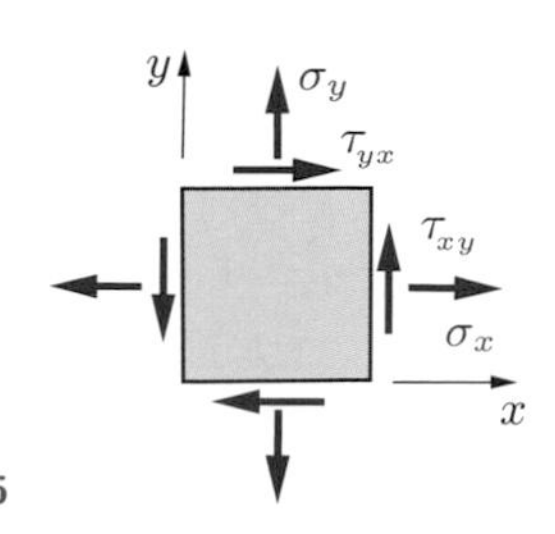

Fig. 2.15

Results: see (**A**) $\sigma_1 = 36.2$ MPa, $\sigma_2 = 13.8$ MPa,

$\varphi_1^* = 58.3°$, $\varphi_2^* = 148.3°$.

E2.5

Example 2.5 A plane stress state is given by the principal stresses $\sigma_1 = 30$ MPa and $\sigma_2 = -10$ MPa (Fig. 2.16).

a) Determine the stress components in a ξ, η-coordinate system which is inclined by 45° with respect to the principal axes.

b) Using Mohr's circle, determine the rotation angle α of an x, y-coordinate system where $\sigma_y = 0$ and $\tau_{xy} < 0$. Calculate σ_x and τ_{xy}.

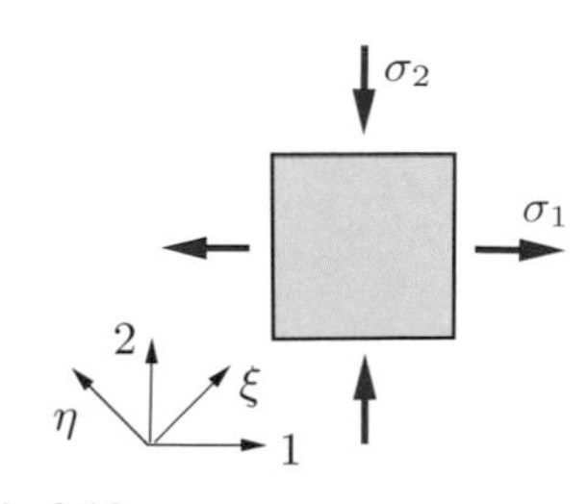

Fig. 2.16

Results: see (**B**)

a) $\sigma_\xi = 10$ MPa, $\sigma_\eta = 10$ MPa, $\tau_{\xi\eta} = -20$ MPa.

b) $\alpha = 30°$, $\sigma_x = 20$ MPa, $\tau_{xy} = -17.3$ MPa.

E2.6

Example 2.6 A thin-walled tube is subjected to bending and torsion such that the following stresses act at points A and B:

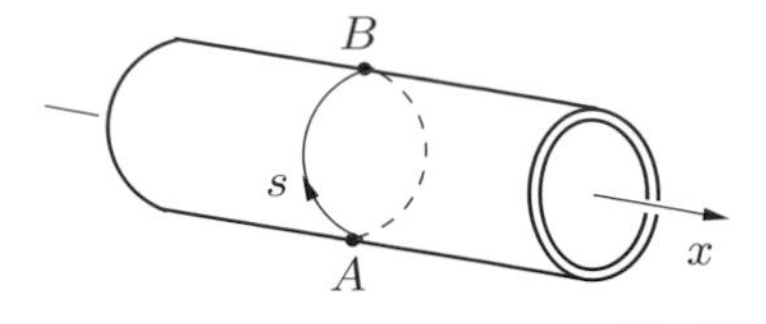

Fig. 2.17

$$\sigma_x^{A,B} = \pm 25\,\text{MPa}, \quad \sigma_s^{A,B} = 50\,\text{MPa}, \quad \tau_{xs}^{A,B} = 50\,\text{MPa}\,.$$

Determine the principal stresses and their directions at A and B.

Results: see (**A**) Point A

$\sigma_1 = 89.0$ MPa, $\sigma_2 = -14.0$ MPa, $\varphi_1^* = 52.0°$, $\varphi_2^* = -38.0°$.

Point B

$\sigma_1 = 75.0$ MPa, $\sigma_2 = -50.0$ MPa, $\varphi_1^* = 63.4°$, $\varphi_2^* = -26.6°$.

E2.7

Example 2.7 A thin-walled bathysphere (radius $r = 500$ mm, wall-thickness $t = 12.5$ mm) is lowered to a depth of 500 m under the water surface (pressure $p = 5$ MPa).

Determine the stresses in the wall.

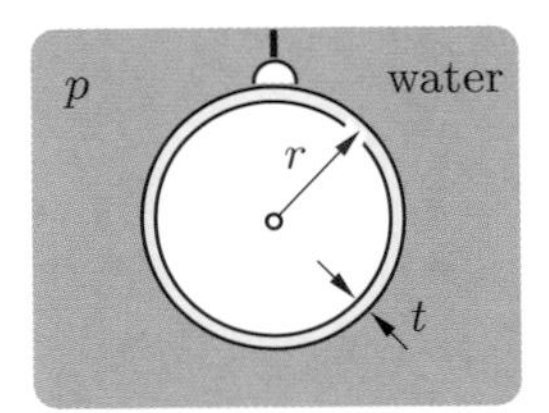

Fig. 2.18

Result: see (**A**) $\sigma_t = -100$ MPa (in any section).

E2.8

Example 2.8 A thin-walled cylindrical vessel has the radius $r = 1$ m and wall-thickness $t = 10$ mm.

Determine the maximum internal pressure $p_{\max}$ so that the maximum stress in the wall does not exceed the allowable stress $\sigma_{\text{allow}} = 150$ MPa.

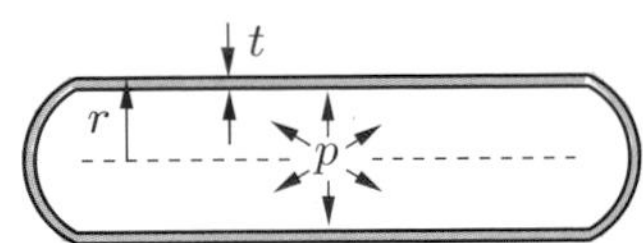

Fig. 2.19

Result: see (**A**) $p_{\max} = 1.5$ MPa.

2.5 Summary

- The stress state at a point of a body is determined by the stress tensor $\boldsymbol{\sigma}$. In the spatial case it has 3×3 components (note the symmetry). In the plane stress state it reduces to

$$\boldsymbol{\sigma} = \begin{bmatrix} \sigma_x & \tau_{xy} \\ \tau_{yx} & \sigma_y \end{bmatrix} \quad \text{where} \quad \tau_{xy} = \tau_{yx} \, .$$

- Sign convention: positive stresses at a positive (negative) face point in positive (negative) directions of the coordinates.
- Transformation relations (plane stress):

$$\begin{aligned} \sigma_\xi &= \tfrac{1}{2}(\sigma_x + \sigma_y) + \tfrac{1}{2}(\sigma_x - \sigma_y)\cos 2\varphi + \tau_{xy}\sin 2\varphi \, , \\ \sigma_\eta &= \tfrac{1}{2}(\sigma_x + \sigma_y) - \tfrac{1}{2}(\sigma_x - \sigma_y)\cos 2\varphi - \tau_{xy}\sin 2\varphi \, , \\ \tau_{\xi\eta} &= -\tfrac{1}{2}(\sigma_x - \sigma_y)\sin 2\varphi + \tau_{xy}\cos 2\varphi \, . \end{aligned}$$

  The axes ξ, η are rotated with respect to x, y by the angle φ.
- Principal stresses and directions (plane stress):

$$\sigma_{1,2} = \tfrac{1}{2}(\sigma_x + \sigma_y) \pm \sqrt{\tfrac{1}{4}(\sigma_x - \sigma_y)^2 + \tau_{xy}^2} \, ,$$

$$\tan 2\varphi^* = \frac{2\tau_{xy}}{\sigma_x - \sigma_y} \quad \rightarrow \quad \varphi_1^*, \ \varphi_2^* = \varphi_1^* \pm \pi/2 \, .$$

  Principal stresses are extreme stresses; the shear stresses vanish in the corresponding sections.
- Maximum shear stresses and their directions (plane stress):

$$\tau_{\max} = \sqrt{\tfrac{1}{4}(\sigma_x - \sigma_y)^2 + \tau_{xy}^2} \, , \qquad \varphi^{**} = \varphi^* \pm \pi/4 \, .$$

- Mohr's circle allows the geometric representation of the coordinate transformation.
- Equilibrium conditions for the stresses (plane stress):

$$\frac{\partial \sigma_x}{\partial x} + \frac{\partial \tau_{yx}}{\partial y} + f_x = 0 \, , \qquad \frac{\partial \tau_{xy}}{\partial x} + \frac{\partial \sigma_y}{\partial y} + f_y = 0 \, .$$

  In the spatial case there are three equilibrium conditions.

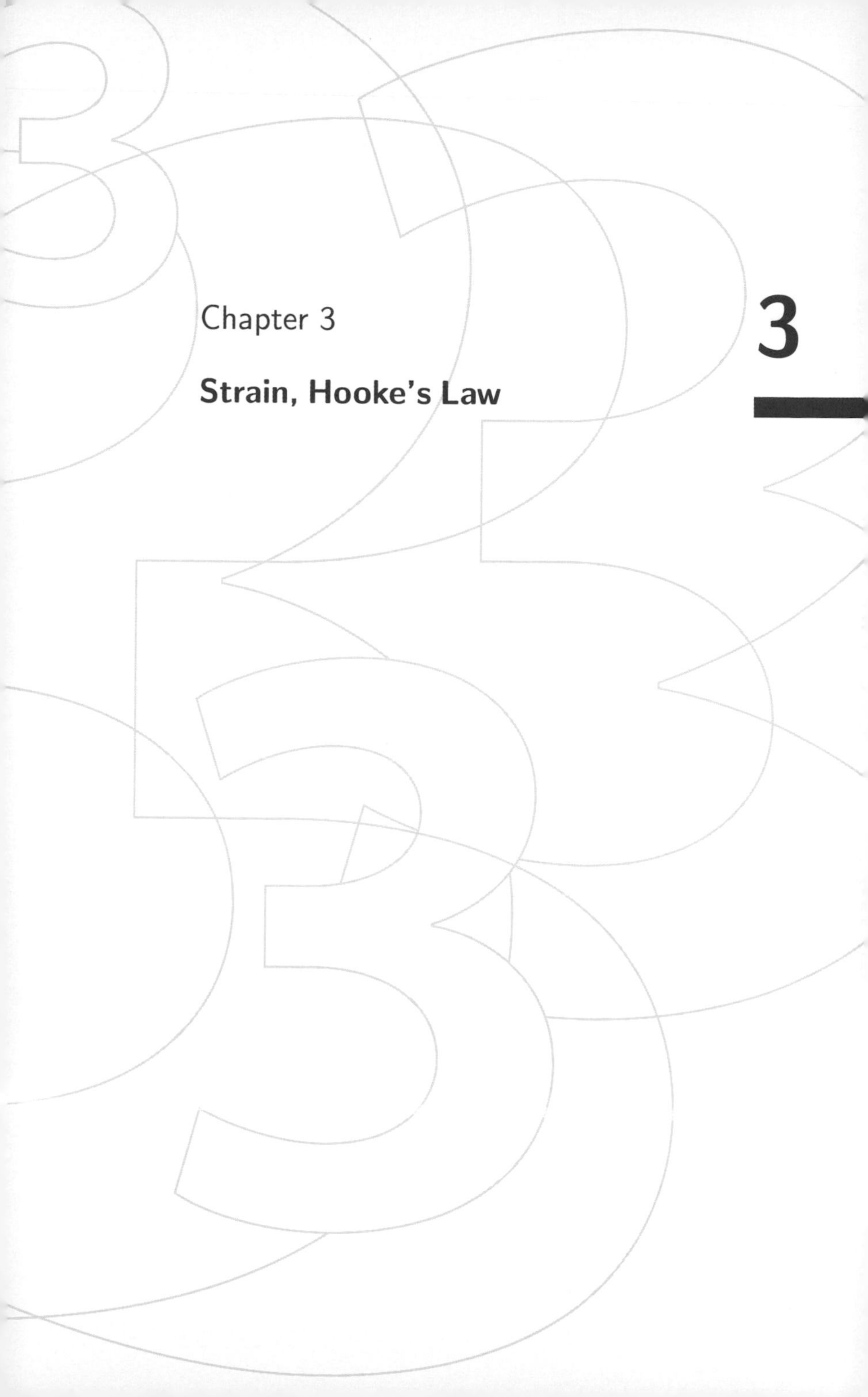

Chapter 3

Strain, Hooke's Law

3

3 Strain, Hooke's Law

3.1 State of Strain 79
3.2 Hooke's Law 84
3.3 Strength Hypotheses 90
3.4 Supplementary Examples 92
3.5 Summary 95

Objectives: In Chapter 1 the deformation of a bar has been characterized by the strain and the displacement. We will now generalize these kinematic quantities to the plane and the spatial cases. For this purpose, we introduce the displacement vector and the *strain tensor*, the latter describing length and angle changes. In addition, we will extend the already known Hooke's law from the uniaxial case to the two and three-dimensional cases. Finally, we will discuss the so-called strength hypotheses in order to assess the exertion of the material under multiaxial stress. The students shall learn how to calculate the stresses from the strains or displacements and vice versa.

3.1 State of Strain

To characterize the uniaxial deformation of a tension bar, the kinematic quantities displacement u and strain $\varepsilon = \mathrm{d}u/\mathrm{d}x$ have been introduced (Section 1.2). We will now explore how the deformation of a plane or spatial structure can be described. Here, we first restrict our attention to the deformation in a plane. To this end we consider a disk where two squares ① and ② are marked such that they are tilted against each other (Fig. 3.1). When the disk is loaded, e.g. by a normal stress σ, point P experiences a displacement $\boldsymbol{u}$ from its initial position to the new position P'. Since the *displacement vector* $\boldsymbol{u}$ depends on the location, the side lengths (square ①) and the side lengths and angles (square ②), respectively, are changed during the deformation.

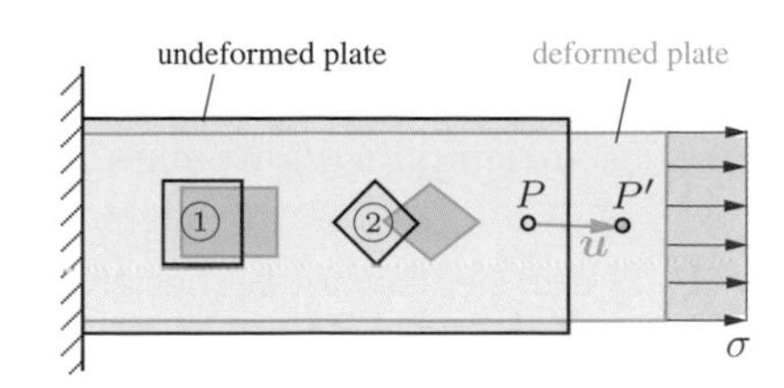

Fig. 3.1

In the discussions that follow we consider the changes of side lengths and angles under the assumption of *small* deformations. Figure 3.2 shows an infinitesimal rectangle $PQRS$ of side lengths $\mathrm{d}x$ and $\mathrm{d}y$ in the undeformed state. During the deformation it is transformed into the new position $P'Q'R'S'$. The displacement vector $\boldsymbol{u}(x,\, y)$ of point $P(x,\, y)$ has the components $u(x,\, y)$ and $v(x,\, y)$ in x- and in y-direction, respectively. The displacement of a point, adjacent to P, can be described with the help of Taylor-expansions. For the functions u and v, which depend on the two variables x and y, we obtain

$$u(x + \mathrm{d}x,\, y + \mathrm{d}y) = u(x,\, y) + \frac{\partial u(x,\, y)}{\partial x}\mathrm{d}x + \frac{\partial u(x,\, y)}{\partial y}\mathrm{d}y + \ldots,$$

$$v(x + \mathrm{d}x,\, y + \mathrm{d}y) = v(x,\, y) + \frac{\partial v(x,\, y)}{\partial x}\mathrm{d}x + \frac{\partial v(x,\, y)}{\partial y}\mathrm{d}y + \ldots.$$

Here $\partial/\partial x$ and $\partial/\partial y$ denote partial derivatives with respect to x and y, respectively.

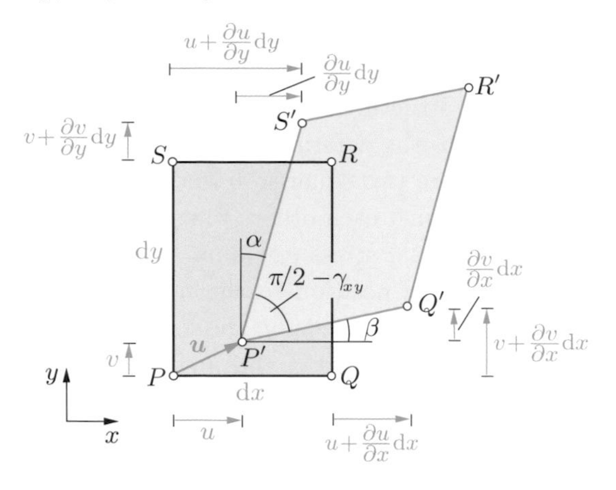

Fig. 3.2

The series are simplified for points Q and S. Taking into account that the y-coordinate does not change ($\mathrm{d}y = 0$) when advancing from P to Q and neglecting higher order terms, the displacements of Q in x- and in y-direction are given by $u + \partial u/\partial x\,\mathrm{d}x$ and $v + \partial v/\partial x\,\mathrm{d}x$, respectively (Fig. 3.2). Similarly, because of $\mathrm{d}x = 0$, the displacement components for point S are $u + \partial u/\partial y\;\mathrm{d}y$ and $v + \partial v/\partial y\,\mathrm{d}y$.

During the deformation, the line $\overline{PQ}$ is transformed into line $\overline{P'Q'}$. Since we assume small deformations ($\beta \ll 1$), the lengths of $\overline{P'Q'}$ and of its projection onto the x-axis are approximately the same (Fig. 3.2):

$$\overline{P'Q'} \approx \mathrm{d}x + \left(u + \frac{\partial u}{\partial x}\,\mathrm{d}x\right) - u = \mathrm{d}x + \frac{\partial u}{\partial x}\,\mathrm{d}x.$$

If we introduce, analogously to Section 1.2, the *normal strain* ε_x in the x-direction as the ratio of length increment to initial length, we obtain

$$\varepsilon_x = \frac{\overline{P'Q'} - \overline{PQ}}{\overline{PQ}} = \frac{\left(\mathrm{d}x + \dfrac{\partial u}{\partial x}\mathrm{d}x\right) - \mathrm{d}x}{\mathrm{d}x} = \frac{\partial u}{\partial x}\,.$$

Similarly, the line $\overline{PS}$ is transformed into line

$$\overline{P'S'} \approx \mathrm{d}y + \left(v + \frac{\partial v}{\partial y}\mathrm{d}y\right) - v = \mathrm{d}y + \frac{\partial v}{\partial y}\mathrm{d}y\,.$$

Thus, the strain ε_y in y-direction is given by

$$\varepsilon_y = \frac{\overline{P'S'} - \overline{PS}}{\overline{PS}} = \frac{\left(\mathrm{d}y + \frac{\partial v}{\partial y}\mathrm{d}y\right) - \mathrm{d}y}{\mathrm{d}y} = \frac{\partial v}{\partial y}\,.$$

Therefore, the two normal strains

$$\varepsilon_x = \frac{\partial u}{\partial x}\,, \qquad \varepsilon_y = \frac{\partial v}{\partial y} \tag{3.1}$$

exist in a plane region.

According to Fig. 3.2, the change of the initially right angle during the deformation is described by α and β. The following geometric relations hold:

$$\tan\alpha = \frac{\frac{\partial u}{\partial y}\mathrm{d}y}{\mathrm{d}y + \frac{\partial v}{\partial y}\mathrm{d}y}\,, \qquad \tan\beta = \frac{\frac{\partial v}{\partial x}\mathrm{d}x}{\mathrm{d}x + \frac{\partial u}{\partial x}\mathrm{d}x}\,.$$

They can be simplified by taking into account the assumption of small deformations, i.e. $\alpha, \beta \ll 1$, $\varepsilon_x, \varepsilon_y \ll 1$, which leads to

$$\alpha = \frac{\partial u}{\partial y}\,, \qquad \beta = \frac{\partial v}{\partial x}\,.$$

Denoting the total change of the angle as γ_{xy}, we obtain

$$\gamma_{xy} = \alpha + \beta \qquad \rightarrow \qquad \gamma_{xy} = \frac{\partial u}{\partial y} + \frac{\partial v}{\partial x}\,. \tag{3.2}$$

The quantity γ is called *shear strain*. The subscripts x and y indicate that γ_{xy} describes the angle increment in the x, y-plane. If x and y as well as u and v are interchanged, we find $\gamma_{yx} = \gamma_{xy}$.

The notion of *strains* comprises the normal strains as well as the shear strains. They are connected with the displacements through the *kinematic relations* (3.1) and (3.2). When the displacements are known, the strains can be determined by differentiation from (3.1) and (3.2).

The values of ε_x, ε_y and γ_{xy} determine the *plane strain state* at point P. It can be shown that the normal strains ε_x, ε_y and *half* the angle change $\varepsilon_{xy} = \gamma_{xy}/2$ are components of a symmetric tensor $\boldsymbol{\varepsilon}$. This tensor is called *strain tensor*. In matrix form it can be written as

$$\boldsymbol{\varepsilon} = \begin{bmatrix} \varepsilon_x & \varepsilon_{xy} \\ \varepsilon_{yx} & \varepsilon_y \end{bmatrix} = \begin{bmatrix} \varepsilon_x & \frac{1}{2}\gamma_{xy} \\ \frac{1}{2}\gamma_{xy} & \varepsilon_y \end{bmatrix} .$$

The main diagonal contains the normal strains whereas the entries of the secondary diagonal are half the shear strains.

The properties of the stress tensor in the case of plane stress as described in Section 2.2 may analogously be transferred to the strain tensor in the case of plane strain. The components ε_ξ, ε_η and $\varepsilon_{\xi\eta} = \gamma_{\xi\eta} / 2$ in a ξ, η-coordinate system, rotated by the angle φ (positive counterclockwise) against the x, y-system, are obtained from the components ε_x, ε_y and $\gamma_{xy} / 2$ by using the transformation relations (2.6). The stresses must simply be replaced by the strains:

$$\begin{aligned} \varepsilon_\xi &= \tfrac{1}{2}(\varepsilon_x + \varepsilon_y) + \tfrac{1}{2}(\varepsilon_x - \varepsilon_y)\cos 2\varphi + \tfrac{1}{2}\gamma_{xy}\sin 2\varphi \,, \\ \varepsilon_\eta &= \tfrac{1}{2}(\varepsilon_x + \varepsilon_y) - \tfrac{1}{2}(\varepsilon_x - \varepsilon_y)\cos 2\varphi - \tfrac{1}{2}\gamma_{xy}\sin 2\varphi \,, \\ \tfrac{1}{2}\gamma_{\xi\eta} &= \qquad\qquad\quad - \tfrac{1}{2}(\varepsilon_x - \varepsilon_y)\sin 2\varphi + \tfrac{1}{2}\gamma_{xy}\cos 2\varphi \,. \end{aligned} \tag{3.3}$$

The strain tensor (as the stress tensor) has two principal directions, perpendicular to each other, which in analogy to (2.8) can be determined from the equation

$$\tan 2\,\varphi^* = \frac{\gamma_{xy}}{\varepsilon_x - \varepsilon_y} \,. \tag{3.4}$$

The principal strains ε_1 and ε_2 are given by (cf. (2.10))

$$\varepsilon_{1,2} = \frac{\varepsilon_x + \varepsilon_y}{2} \pm \sqrt{\left(\frac{\varepsilon_x - \varepsilon_y}{2}\right)^2 + \left(\frac{1}{2}\gamma_{xy}\right)^2} . \tag{3.5}$$

In analogy to Mohr's stress circle, a Mohr's *strain circle* may be drawn. Here, the stresses σ and τ must be replaced by the strains ε and $\gamma/2$.

A *spatial* deformation state can be described by the changes of the edge lengths and angles of infinitesimal cubes. The displacement vector $\boldsymbol{u}$ in space has the components u, v and w which depend on the three coordinates x, y and z. From these, the normal strains

$$\varepsilon_x = \frac{\partial u}{\partial x}, \quad \varepsilon_y = \frac{\partial v}{\partial y}, \quad \varepsilon_z = \frac{\partial w}{\partial z} \tag{3.6a}$$

and the shear strains

$$\gamma_{xy} = \frac{\partial u}{\partial y} + \frac{\partial v}{\partial x}, \quad \gamma_{xz} = \frac{\partial u}{\partial z} + \frac{\partial w}{\partial x}, \quad \gamma_{yz} = \frac{\partial v}{\partial z} + \frac{\partial w}{\partial y} \tag{3.6b}$$

can be derived. They are the components of the symmetric strain tensor $\boldsymbol{\varepsilon}$ which, as the stress tensor (2.4), can be arranged in a matrix:

$$\boldsymbol{\varepsilon} = \begin{bmatrix} \varepsilon_x & \varepsilon_{xy} & \varepsilon_{xz} \\ \varepsilon_{yx} & \varepsilon_y & \varepsilon_{yz} \\ \varepsilon_{zx} & \varepsilon_{zy} & \varepsilon_z \end{bmatrix} = \begin{bmatrix} \varepsilon_x & \frac{1}{2}\gamma_{xy} & \frac{1}{2}\gamma_{xz} \\ \frac{1}{2}\gamma_{xy} & \varepsilon_y & \frac{1}{2}\gamma_{yz} \\ \frac{1}{2}\gamma_{xz} & \frac{1}{2}\gamma_{yz} & \varepsilon_z \end{bmatrix} . \tag{3.7}$$

The main diagonal again contains the normal strains and the remaining elements are half the angle changes.

It should be mentioned that the second and third equations in (3.6a) and (3.6b) can be obtained from the respective first equation also by *cyclic permutation*. Here x is replaced by y, y by z, z by x and u by v, v by w, w by u.

3.2

3.2 Hooke's Law

The strains in a structural member depend on the external loading and therefore on the stresses. According to Chapter 1, stresses and strains are connected by Hooke's law. In the uniaxial case (bar) it takes the form $\sigma = E\,\varepsilon$ where E is Young's modulus.

We will now formulate Hooke's law for the plane stress state. Here we restrict our attention to materials which are *homogeneous* and *isotropic*. A homogeneous material has the same properties at each material point. For an isotropic material these properties are independent of the direction. In contrast, for an anisotropic material the properties depend on the direction. An example of an anisotropic material is wood: on account of its fiber structure the stiffness in the direction of the fibers is different from the stiffness perpendicular to the fibers.

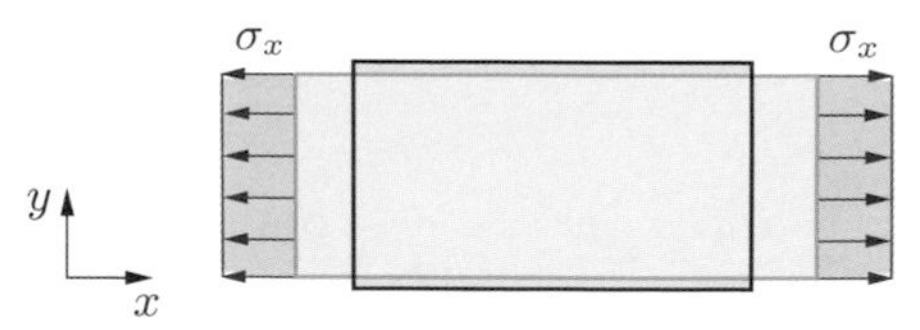

Fig. 3.3

To derive Hooke's law for plane stress we consider a rectangular domain (Fig. 3.3) which is cut out from a disk and which is loaded only by the normal stress σ_x. Then, according to (1.8)

$$\varepsilon_x = \frac{1}{E}\,\sigma_x$$

holds.

Experiments show that the tensile stress σ_x causes not only an increase of length but also a reduction of width of the rectangle. Thus, also a strain ε_y in y-direction appears. This phenomenon is called *lateral contraction* or *Poisson effect* (Siméon Denis Poisson, 1781–1840). The lateral strain ε_y is proportional to the axial strain ε_x and can be written as

$$\varepsilon_y = -\,\nu\,\varepsilon_x\,. \tag{3.8}$$

The dimensionless parameter ν is called *Poisson's ratio*. It is a material constant which is determined from experiments. Most metallic materials exhibit values about $\nu \approx 0.3$. Generally it can be shown that Poisson's ratio must be in the range of $-1 \leq \nu \leq \frac{1}{2}$.

As just discussed, the stress σ_x induces the strains $\varepsilon_x = \sigma_x/E$ and $\varepsilon_y = -\nu\,\sigma_x/E$. Analogously, a stress σ_y induces the strains $\varepsilon_x = -\nu\,\sigma_y/E$ and $\varepsilon_y = \sigma_y/E$. When both stresses, σ_x as well as σ_y, are acting, the total strains are obtained by superposition:

$$\varepsilon_x = \frac{1}{E}(\sigma_x - \nu\,\sigma_y)\,, \qquad \varepsilon_y = \frac{1}{E}(\sigma_y - \nu\,\sigma_x)\,. \tag{3.9}$$

It should be noted that the stresses σ_x and σ_y also induce a lateral contraction in the z-direction:

$$\varepsilon_z = -\,\frac{\nu}{E}\,\sigma_x - \frac{\nu}{E}\,\sigma_y = -\,\frac{\nu}{E}\,(\sigma_x + \sigma_y)\,.$$

Thus, a plane stress state leads to a spatial strain state. Since in this section we are only interested in studying the *in-plane* deformation, we will not consider further these strains in z-direction.

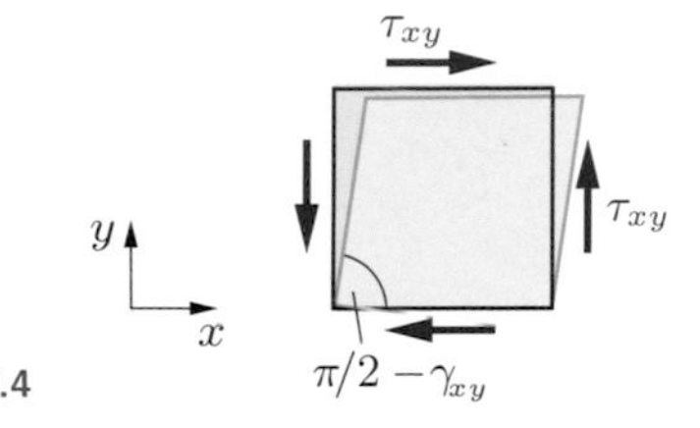

Fig. 3.4

If a disk (Fig. 3.4) is loaded solely by shear stresses τ_{xy} (pure shear), a linear relationship between the angle change γ_{xy} and the shear stress τ_{xy} is experimentally observed:

$$\tau_{xy} = G\,\gamma_{xy}\,. \tag{3.10}$$

The proportionality constant G is called *shear modulus*. It is a material parameter which can be experimentally determined in a shear test or a torsion test. The shear modulus G has the same dimension as Young's modulus E, i.e. force/area, and it is usually expressed in MPa or GPa (1 GPa $= 10^3$ MPa $= 10^6$ N/mm^2). It can be shown that there exist only *two independent* material constants for isotropic, linear elastic materials. The following relationship holds between the three constants E, G and ν:

$$G = \frac{E}{2(1+\nu)} \,. \tag{3.11}$$

The relationships (3.9) and (3.10) are *Hooke's law* for the plane stress state:

$$\begin{aligned} \varepsilon_x &= \frac{1}{E}(\sigma_x - \nu\,\sigma_y)\,, \\ \varepsilon_y &= \frac{1}{E}(\sigma_y - \nu\,\sigma_x)\,, \\ \gamma_{xy} &= \frac{1}{G}\,\tau_{xy}\,. \end{aligned} \tag{3.12}$$

If we introduce (3.12) in conjunction with (3.11) into (3.4) to determine the principal directions of the strain tensor, we obtain

$$\begin{aligned} \tan 2\,\varphi^* = \frac{\frac{1}{G}\,\tau_{xy}}{\frac{1}{E}(\sigma_x - \nu\,\sigma_y) - \frac{1}{E}(\sigma_y - \nu\,\sigma_x)} &= \frac{E\,\tau_{xy}}{G(1+\nu)(\sigma_x - \sigma_y)} \\ &= \frac{2\,\tau_{xy}}{\sigma_x - \sigma_y}\,. \end{aligned}$$

By comparison with (2.8) it can be recognized that (for an isotropic elastic material) the principal directions of the strain tensor coincide with those of the stress tensor.

Hooke's law (3.12) is valid in any arbitrary cartesian coordinate system. Specifically in the principal coordinate system it takes the form

$$\varepsilon_1 = \frac{1}{E}(\sigma_1 - \nu\,\sigma_2)\,, \qquad \varepsilon_2 = \frac{1}{E}(\sigma_2 - \nu\,\sigma_1). \tag{3.13}$$

We finally formulate Hooke's law in the three-dimensional space without going into the details of its derivation. Here, also temperature changes will be taken into account. Experiments show that a temperature change ΔT in an isotropic material causes only normal strains which are equal in all directions, i.e.

$$\varepsilon_{xT} = \varepsilon_{yT} = \varepsilon_{zT} = \alpha_T \Delta T$$

where α_T is the coefficient of thermal expansion (see Chapter 1). No shear strains are induced by ΔT. Then, as a generalization of (3.12), Hooke's law can be written as

$$\begin{aligned}
\varepsilon_x &= \frac{1}{E}[\sigma_x - \nu(\sigma_y + \sigma_z)] + \alpha_T \Delta T\,, \\
\varepsilon_y &= \frac{1}{E}[\sigma_y - \nu(\sigma_z + \sigma_x)] + \alpha_T \Delta T\,, \\
\varepsilon_z &= \frac{1}{E}[\sigma_z - \nu(\sigma_x + \sigma_y)] + \alpha_T \Delta T\,, \\
\gamma_{xy} &= \frac{1}{G}\tau_{xy}\,, \quad \gamma_{xz} = \frac{1}{G}\tau_{xz}\,, \quad \gamma_{yz} = \frac{1}{G}\tau_{yz}\,.
\end{aligned} \tag{3.14}$$

E3.1

Example 3.1 By using a strain gage rosette, the strains $\varepsilon_a = 12 \cdot 10^{-4}$, $\varepsilon_b = 2 \cdot 10^{-4}$ and $\varepsilon_c = -2 \cdot 10^{-4}$ have been measured in a steel sheet in the directions a, b and c (Fig. 3.5a).

Calculate the principal strains, the principal stresses and the principal directions.

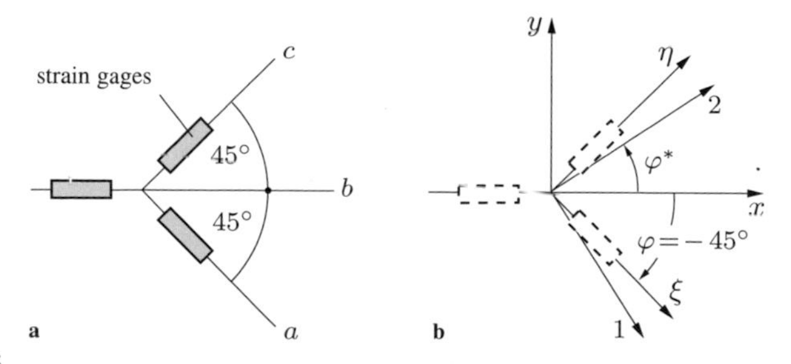

Fig. 3.5

Solution We introduce the two coordinate systems x, y and ξ, η according to Fig. 3.5b. Inserting the angle $\varphi = -45°$ into the first and second transformation equation (3.3), we obtain

$$\varepsilon_\xi = \frac{1}{2}(\varepsilon_x + \varepsilon_y) - \frac{1}{2}\gamma_{xy}, \qquad \varepsilon_\eta = \frac{1}{2}(\varepsilon_x + \varepsilon_y) + \frac{1}{2}\gamma_{xy}\,.$$

Addition and subtraction, respectively, yields

$$\varepsilon_\xi + \varepsilon_\eta = \varepsilon_x + \varepsilon_y\,, \qquad \varepsilon_\eta - \varepsilon_\xi = \gamma_{xy}\,.$$

With $\varepsilon_\xi = \varepsilon_a$, $\varepsilon_\eta = \varepsilon_c$ and $\varepsilon_x = \varepsilon_b$ we get

$$\varepsilon_y = \varepsilon_a + \varepsilon_c - \varepsilon_b = 8 \cdot 10^{-4}\,, \qquad \gamma_{xy} = \varepsilon_c - \varepsilon_a = -14 \cdot 10^{-4}\,.$$

The principal strains and principal directions are determined according to (3.5) and (3.4):

$$\varepsilon_{1,2} = (5 \pm \sqrt{9+49}) \cdot 10^{-4} \quad \rightarrow \quad \underline{\underline{\varepsilon_1 = 12.6 \cdot 10^{-4}}}\,, \quad \underline{\underline{\varepsilon_2 = -2.6 \cdot 10^{-4}}}\,,$$

$$\tan 2\,\varphi^* = \frac{-14}{2-8} = 2.33 \quad \rightarrow \quad \underline{\underline{\varphi^* = 33.4^\circ}}\,.$$

Introducing the angle φ^* into (3.3) shows that it is associated with the principal strain ε_2. The principal directions 1 and 2 are plotted in Fig. 3.5b.

Solving (3.13) for the stresses yields

$$\sigma_1 = \frac{E}{1-\nu^2}(\varepsilon_1 + \nu\,\varepsilon_2)\,, \qquad \sigma_2 = \frac{E}{1-\nu^2}(\varepsilon_2 + \nu\,\varepsilon_1)\,.$$

With $E = 2.1 \cdot 10^2$ GPa and $\nu = 0.3$ we obtain

$$\underline{\underline{\sigma_1 = 273\ \text{MPa}}}\,, \qquad \underline{\underline{\sigma_2 = 27\ \text{MPa}}}\,.$$

E3.2

Example 3.2 A steel cuboid with a quadratic base area ($h = 60$ mm, $a = 40$ mm) fits in the unloaded state exactly into an opening with rigid walls (Fig. 3.6a).

Determine the change of height of the cuboid when it is

a) loaded by the force $F = 160$ kN or
b) heated uniformly by the temperature $\Delta T = 100^\circ$ C.

Assume that the force F is uniformly distributed across the top surface and that the cuboid can slide without friction along the contact faces.

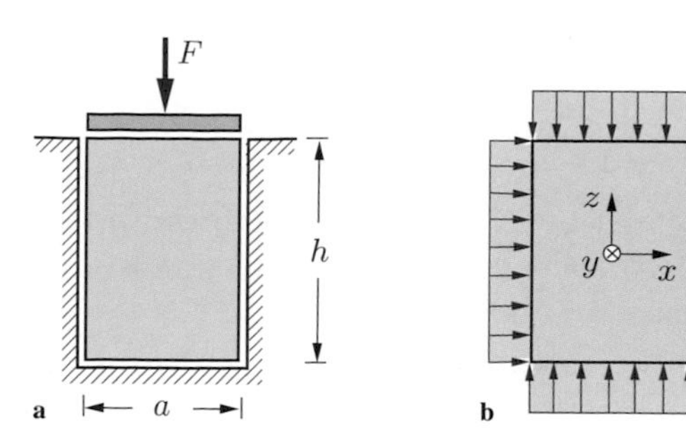

Fig. 3.6

Solution a) A homogeneous spatial stress state acts in the cuboid. The stress σ_z in the vertical direction, generated by the compressive force F (cf. Fig. 3.6b), is known:

$$\sigma_z = -\frac{F}{a^2}\,.$$

Since the body cannot undergo strains in x- and in y-direction on account of the deformation constraints, the conditions

$$\varepsilon_x = 0\,, \quad \varepsilon_y = 0$$

must be fulfilled. Introducing them into the first and second equation of Hooke's law (3.14) yields with $\Delta T = 0$

$$\begin{aligned} \sigma_x - \nu\,(\sigma_y + \sigma_z) &= 0 \\ \sigma_y - \nu\,(\sigma_z + \sigma_x) &= 0 \end{aligned} \quad \rightarrow \quad \sigma_x = \sigma_y = \frac{\nu}{1-\nu}\,\sigma_z\,.$$

Thus, the strain in the vertical direction is obtained from the third equation (3.14):

$$\begin{aligned} \varepsilon_z = \frac{1}{E}[\sigma_z - \nu(\sigma_x + \sigma_y)] &= \frac{\sigma_z}{E}\left(1 - \frac{2\,\nu^2}{1-\nu}\right) \\ &= -\frac{F}{a^2 E}\,\frac{(1+\nu)(1-2\,\nu)}{1-\nu}\,. \end{aligned}$$

Since the strain ε_z is constant, it can also be written as the ratio of height change Δh versus height h (cf. Section 1.2): $\varepsilon_z = \Delta h/h$. From that, with $E = 2.1 \cdot 10^5$ MPa and $\nu = 0.3$, the change of height is calculated:

$$\underline{\underline{\Delta h}} = \varepsilon_z h = -\frac{F\,h}{a^2\,E}\frac{(1+\nu)(1-2\,\nu)}{1-\nu} = \underline{\underline{-0.02\text{ mm}}}\,.$$

b) Now the cuboid is heated by ΔT while no compressive force is acting on the top surface ($F = 0$). Therefore, the stress in the vertical direction vanishes:

$$\sigma_z = 0\,.$$

Since the cuboid again cannot deform in x- and in y-direction, the conditions $\varepsilon_x = 0$, $\varepsilon_y = 0$ still hold. The first and the second equation of Hooke's law (3.14) now yield

$$\begin{aligned} \sigma_x - \nu\,\sigma_y + E\,\alpha_T\,\Delta T &= 0 \\ \sigma_y - \nu\,\sigma_x + E\,\alpha_T\,\Delta T &= 0 \end{aligned} \quad \rightarrow \quad \sigma_x = \sigma_y = -\frac{E\,\alpha_T\,\Delta T}{1-\nu}\,.$$

Thus, from the third equation (3.14), the strain in the vertical direction is given as

$$\varepsilon_z = -\frac{\nu}{E}(\sigma_x + \sigma_y) + \alpha_T\,\Delta T = \frac{1+\nu}{1-\nu}\,\alpha_T\,\Delta T\,.$$

With $\alpha_T = 1.2 \cdot 10^{-5}/^\circ\text{C}$ this leads to the change of height

$$\underline{\underline{\Delta h}} = \varepsilon_z\,h = \frac{1+\nu}{1-\nu}\,\alpha_T\,\Delta T\,h = \underline{\underline{0.13\text{ mm}}}\,.$$

3.3

3.3 Strength Hypotheses

For a bar under tensile loading one can conclude from the stress-strain diagram at which stress (or strain) failure will occur, e.g. on account of plastic flow or fracture. To prevent such a failure, an *allowable stress* σ_{allow} is introduced and it is postulated that the stress σ in the bar must not exceed σ_{allow}, i.e.: $\sigma \leq \sigma_{\text{allow}}$ (cf. Chapter 1).

In an arbitrary structural member, a spatial stress state is present. Also here it is necessary to determine the circumstances under which the load carrying capacity is lost and the material starts to fail. Since there exists no experimental setup which can provide a general answer, hypotheses on the basis of specific experiments are used. These so-called *strength hypotheses* allow us to calculate, according to a specific rule, an *equivalent stress* σ_e from the normal and shear stresses (or strains). It is assumed that the stress σ_e, when applied to the uniaxial case of a bar, has the same effect regarding failure through plastic flow or fracture as the given spatial stress state in the body under consideration. Since the stress states in the body and in a tensile bar are then said to be *equivalent*, the stress σ_e is called *equivalent stress*. Therefore, if a structural element shall not lose its load carrying capacity, the equivalent stress must not exceed the allowable stress:

$$\sigma_e \leq \sigma_{\text{allow}} \,. \tag{3.15}$$

In the following we will present three different strength hypotheses where we restrict ourselves to *plane* stress states.

1) *Maximum-normal-stress hypothesis:* It is assumed that the material starts to fail when the largest principal stress reaches a critical value. Thus,

$$\sigma_e = \sigma_1 \,. \tag{3.16}$$

This hypothesis may be applied to brittle materials under a predominating tensile stress state.

2) *Maximum-shear-stress hypothesis:* This hypothesis is based on the assumption that failure occurs when the maximum shear stress reaches a critical value. According to (2.12b), the maximum shear stress in a plane stress state is given by $\tau_{\max} = \frac{1}{2}(\sigma_1 - \sigma_2)$ whereas in a bar under tensile stress σ_e, the maximum shear stress is $\tau_{\max} = \frac{1}{2}\sigma_e$ (see (1.3)). Equating yields

$$\tau_{\max} = \frac{1}{2}(\sigma_1 - \sigma_2) = \frac{1}{2}\sigma_e \qquad \rightarrow \qquad \sigma_e = \sigma_1 - \sigma_2 \,.$$

Introducing (2.10) we obtain

$$\sigma_e = \sqrt{(\sigma_x - \sigma_y)^2 + 4\,\tau_{xy}^2} \,. \tag{3.17}$$

This hypothesis is applicable to materials where failure occurs predominantly through plastic flow. It was first formulated 1864 by Henri Édouard Tresca (1814–1885) and is usually named after him (note: the relation $\sigma_e = \sigma_1 - \sigma_2$ for plane stress is only valid if both stresses have different algebraic signs. Otherwise the maximum absolute stress value, σ_1 or σ_2, must be chosen for σ_e).

3) *Von Mises hypothesis (maximum-distortion-energy hypothesis)*: Here it is assumed that the material state becomes critical when the energy needed for the "distortion" of a material element (volume remains unchanged) reaches a critical value. Without going into the details of the derivation, we state that the equivalent stress is given by

$$\sigma_e = \sqrt{\sigma_1^2 + \sigma_2^2 - \sigma_1\sigma_2}$$

or, after introducing (2.10),

$$\sigma_e = \sqrt{\sigma_x^2 + \sigma_y^2 - \sigma_x\sigma_y + 3\tau_{xy}^2}\,. \tag{3.18}$$

It is usually called *von Mises stress* (Richard von Mises, 1883–1953). This hypothesis was developed independently also by Maxymilian Tytus Huber (1872–1950) and Heinrich Hencky (1885–1951). Since it agrees well with experiments on ductile materials, it is preferably used to characterize the onset of plastic flow.

In Example 5.3 the von Mises hypothesis is applied to dimension a shaft subjected to bending and torsion.

3.4

3.4 Supplementary Examples

Detailed solutions to the following examples are given in (**A**) D. Gross et al. *Formeln und Aufgaben zur Technischen Mechanik 2*, Springer, Berlin 2010 or (**B**) W. Hauger et al. *Aufgaben zur Technischen Mechanik 1-3*, Springer, Berlin 2008.

E3.3

Example 3.3 In a structure the following plane displacement field has been found from experiments:

$$u(x,y) = u_0 + 7\cdot 10^{-3}x + 4\cdot 10^{-3}y\,,$$
$$v(x,y) = v_0 + 2\cdot 10^{-3}x - 1\cdot 10^{-3}y\,.$$

a) Determine the strains.
b) Calculate the principal strains and their directions with respect to x.
c) Determine the maximum shear strain $\gamma_{\max}$.

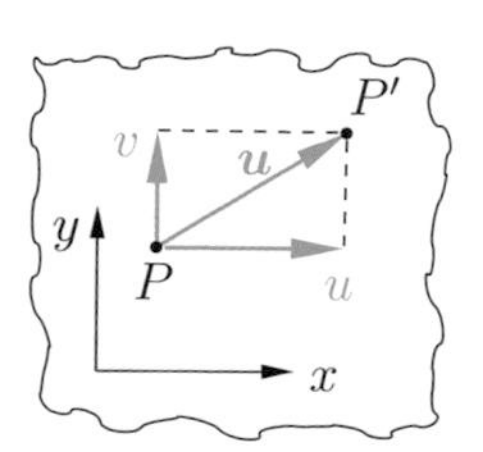

Fig. 3.7

Results: see (**A**)

a) $\varepsilon_x = 7\cdot 10^{-3}$, $\varepsilon_y = -1\cdot 10^{-3}$, $\gamma_{xy} = 6\cdot 10^{-3}$.

b) $\varepsilon_1 = 8\cdot 10^{-3}$, $\varepsilon_2 = -2\cdot 10^{-3}$, $\varphi_1^* = 36.9°$, $\varphi_2^* = 126.9°$.

c) $\gamma_{\max} = 1\cdot 10^{-2}$.

E3.4

Example 3.4 Two quadratic plates made of different materials (side lengths a) are pressed into a rigid opening whose width l is smaller than $2a$ (Fig. 3.8).

Determine the stresses and the changes of the side lengths. Assume that the contact faces can slide without friction.

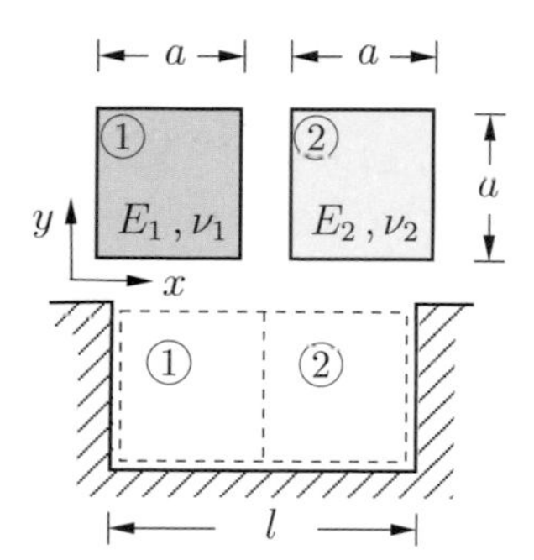

Fig. 3.8

Results: see (**A**)

$$\sigma_x = -\frac{2a-l}{a}\,\frac{E_1E_2}{E_1+E_2}\,,\qquad \sigma_y = 0\,,\qquad \tau_{xy} = 0\,,$$

$$\Delta u_1 = -(2a-l)\frac{E_2}{E_1+E_2}\,,\qquad \Delta u_2 = -(2a-l)\frac{E_1}{E_1+E_2}\,,$$

$$\Delta v_1 = -\nu_1\Delta u_1\,,\qquad \Delta v_2 = -\nu_2\Delta u_2\,.$$

E3.5

Example 3.5 The stresses $\sigma_x = 30$ MPa and $\tau_{xy} = 15$ MPa act on a rectangular metal sheet (Young's modulus $E = 2.1 \cdot 10^2$ GPa, Poisson's ratio $\nu = 0.3$) as shown in Fig. 3.9.

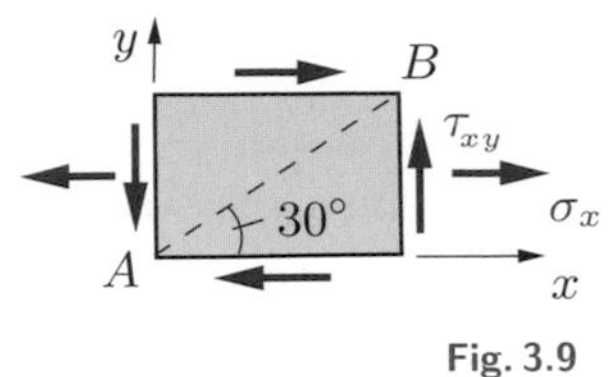

Fig. 3.9

Calculate the strain ε_{AB} in the direction of the diagonal.

Result: see (**B**) $\varepsilon_{AB} = 1.8 \cdot 10^{-4}$.

E3.6

Example 3.6 A strain gage rosette is bonded to the surface of a wrench (Young's modulus $E = 2.1 \cdot 10^2$ GPa, Posson's ratio $\nu = 0.3$) as shown in Fig. 3.10. The strain gage readings give the normal strains $\varepsilon_q = 5.8 \cdot 10^{-4}$, $\varepsilon_r = -1.0 \cdot 10^{-4}$ and $\varepsilon_s = 1.2 \cdot 10^{-4}$ in the directions q, r, s.

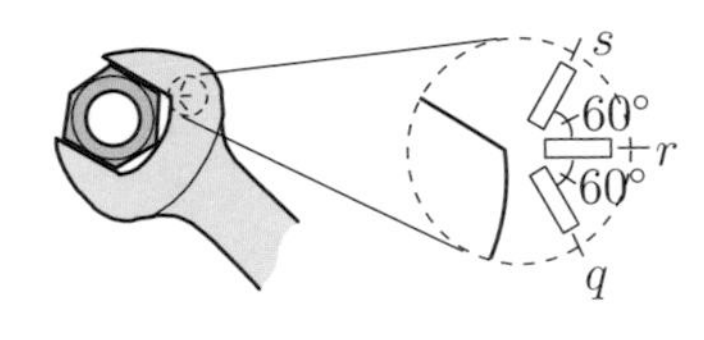

Fig. 3.10

Assume a plane stress state and determine the maximum normal stress at the location of the strain rosette.

Result: see (**B**) $\sigma_{\max} = 125$ MPa.

E3.7

Example 3.7 A thin-walled spherical vessel is heated by hot gas ($\Delta T = 200°$ C) and loaded by a gage pressure ($p = 1$ MPa).

Vessel data: $r = 2$ m, $t = 10$ mm, $E = 2.1 \cdot 10^2$ GPa, $\nu = 0.3$, $\alpha_T = 12 \cdot 10^{-}6\,/°$C.

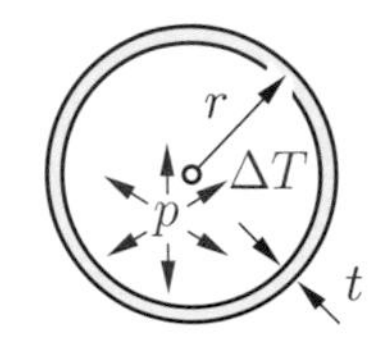

Fig. 3.11

Determine the change in radius of the vessel.

Result: see (**A**) $\Delta r = 5.5$ mm.

3.5 Summary

- The deformation state at a point of a body is described by the displacement vector $\boldsymbol{u}$ and the strain tensor $\boldsymbol{\varepsilon}$. In the spatial case the strain tensor has 3×3 components (note the symmetry). In the plane strain state it reduces to

$$\boldsymbol{\varepsilon} = \begin{bmatrix} \varepsilon_x & \varepsilon_{xy} \\ \varepsilon_{yx} & \varepsilon_y \end{bmatrix} = \begin{bmatrix} \varepsilon_x & \frac{1}{2}\gamma_{xy} \\ \frac{1}{2}\gamma_{yx} & \varepsilon_y \end{bmatrix} \quad \text{where} \quad \varepsilon_{xy} = \varepsilon_{yx} \, .$$

- Relationships between displacements u, v, normal strains ε_x, ε_y and shear strain γ_{xy}:

$$\varepsilon_x = \frac{\partial u}{\partial x}, \quad \varepsilon_y = \frac{\partial v}{\partial y}, \quad \gamma_{xy} = \frac{\partial u}{\partial y} + \frac{\partial v}{\partial x} \, .$$

- The transformation relations as well as the equations to determine the principal strains and principal directions are analogous to those of the stresses. The same is valid for Mohr's strain circle.
- The principal directions for stresses and strains coincide in the case of an isotropic elastic material.
- Hooke's law (spatial case):

$$E\varepsilon_x = \sigma_x - \nu(\sigma_y + \sigma_z) + E\alpha_T \Delta T \, , \qquad G\gamma_{xy} = \tau_{xy} \, .$$

  Respective further equations are obtained by cyclic permutation of the coordinate indices.
- Relation between G, E and ν for isotropic materials:

$$G = \frac{E}{2(1+\nu)} \, .$$

- To assess the stressing of the material, i.e. the criticality against material failure in the two- and three-dimensional cases, an equivalent stress σ_e, based on a strength hypothesis, is introduced.
  Example: von Mises hypothesis in a plane stress state

$$\sigma_e = \sqrt{\sigma_x^2 + \sigma_y^2 - \sigma_x \sigma_y + 3\tau_{xy}^2} \, .$$

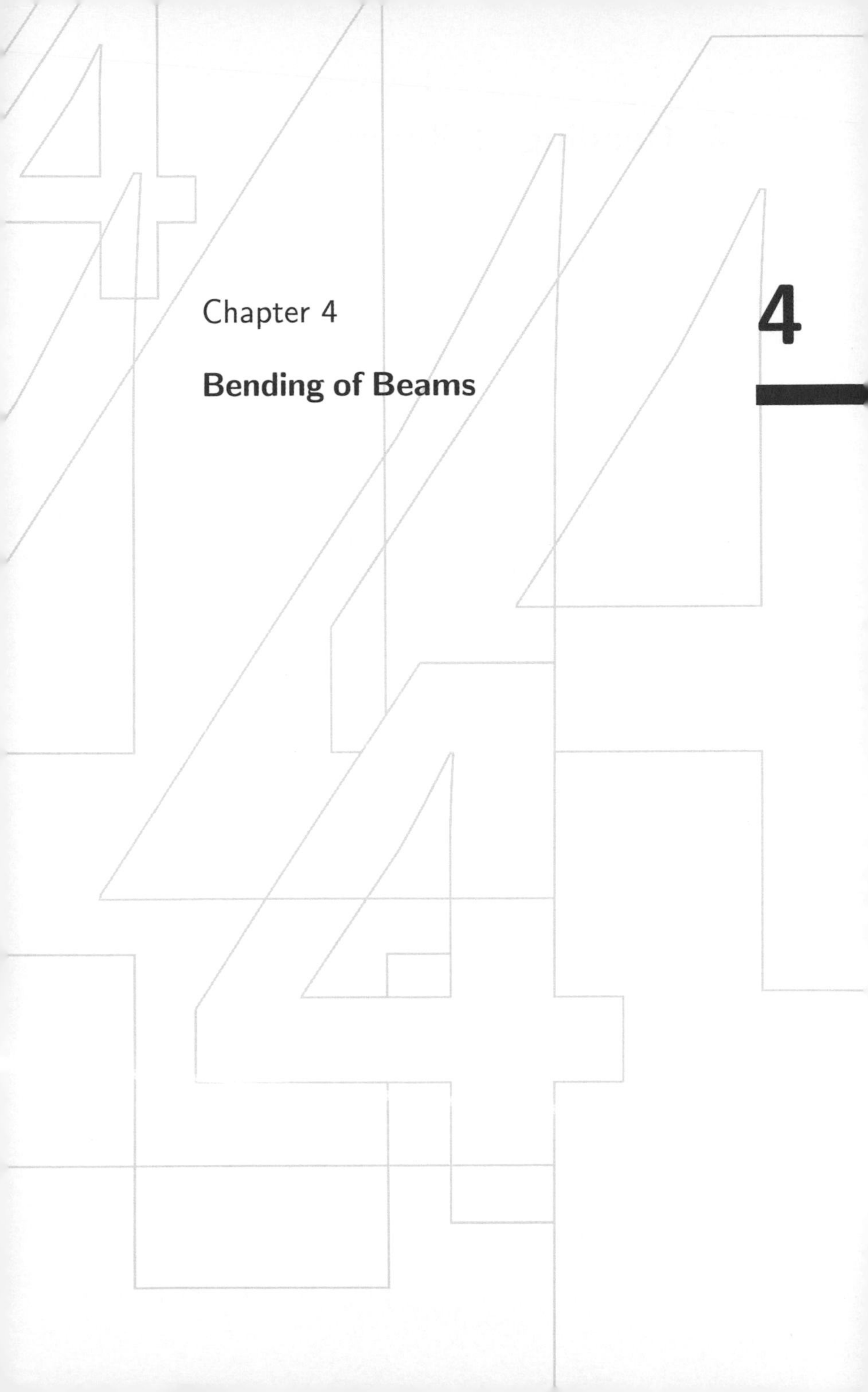

Chapter 4

Bending of Beams

4

4 Bending of Beams

4.1	Introduction	99
4.2	Second Moments of Area	101
4.2.1	Definitions	101
4.2.2	Parallel-Axis Theorem	108
4.2.3	Rotation of the Coordinate System, Principal Moments of Inertia	113
4.3	Basic Equations of Ordinary Bending Theory	117
4.4	Normal Stresses	121
4.5	Deflection Curve	125
4.5.1	Differential Equation of the Deflection Curve	125
4.5.2	Beams with one Region of Integration	129
4.5.3	Beams with several Regions of Integration	138
4.5.4	Method of Superposition	140
4.6	Influence of Shear	151
4.6.1	Shear Stresses	151
4.6.2	Deflection due to Shear	161
4.7	Unsymmetric Bending	162
4.8	Bending and Tension/Compression	171
4.9	Core of the Cross Section	174
4.10	Thermal Bending	176
4.11	Supplementary Examples	180
4.12	Summary	187

Objectives: In this chapter the bending of beams is investigated. We will derive the equations which enable us to determine the stresses and the deformations during bending. The theory also makes it possible to analyse statically indeterminate systems. The students will learn how to apply the equations to specific problems.

4.1 Introduction

Beams are among the most important elements in structural engineering. A beam has the geometrical shape of a straight bar, i.e., the dimensions of its cross-sectional area are much smaller than its length. However, in contrast to the members of a truss it is loaded by forces which are *perpendicular* to its axis. Due to the applied loads, the originally straight beam deforms (Fig. 4.1a). This is referred to as the *bending* of the beam. As a consequence, internal forces (= stresses) are generated in the beam, the resultants of which are the shear force V and the bending moment M (see Volume 1). It is the aim of the *bending theory* to derive equations that allow the determination of the stresses and the deformations.

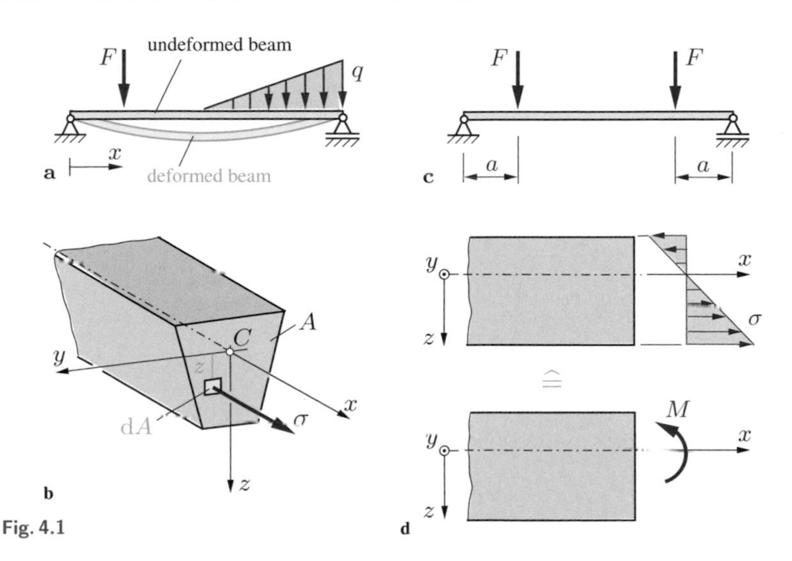

Fig. 4.1

Let us consider the beam in Fig. 4.1b where the z-axis is an axis of symmetry. In accordance with Volume 1, the x-axis coincides with the axis of the beam and passes through the centroid C of the cross-sectional area A. The reason for this particular choice of the location of the x-axis will become apparent in Section 4.3. The z-axis points downward and the orientation of the y-axis is chosen in such a way that the axes x, y and z represent a right-hand system.

In the following discussion, the load is assumed to cause a bending moment M only (no shear force and no normal force). The beam is then said to be in a state of *pure bending*. This is the case, for example, in the region between the two forces F acting at the beam in Fig. 4.1c. In pure bending there are only normal stresses σ which act in the direction of the x-axis in the cross-sectional areas (Figs. 4.1b, d). They are independent of the y-coordinate and they are linearly distributed in the z-direction (this will be explained in Sections 4.3 and 4.4). Therefore, the normal stresses may be written as

$$\sigma(z) = c\,z \tag{4.1}$$

where c is as yet an unknown factor of proportionality.

The bending moment M is statically equivalent to the sum of the moments of the distributed normal stresses with respect to the y-axis (Fig. 4.1d). The infinitesimal force $\mathrm{d}F = \sigma\,\mathrm{d}A$ causes the infinitesimal moment $\mathrm{d}M = z\,\mathrm{d}F = z\,\sigma\,\mathrm{d}A$ (Fig. 4.1b). Integration leads to

$$M = \int z\,\sigma\,\mathrm{d}A\,. \tag{4.2}$$

Inserting of (4.1) yields

$$M = c\int z^2\,\mathrm{d}A\,.$$

The quantity

$$I = \int z^2\,\mathrm{d}A \tag{4.3}$$

is called *moment of inertia*. With (4.3), the above equation can be written as $c = M/I$, and (4.1) yields a relation between the stresses and the bending moment:

$$\sigma = \frac{M}{I}\,z\,. \tag{4.4}$$

Hence, the stress at a fixed value of z depends not only on the bending moment M but also on the moment of inertia I. The term "inertia" is used since there is a similarity between the definition of the moments of inertia of an area and the definition of the mass moments of inertia (see Volume 3). The moment of inertia is a purely geometrical quantity which is related to the shape of the area. It plays an important role in the bending of beams. In the following section, the properties of moments of inertia will be discussed in detail.

4.2 Second Moments of Area

4.2

4.2.1 Definitions

Fig. 4.2 shows an area A in the y, z-plane. The axes and their directions (z downward, y to the left) are chosen in accordance with the axes in the cross section of a beam. The origin 0 is located at an arbitrary position.

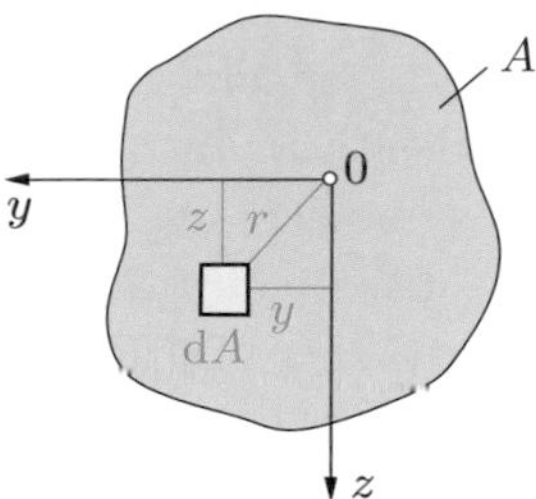

Fig. 4.2

The coordinates of the centroid C of an area may be obtained from $y_c = \frac{1}{A}\int y\,\mathrm{d}A$ and $z_c = \frac{1}{A}\int z\,\mathrm{d}A$ (see Volume 1, Section 4.3). The integrals

$$S_y = \int z\,\mathrm{d}A, \qquad S_z = \int y\,\mathrm{d}A \tag{4.5}$$

are called *first moments of area* since the distances y and z, respectively, appear linearly in (4.5).

Integrals which contain the square of the distances of the element $\mathrm{d}A$ or the distances as a product are called *second moments of area*. They are also referred to as *moments of inertia* of the area. These integrals are purely geometrical quantities without physical significance. They are defined as follows:

$$I_y = \int z^2 \, \mathrm{d}A, \quad I_z = \int y^2 \, \mathrm{d}A, \tag{4.6a}$$

$$I_{yz} = \int I_{zy} = -\int y\, z \, \mathrm{d}A, \tag{4.6b}$$

$$I_p = \int r^2 \, \mathrm{d}A = \int (z^2 + y^2) \, \mathrm{d}A = I_y + I_z \,. \tag{4.6c}$$

The integrals I_y and I_z in (4.6a) are referred to as *rectangular moments of inertia*, I_{yz} is called the *product of inertia* and I_p is the *polar moment of inertia*. The second moments of area have the dimension length^4; they are given, for example, in the unit cm^4.

The magnitude of a moment of inertia depends on the location of the origin and on the orientation of the axes. Whereas I_y, I_z and I_p are always positive (the integrals involve the squares of the distances), the product of inertia I_{yz} may be positive, negative or zero (the integrals contain the product of y and z which need not be positive). In particular, $I_{yz} = 0$ if one of the axes is an axis of symmetry of the area A. Let, for example, the z-axis be an axis of symmetry (Fig. 4.3a). Then for every infinitesimal area $\mathrm{d}A$ located at a positive distance y, there exists a corresponding element located at a negative distance. Therefore, the integral (4.6b) is zero.

In some cases it is practical to use the *radii of gyration* instead of the second moments of area. They have the dimension "length"

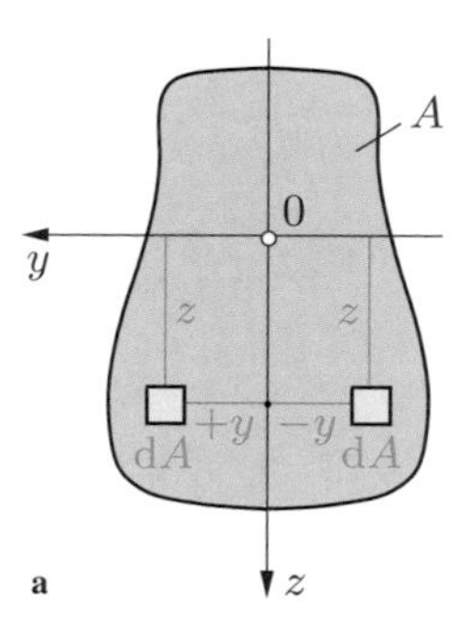

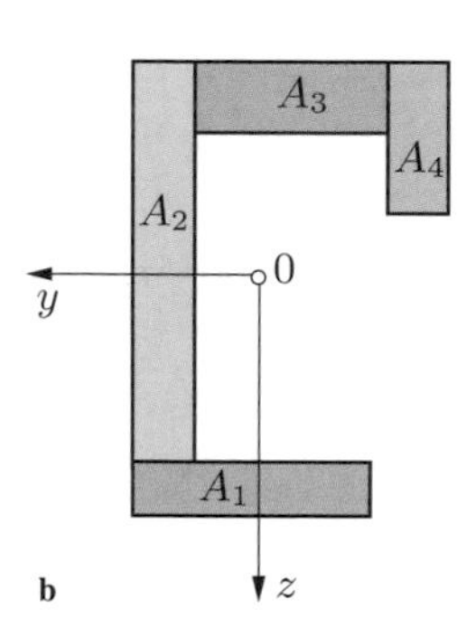

Fig. 4.3

and are defined by

$$r_{gy} = \sqrt{\frac{I_y}{A}}, \qquad r_{gz} = \sqrt{\frac{I_z}{A}}, \qquad r_{gp} = \sqrt{\frac{I_p}{A}}\,. \tag{4.7}$$

These equations yield, for example, the relation $I_y = r_{gy}^2 A$. Therefore, one may interpret the radius r_{gy} as the distance from the y-axis at which the area A can be imagined to be "concentrated" in order to have the moment of inertia I_y.

Frequently, an area A is composed of several parts A_i, the moments of inertia of which are known (Fig. 4.3b). In this case, the moment of inertia about the y-axis, for example, is obtained as the sum of the moments of inertia I_{y_i} of the individual parts about the *same axis*:

$$I_y = \int\limits_A z^2 \,\mathrm{d}A = \int\limits_{A_1} z^2 \,\mathrm{d}A + \int\limits_{A_2} z^2 \,\mathrm{d}A + \ldots = \sum I_{y_i}\,.$$

The other moments of inertia can also be found by summation from similar expressions:

$$I_z = \sum I_{z_i}, \qquad I_{yz} = \sum I_{yz_i}\,.$$

The methods to calculate the second moments of area either by integration (in particular, the considerations of how to choose an infinitesimal area $\mathrm{d}A$) or by summation (in the case of a composite area) are quite analogous to those discussed in Volume 1, Chapter 4, for the determination of centroids, i.e., first moments of area.

As a first example we consider a rectangular area (width b, height h). The coordinate system with the origin at the centroid C is given; the y-axis is parallel to the base (Fig. 4.4a). In order to determine I_y, we select an infinitesimal area $\mathrm{d}A = b\mathrm{d}z$ according to Fig. 4.4b. Then every point of the element has the same distance z from the y-axis. Thus, we obtain

$$I_y = \int z^2 \,\mathrm{d}A = \int\limits_{-h/2}^{+h/2} z^2 \,(b\,\mathrm{d}z) = \frac{b\,z^3}{3}\Big|_{-h/2}^{+h/2} = \frac{b\,h^3}{12}\,. \tag{4.8a}$$

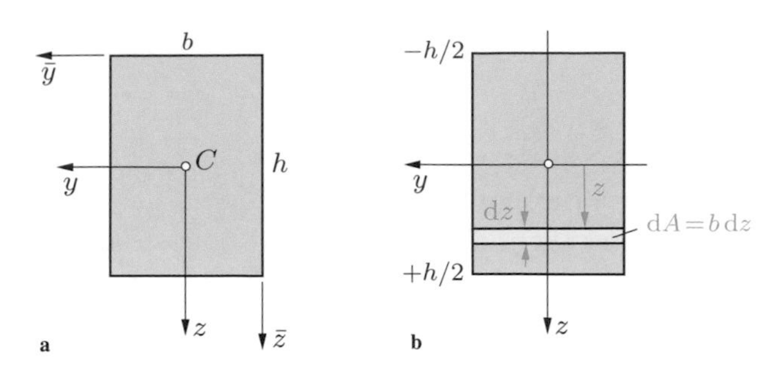

Fig. 4.4

Exchanging b and h yields

$$I_z = \frac{h\,b^3}{12}\,. \tag{4.8b}$$

Since the z-axis is an axis of symmetry, the product of inertia I_{yz} is zero:

$$I_{yz} = 0\,. \tag{4.8c}$$

(Note that in this example the y-axis is also an axis of symmetry.) The polar moment of inertia is calculated with the aid of (4.6c) and the already known quantities I_y and I_z:

$$I_p = I_y + I_z = \frac{b\,h^3}{12} + \frac{h\,b^3}{12} = \frac{b\,h}{12}(h^2 + b^2). \tag{4.8d}$$

The radii of gyration follow with the area $A = b\,h$ and the length $d = \sqrt{b^2 + h^2}$ of the diagonal of the rectangular area from (4.7):

$$r_{gy} = \frac{h}{2\sqrt{3}}, \qquad r_{gz} = \frac{b}{2\sqrt{3}}, \qquad r_{gp} = \frac{d}{2\sqrt{3}}\,. \tag{4.8e}$$

In a second example we calculate the moments of inertia and the radii of gyration of a circular area (radius R), the origin of the coordinate system being at the centroid C (Fig. 4.5a). Due to the symmetry of the problem, the moments of inertia about every axis through C are equal. Therefore, according to (4.6c),

$$I_y = I_z = \frac{1}{2}\,I_p\,. \tag{4.9}$$

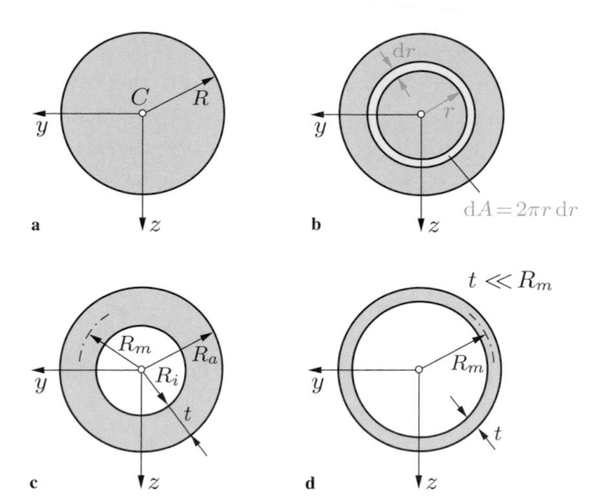

Fig. 4.5

The product of inertia I_{yz} is zero (symmetry). To determine the polar moment I_p, we choose an infinitesimal circular ring with the area $\mathrm{d}A = 2\pi\, r\, \mathrm{d}r$ (every point of this ring has the same distance r from the center C), see Fig. 4.5b. Then we obtain

$$I_p = \int r^2 \mathrm{d}A = \int_0^R r^2 (2\,\pi\, r\, \mathrm{d}r) = \frac{\pi}{2}\, R^4, \tag{4.10a}$$

and (4.9) yields

$$I_y = I_z = \frac{\pi}{4}\, R^4\,. \tag{4.10b}$$

With the area $A = \pi R^2$, the radii of gyration follow as

$$r_{gy} = r_{gz} = \frac{R}{2}, \qquad r_{gp} = \frac{R}{\sqrt{2}}\,. \tag{4.10c}$$

The results for the circular area can be used to find the moments of inertia of a circular ring with outer radius R_a and inner radius R_i (Fig. 4.5c). Subtraction leads to

$$\begin{aligned} I_p &= \frac{\pi}{2}\, R_a^4 - \frac{\pi}{2}\, R_i^4 = \frac{\pi}{2}(R_a^4 - R_i^4)\,, \\ I_y = I_z &= \frac{\pi}{4}\, R_a^4 - \frac{\pi}{4}\, R_i^4 = \frac{\pi}{4}(R_a^4 - R_i^4). \end{aligned} \tag{4.11}$$

If we introduce the radius $R_m = \frac{1}{2}(R_a + R_i)$ (= arithmetic mean value of the radii R_i and R_a) and the thickness $t = R_a - R_i$ of the ring, we can write the terms in the parentheses in (4.11) in the form $R_a^4 - R_i^4 = 4R_m^3 t(1 + t^2/(4R_m^2))$. If the thickness t is small as compared with the radius R_m (i.e., $t \ll R_m$), the term $t^2/(4R_m^2)$ may be neglected. Therefore, the moments of inertia of a *thin circular ring* (Fig. 4.5d) are

$$I_p \approx 2\,\pi\,R_m^3\,t, \quad I_y = I_z \approx \pi\,R_m^3\,t\,. \tag{4.12}$$

Additional moments of inertia for a number of typical areas can be found in Table 4.1

E4.1

Example 4.1 Determine the second moments of area and the radii of gyration of an ellipse for the coordinate system as given in Fig. 4.6a.

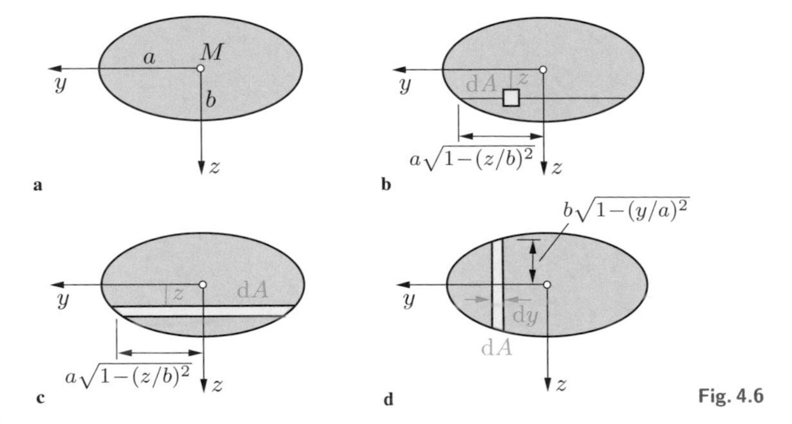

Fig. 4.6

Solution The area is defined by the equation $(y/a)^2 + (z/b)^2 = 1$ of the ellipse. First we determine I_y and use three different ways to choose the infinitesimal area $\mathrm{d}A$.

a) If we choose the element $\mathrm{d}A = \mathrm{d}y\,\mathrm{d}z$ as shown in Fig. 4.6b, we have to integrate over the two variables y and z (double integral). First we integrate over y (the variable boundaries $\pm a\sqrt{1-(z/b)^2}$ follow from the equation of the ellipse) and then we integrate over z (with the fixed boundaries $\pm b$):

$$I_y = \int z^2 \, \mathrm{d}A = \int_{-b}^{+b} z^2 \left\{ \int_{-a\sqrt{1-(z/b)^2}}^{+a\sqrt{1-(z/b)^2}} \mathrm{d}y \right\} \mathrm{d}z = 2a \int_{-b}^{+b} z^2 \sqrt{1-(z/b)^2} \, \mathrm{d}z \, .$$

Using the substitution $z = b \sin \frac{\varphi}{2}$, we obtain

$$\underline{\underline{I_y}} = a\, b^3 \int_{-\pi}^{+\pi} \sin^2 \frac{\varphi}{2} \cos^2 \frac{\varphi}{2} \mathrm{d}\varphi = \underline{\underline{\frac{\pi}{4} a\, b^3}} \, . \qquad \text{(a)}$$

b) The integration over y may be avoided if the infinitesimal strip according to Fig. 4.6c is used. Every point of this element has the same distance z from the y-axis. With $\mathrm{d}A = 2a\sqrt{1-(z/b)^2}\,\mathrm{d}z$ we immediately obtain the integral

$$I_y = \int z^2 \, \mathrm{d}A = 2\, a \int_{-b}^{+b} z^2 \sqrt{1-(z/b)^2} \, \mathrm{d}z \, ,$$

which as before leads to the result (a).

c) We will now consider the ellipse to be composed of infinitesimal rectangular areas as shown in Fig. 4.6d. Each area element has the width $\mathrm{d}y$ and the height $2b\sqrt{1-(y/a)^2}$. Therefore, according to (4.8a), its moment of inertia is given by

$$\mathrm{d}I_y = \frac{1}{12} 8\, b^3 (1 - y^2/a^2)^{3/2} \, \mathrm{d}y \, .$$

The moment of inertia of the ellipse follows from the summation (= integration) of the infinitesimal moments of inertia (substitution: $y = a \sin \psi$):

$$I_y = \int \mathrm{d}I_y = \frac{8}{12} b^3 \int_{-a}^{+a} (1 - y^2/a^2)^{3/2} \mathrm{d}y$$

$$= \frac{2}{3} b^3 a \int_{-\pi/2}^{+\pi/2} \cos^4 \psi \, \mathrm{d}\psi = \frac{\pi}{4} a\, b^3 \, .$$

The moment of inertia I_z is obtained from I_y by exchanging a and b:

$$\underline{\underline{I_z = \frac{\pi}{4}\, b\, a^3.}} \tag{b}$$

The coordinate axes y and z are axes of symmetry. Therefore, $I_{yz} = 0$. The polar moment of inertia is determined from (4.6c):

$$\underline{\underline{I_p}} = I_y + I_z = \underline{\underline{\frac{\pi}{4}\, a\, b(a^2 + b^2)}}\,.$$

The area of an ellipse is given by $A = \pi\, a\, b$. Thus, the Equations (4.7) yield the radii of gyration:

$$\underline{\underline{r_{gy} = \frac{b}{2}}}, \qquad \underline{\underline{r_{gz} = \frac{a}{2}}}, \qquad \underline{\underline{r_{gp} = \frac{1}{2}\sqrt{a^2 + b^2}}}\,.$$

4.2.2 Parallel-Axis Theorem

Let us now consider two different coordinate systems, namely, y, z and $\bar{y}, \bar{z}$, as shown in Fig. 4.7. The axes of the two systems are assumed to be parallel and the origin of the y, z-system is the centroid C of the area. In the following we shall investigate how the second moments of area with respect to the different coordinate systems are related.

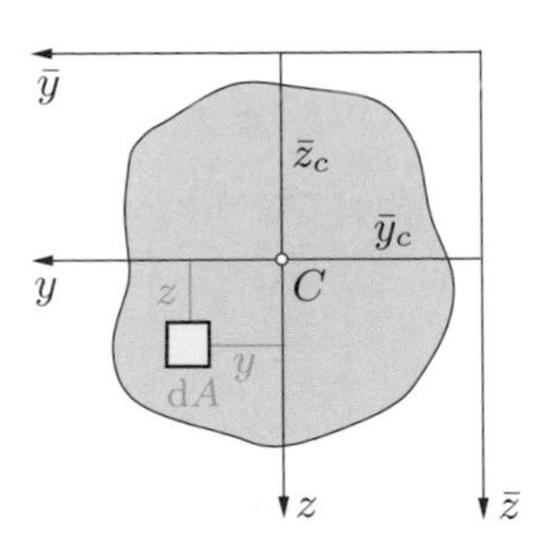

Fig. 4.7

From Fig. 4.7 we take the relations

$$\bar{y} = y + \bar{y}_c, \qquad \bar{z} = z + \bar{z}_c\,.$$

Then the second moments of area with respect to the $\bar{y}$, $\bar{z}$-system are given by

$$I_{\bar{y}} = \int \bar{z}^2\, \mathrm{d}A = \int (z+\bar{z}_c)^2\, \mathrm{d}A = \int z^2\, \mathrm{d}A + 2\,\bar{z}_c \int z\, \mathrm{d}A + \bar{z}_c^2 \int \mathrm{d}A,$$

$$I_{\bar{z}} = \int \bar{y}^2\, \mathrm{d}A = \int (y+\bar{y}_c)^2\, \mathrm{d}A = \int y^2\, \mathrm{d}A + 2\,\bar{y}_c \int y\, \mathrm{d}A + \bar{y}_c^2 \int \mathrm{d}A,$$

$$I_{\bar{y}\bar{z}} = -\int \bar{y}\bar{z} \mathrm{d}A = -\int (y+\bar{y}_c)(z+\bar{z}_c) \mathrm{d}A = -\int y\, z\, \mathrm{d}A - \bar{y}_c \int z\, \mathrm{d}A$$
$$- \bar{z}_c \int y\, \mathrm{d}A - \bar{y}_c \bar{z}_c \int \mathrm{d}A\,.$$

The first moments of area $\int z\, \mathrm{d}A$ and $\int y\, \mathrm{d}A$ about the axes y and z through the centroid C are zero (see Volume 1, Chapter 4). With $A = \int \mathrm{d}A$ and $I_y = \int z^2\, \mathrm{d}A$ etc we thus obtain

$$\begin{aligned} I_{\bar{y}} &= I_y + \bar{z}_c^2\, A\,, \\ I_{\bar{z}} &= I_z + \bar{y}_c^2\, A\,, \\ I_{\bar{y}\bar{z}} &= I_{yz} - \bar{y}_c \bar{z}_c\, A\,. \end{aligned} \tag{4.13}$$

These are the relations between the moments of inertia with respect to the axes through the centroid C and the moments of inertia with respect to axes which are parallel to them. The Equations (4.13) are known as the *parallel-axis theorem*. Note that when applying the parallel-axis theorem one of the two axes must be a centroidal axis. The terms $\bar{z}_c^2 A$ and $\bar{y}_c^2 A$ are always positive. Therefore, the moments of inertia with respect to centroidal axes are always smaller than those with respect to axes parallel to them. The term $\bar{y}_c \bar{z}_c A$ which appears in the expression for $I_{\bar{y}\bar{z}}$ may be positive, negative or zero, depending on the location of the respective axes.

As an example we determine the moments of inertia with respect to the axes $\bar{y}, \bar{z}$ for the rectangle shown in Fig. 4.4a. Since the moments of inertia with respect to centroidal axes are given by (4.8), we obtain from (4.13)

$$I_{\bar{y}} = \frac{b\,h^3}{12} + \left(\frac{h}{2}\right)^2 b\,h = \frac{b\,h^3}{3}\,,$$

Table 4.1. Moments of Inertia

Area	I_y	I_z	I_{yz}	I_p	$I_{\bar{y}}$
Rectangle 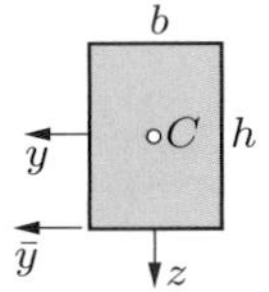	$\dfrac{b\,h^3}{12}$	$\dfrac{h\,b^3}{12}$	0	$\dfrac{b\,h}{12}(h^2+b^2)$	$\dfrac{b\,h^3}{3}$
Square 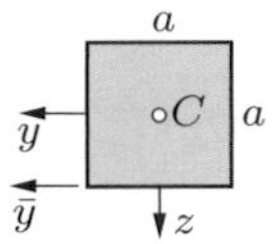	$\dfrac{a^4}{12}$	$\dfrac{a^4}{12}$	0	$\dfrac{a^4}{6}$	$\dfrac{a^4}{3}$
Triangle 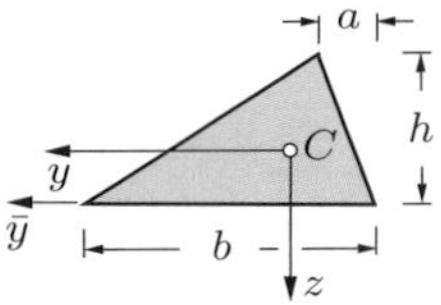	$\dfrac{b\,h^3}{36}$	$\dfrac{b\,h}{36}(b^2-b\,a+a^2)$	$-\dfrac{b\,h^2}{72}(b-2\,a)$	$\dfrac{b\,h}{36}(h^2+b^2-b\,a+a^2)$	$\dfrac{b\,h^3}{12}$

Circle	$\dfrac{\pi R^4}{4}$	$\dfrac{\pi R^4}{4}$	0	$\dfrac{\pi R^4}{2}$	$\dfrac{5\pi}{4} R^4$
Thin Circular Ring $t \ll R_m$	$\pi R_m^3 t$	$\pi R_m^3 t$	0	$2 \pi R_m^3 t$	$3 \pi R_m^3 t$
Semi-Circle	$\dfrac{R^4}{72 \pi}(9 \pi^2 - 64)$	$\dfrac{\pi R^4}{8}$	0	$\dfrac{R^4}{36 \pi}(9 \pi^2 - 32)$	$\dfrac{\pi R^4}{8}$
Ellipse	$\dfrac{\pi}{4} a b^3$	$\dfrac{\pi}{4} b a^3$	0	$\dfrac{\pi a b}{4}(a^2 + b^2)$	$\dfrac{5\pi}{4} a b^3$

$$I_{\bar{z}} = \frac{h\,b^3}{12} + \left(\frac{b}{2}\right)^2 b\,h = \frac{h\,b^3}{3}\,,$$

$$I_{\bar{y}\bar{z}} = 0 - \frac{b}{2}\frac{h}{2}\,b\,h = -\frac{b^2h^2}{4}\,.$$

E4.2

Example 4.2 Determine the moments of inertia for the I-profile shown in Fig. 4.8a. Simplify the results for $d, t \ll b, h$.

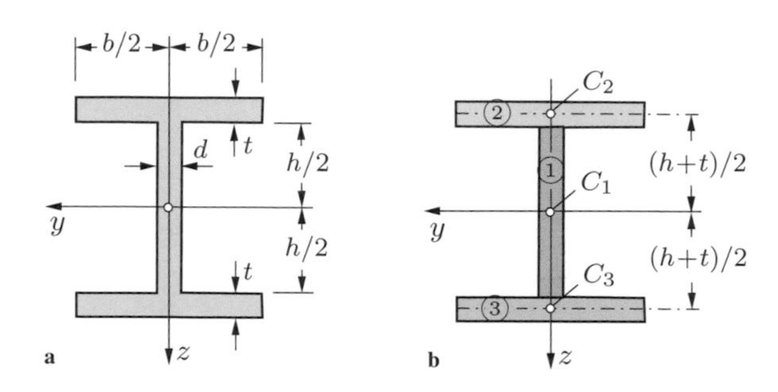

Fig. 4.8

Solution We consider the area to be composed of three rectangles (Fig. 4.8b). According to (4.13), the moments of inertia of each part consist of the moments of inertia with respect to the centroidal axes (see (4.8)) and the corresponding additional terms:

$$\underline{\underline{I_y}} = \frac{d\,h^3}{12} + 2\left[\frac{b\,t^3}{12} + \left(\frac{h}{2} + \frac{t}{2}\right)^2 b\,t\right] = \frac{d\,h^3}{12} + \frac{b\,t^3}{6} + \frac{h^2 b\,t}{2}$$

$$+\,t^2 h\,b + \frac{b\,t^3}{2} = \underline{\underline{\frac{d\,h^3}{12} + \frac{2\,b\,t^3}{3} + \frac{h^2 b\,t}{2} + t^2 h\,b}}\,,$$

$$\underline{\underline{I_z = \frac{h\,d^3}{12} + 2\,\frac{t\,b^3}{12}}}\,.$$

In the case of $d, t \ll b, h$ the terms which contain d, t quadratically or to the third power may be neglected as compared with the terms that are linear in d and t:

$$\underline{\underline{I_y \approx \frac{d\,h^3}{12} + \frac{h^2 b\,t}{2}}}\,, \qquad \underline{\underline{I_z \approx \frac{t\,b^3}{6}}}\,.$$

One can see that in the case of a thin-walled profile the moments of inertia $2bt^3/12$ of the flanges about their respective centroidal axis may be neglected when calculating I_y. Calculating I_z, we may neglect the moment of inertia $h\,d^3/12$ of the web.

4.2.3 Rotation of the Coordinate System, Principal Moments of Inertia

Let us now consider two coordinate systems y, z and η, ζ (Fig. 4.9). Their relative orientation is given by the angle φ. In order to derive relations between the respective moments of inertia, we use the geometrical relations

$$\eta = y\cos\varphi + z\sin\varphi, \qquad \zeta = -y\sin\varphi + z\cos\varphi$$

(see Fig. 4.9). Then the moments of inertia with respect to the axes η, ζ are given by

$$\begin{aligned} I_\eta &= \int \zeta^2\,\mathrm{d}A = \sin^2\varphi \int y^2\,\mathrm{d}A + \cos^2\varphi \int z^2\,\mathrm{d}A \\ &\quad - 2\sin\varphi\cos\varphi \int y\,z\,\mathrm{d}A, \\ I_\zeta &= \int \eta^2\,\mathrm{d}A = \cos^2\varphi \int y^2\,\mathrm{d}A + \sin^2\varphi \int z^2\,\mathrm{d}A \\ &\quad + 2\sin\varphi\cos\varphi \int y\,z\,\mathrm{d}A, \\ I_{\eta\zeta} &= -\int \eta\,\zeta\,\mathrm{d}A = \sin\varphi\cos\varphi \int y^2\,\mathrm{d}A - \cos^2\varphi \int y\,z\,\mathrm{d}A \\ &\qquad + \sin^2\varphi \int y\,z\,\mathrm{d}A - \sin\varphi\cos\varphi \int z^2\,\mathrm{d}A\,. \end{aligned}$$

With the moments of inertia about the axes y,z according to (4.6) and the trigonometrical relations $\sin^2\varphi = \frac{1}{2}(1-\cos 2\varphi)$,

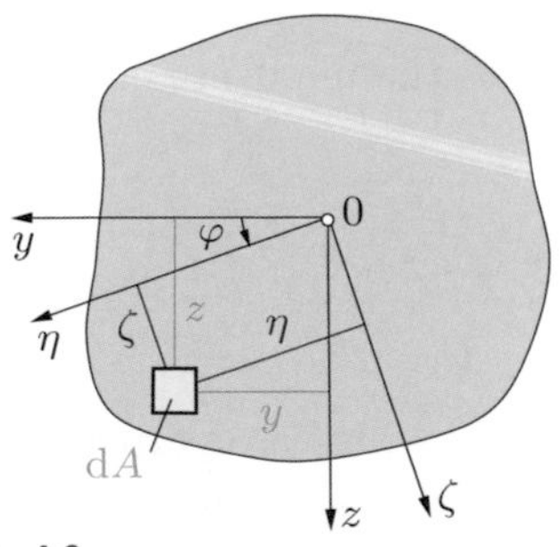

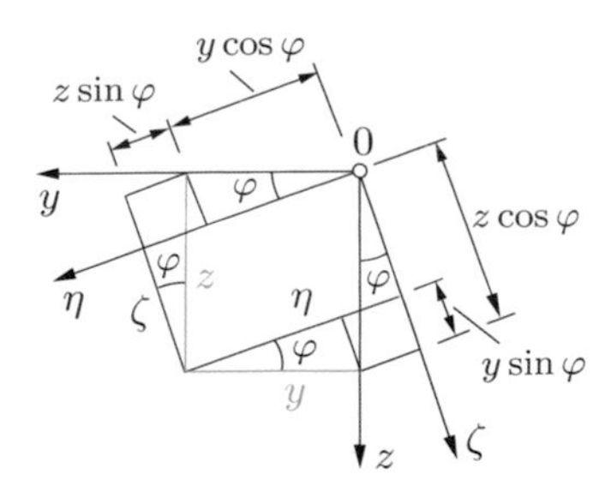

Fig. 4.9

$\cos^2\varphi = \frac{1}{2}(1+\cos 2\varphi)$ and $2\sin\varphi\cos\varphi = \sin 2\varphi$ we obtain the *transformation equations*

$$
\begin{aligned}
I_\eta &= \tfrac{1}{2}(I_y + I_z) + \tfrac{1}{2}(I_y - I_z)\cos 2\varphi + I_{yz}\sin 2\varphi\,,\\
I_\zeta &= \tfrac{1}{2}(I_y + I_z) - \tfrac{1}{2}(I_y - I_z)\cos 2\varphi - I_{yz}\sin 2\varphi\,,\\
I_{\eta\zeta} &= \qquad\qquad\quad - \tfrac{1}{2}(I_y - I_z)\sin 2\varphi + I_{yz}\cos 2\varphi\,.
\end{aligned}
\tag{4.14}
$$

If the moments of inertia with respect to the y, z-coordinate system are known, the moments of inertia with respect to the inclined system η, ζ can be determined from (4.14).

If we add the first two equations in (4.14) and use (4.6c), we obtain

$$
I_\eta + I_\zeta = I_y + I_z = I_p\,. \tag{4.15}
$$

Hence, the sum of the moments of inertia I_y and I_z (= polar moment of inertia) is independent of the angle φ. Therefore, $I_\eta + I_\zeta$ is referred to as an *invariant* of the coordinate transformation (= rotation of the coordinate system). One may verify that another invariant is given by $[\frac{1}{2}(I_\eta - I_\zeta)]^2 + I_{\eta\zeta}^2$.

According to (4.14), the magnitude of a moment of inertia depends on the angle φ. The moments I_η or I_ζ have a maximum or a minimum value if the conditions $\mathrm{d}I_\eta/\mathrm{d}\varphi = 0$ or $\mathrm{d}I_\zeta/\mathrm{d}\varphi = 0$ are satisfied. Both conditions lead to the same result:

$$
-\frac{1}{2}(I_y - I_z)\sin 2\varphi + I_{yz}\cos 2\varphi = 0\,.
$$

Therefore, the angle $\varphi = \varphi^*$ which makes the moments of inertia a maximum or a minimum is given by

$$
\tan 2\varphi^* = \frac{2\,I_{yz}}{I_y - I_z}\,. \tag{4.16}
$$

Because of $\tan 2\varphi^* = \tan 2\,(\varphi^* + \pi/2)$ there exist two perpendicular axes with the directions given by the angles φ^* and $\varphi^* + \pi/2$

for which the moments of inertia become an extremum. These axes are called *principal axes*. We obtain the corresponding *principal moments of inertia* if we insert (4.16) into (4.14). Using the trigonometrical relations

$$\cos 2\varphi^* = \frac{1}{\sqrt{1+\tan^2 2\varphi^*}} = \frac{I_y - I_z}{\sqrt{(I_y - I_z)^2 + 4\,I_{yz}^2}}\,,$$

$$\sin 2\varphi^* = \frac{\tan 2\varphi^*}{\sqrt{1+\tan^2 2\varphi^*}} = \frac{2\,I_{yz}}{\sqrt{(I_y - I_z)^2 + 4\,I_{yz}^2}}$$

we find

$$I_{1,2} = \frac{I_y + I_z}{2} \pm \sqrt{\left(\frac{I_y - I_z}{2}\right)^2 + I_{yz}^2}\,. \tag{4.17}$$

The positive (negative) sign corresponds to the maximum (minimum) moment of inertia.

We will now determine the angle for which the product of inertia $I_{\eta\zeta}$ vanishes. If we introduce the condition $I_{\eta\zeta} = 0$ into (4.14) we obtain the same angle φ^* as given by (4.16). Thus, the product of inertia with respect to the principal axes is zero. As already mentioned in Section 4.2.1, the product of inertia is zero with respect to an axis of symmetry. Therefore, an axis of symmetry and the axis perpendicular to it are principal axes.

As an example we consider the rectangular area in Fig. 4.10. Since $I_{yz} = 0$ (see (4.8c)), the axes y and z are principal axes and the moments of inertia $I_y = b\,h^3/12$ and $I_z = h\,b^3/12$ are the

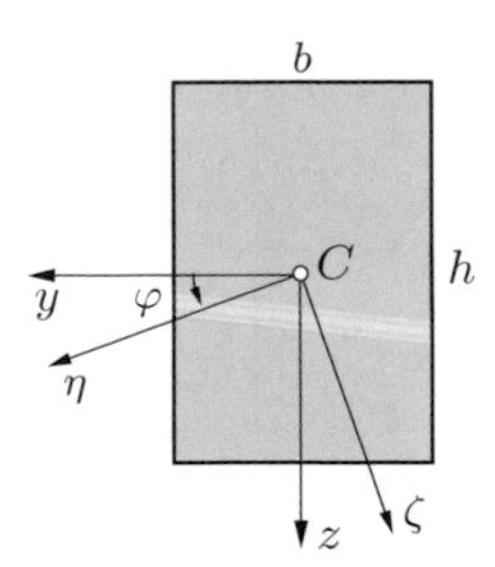

Fig. 4.10

principal moments of inertia. The moments of inertia with respect to the η, ζ-coordinate system are determined from (4.14):

$$I_\eta = \frac{b\,h}{24}[(h^2 + b^2) + (h^2 - b^2)\cos 2\,\varphi]\,,$$

$$I_\zeta = \frac{b\,h}{24}[(h^2 + b^2) - (h^2 - b^2)\cos 2\,\varphi]\,,$$

$$I_{\eta\zeta} = -\frac{b\,h}{24}(h^2 - b^2)\sin 2\,\varphi\,.$$

In the special case of a square ($h = b$) we obtain $I_\eta = I_\zeta = h^4/12$ and $I_{\eta\zeta} = 0$. These results are independent of the angle φ. Hence, in the case of a square each inclined coordinate system represents principal axes.

Note that the moments of inertia are components of a *tensor*. Therefore, the transformation relations (4.14) and the resulting equations (4.15) - (4.17) are analogous to the corresponding equations for the stress tensor (see Section 2.2). In analogy to Mohr's circle for the stresses, a circle for the moments of inertia can be constructed. In doing so, the normal stresses σ_x, σ_y have to be replaced with I_y, I_z and the shear stress τ_{xy} has to be replaced with the product of inertia I_{yz}.

E4.3

Example 4.3 Determine the principal axes and the principal moments of inertia for the thin-walled area with constant thickness $t\,(t \ll a)$ shown in Fig. 4.11a.

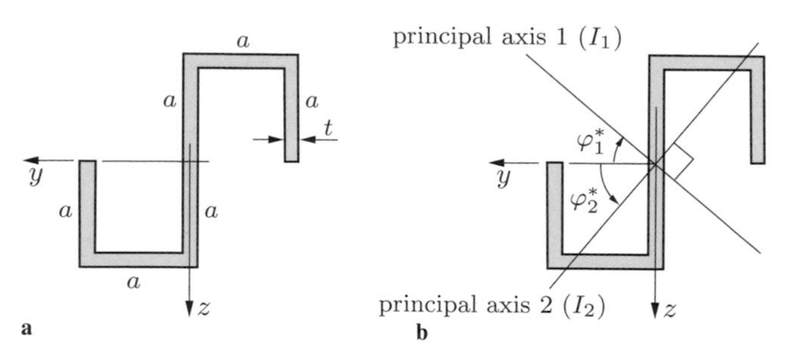

Fig. 4.11

Solution First we determine the moments of inertia about the axes y and z. If we divide the area into several rectangular parts and neglect terms of higher order in the thickness t, we obtain

$$I_y = \frac{1}{12}\, t(2\,a)^3 + 2\left\{\frac{1}{3}\, t\, a^3 + a^2(a\,t)\right\} = \frac{10}{3}\, t\, a^3\,,$$

$$I_z = 2\left\{\frac{1}{3}\, t\, a^3 + a^2(a\,t)\right\} = \frac{8}{3}\, t\, a^3\,,$$

$$I_{yz} = 2\left\{-\left[\frac{a}{2}\, a(a\,t)\right] - \left[a\,\frac{a}{2}(a\,t)\right]\right\} = -\,2\,t\,a^3\,.$$

The orientation of the principal axes follows from (4.16)

$$\tan 2\,\varphi^* = \frac{2\,I_{yz}}{I_y - I_z} = -\,\frac{2\cdot 2\cdot 3}{10-8} = -\,6$$

which leads to

$$\underline{\underline{\varphi_1^* = -\,40.3^\circ}}\,, \qquad \underline{\underline{\varphi_2^*}} = \varphi_1^* + 90^\circ = \underline{\underline{49.7^\circ}}\,. \tag{a}$$

The principal moments of inertia are calculated from (4.17):

$$I_{1,2} = \frac{t\,a^3}{3}\left[\frac{10+8}{2} \pm \sqrt{\left(\frac{10-8}{2}\right)^2 + (-6)^2}\,\right] = \left(3 \pm \frac{\sqrt{37}}{3}\right) t\,a^3$$

$$\rightarrow \quad \underline{\underline{I_1 = 5.03\,t\,a^3}}\,, \qquad \underline{\underline{I_2 = 0.97\,t\,a^3}}\,. \tag{b}$$

The principal axes are shown in Fig. 4.11b. In order to decide which principal moment of inertia (b) corresponds to which direction (a), we may insert (a) into (4.14). However, in this example one can see by inspection that the maximum moment of inertia I_1 belongs to the angle φ_1^* since the distances of the area elements from the corresponding axis are larger than those for the angle φ_2^*.

4.3 Basic Equations of Ordinary Bending Theory

4.3

We will now derive the equations which enable us to determine the stresses and the deformations due to the bending of a beam. In the following we restrict ourselves to *ordinary (uniaxial) bending*,

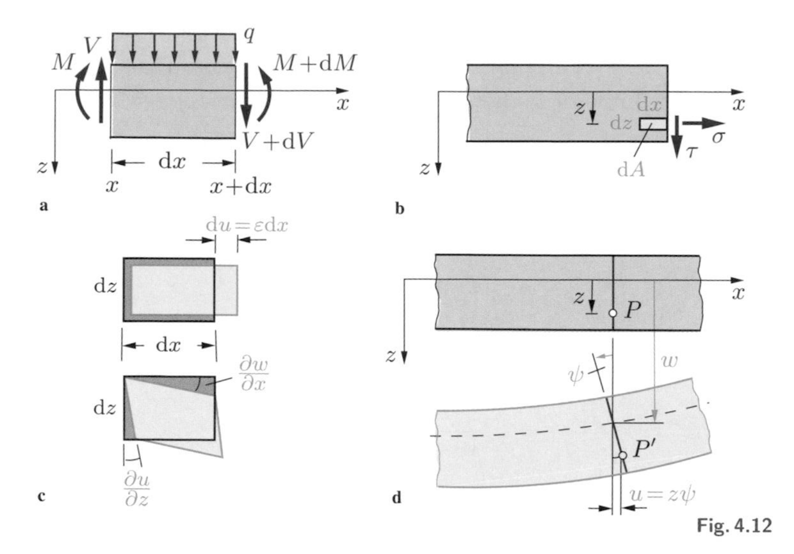

Fig. 4.12

i.e., we assume that the axes y and z are the principal axes of the cross sectional area ($I_{yz} = 0$). In addition, the applied loads are assumed to cause only a shear force V in the z-direction and a bending moment M about the y-axis. These assumptions are satisfied, for example, if the z-axis is an axis of symmetry of the cross section and the applied forces act in the x, z-plane.

We have to use three different types of equations: equations from statics, geometrical (kinematic) relations and Hooke's law. The equations from statics are the equilibrium conditions formulated for a beam element as shown in Fig. 4.12a. They are taken from Volume 1, Section 7.2.2:

$$\frac{\mathrm{d}V}{\mathrm{d}x} = -q, \qquad \frac{\mathrm{d}M}{\mathrm{d}x} = V\,. \tag{4.18}$$

The bending moment M and the shear force V are the resultants of the normal stresses σ (acting in the x-direction) and the shear stresses τ (acting in the z-direction), respectively (see Fig. 4.12b):

$$M = \int z\,\sigma\,\mathrm{d}A, \qquad V = \int \tau\,\mathrm{d}A\,. \tag{4.19a,b}$$

The normal force

$$N = \int \sigma \, \mathrm{d}A \tag{4.19c}$$

is zero due to the assumptions. Since only one normal stress and one shear stress are needed here, we have omitted the subscripts of the stress components: ($\sigma = \sigma_x$, $\tau = \tau_{xz}$).

The kinematic relations between the strain components and the displacements are taken from Section 3.1:

$$\varepsilon = \frac{\partial u}{\partial x}, \qquad \gamma = \frac{\partial w}{\partial x} + \frac{\partial u}{\partial z}. \tag{4.20}$$

Here, u and w are the displacements in the x-direction (axis of the beam) and in the z-direction, respectively. Since no other strain components are needed, the subscripts are also omitted: $\varepsilon = \varepsilon_x$, $\gamma = \gamma_{xz}$. The strain ε and the shear strain γ describe the deformation of an arbitrary element of the beam with length $\mathrm{d}x$ and height $\mathrm{d}z$ (Fig. 4.12b). This is illustrated in Fig. 4.12c.

We assume that the normal stresses σ_y and σ_z in the beam are small as compared with $\sigma = \sigma_x$ and therefore may be neglected. Then Hooke's law is given by (compare Section 3.2)

$$\sigma = E\,\varepsilon, \quad \tau = G\,\gamma. \tag{4.21}$$

It is not possible to uniquely determine the stresses and displacements with the aid of Equations (4.18) - (4.21). Therefore, we have to introduce *additional assumptions*. They concern the displacements of the points of a cross section at an arbitrary position x (Fig. 4.12d):

a) The displacement w is independent of z:

$$w = w(x). \tag{4.22a}$$

Hence, every point of a cross section undergoes the same deflection in the z-direction. This implies that the height of the beam does not change due to bending: $\varepsilon_z = \partial w/\partial z = 0$.

b) Plane cross sections of the beam remain plane during the bending. In addition to the displacement w, a cross section undergoes a rotation. The angle of rotation $\psi = \psi(x)$ is a *small angle*; it is counted as positive if the rotation is counterclockwise. Thus, the displacement u of a point P which is located at a distance z from the x-axis is given by

$$u(x, z) = \psi(x)\, z \,. \tag{4.22b}$$

Experiments show that the assumptions a) and b) lead to results which are sufficiently accurate in the case of a slender beam with a constant cross section or with a slight taper.

Introducing (4.22a, b) and (4.20) into (4.21) yields

$$\sigma = E\,\frac{\partial u}{\partial x} = E\,\psi'\,z\,, \tag{4.23a}$$

$$\tau = G\left(\frac{\partial w}{\partial x} + \frac{\partial u}{\partial z}\right) = G(w' + \psi) \tag{4.23b}$$

where $\mathrm{d}(\ )/\mathrm{d}x = (\ )'$ and w' represents the slope of the deformed axis of the beam. Since $|w'| \ll 1$, the slope is equal to the angle between the deformed axis of the beam and the x-axis. Using (4.23a), the Equations (4.19a) and (4.19c) yield the stress resultants

$$M = E\,\psi' \int z^2\,\mathrm{d}A, \qquad N = E\,\psi' \int z\,\mathrm{d}A\,.$$

It was assumed that the normal force is zero: $N = 0$. Hence, according to the second equation, we obtain $S_y = \int z\,\mathrm{d}A = 0$, which implies that the y-axis has to be a centroidal axis (see Volume 1, Section 4.3). This is the reason for the particular choice of the coordinate system mentioned in Section 4.1. Introducing the second moment of area $I = I_y = \int z^2\,\mathrm{d}A$, the first equation can be written in the form

$$M = EI\,\psi'\,. \tag{4.24}$$

Thus, the change $\mathrm{d}\psi$ of the angle ψ in the x-direction is proportional to the bending moment M. The corresponding deformation of a beam element of length $\mathrm{d}x$ is illustrated in Fig. 4.13a. Equation (4.24) is a *constitutive equation for the bending moment.* The quantity EI is referred to as *flexural rigidity* or *bending stiffness.*

Equation (4.23b) represents a shear stress τ which is constant in the cross section. This result is due to the simplifying assumptions a) and b). In reality, however, the shear stress is not evenly distributed as will be shown in Section 4.6.1. In particular, $\tau = 0$ at the outer fibers of the cross section. This can easily be shown with the fact that, according to (2.3), the shear stresses in two perpendicular planes are equal (complementary shear stresses). Note that there are no shear stresses at the upper and the lower surface of the beam acting in the direction of the beam axis (no applied loads in this direction). Therefore, the shear stress also has to be zero at the extreme fibers of a cross section which is orthogonal to the surfaces of the beam. The actual distribution of the shear stress τ may approximately be taken into account by introducing a factor $\varkappa$, called the *shear correction factor*, when (4.23b) is inserted into (4.19b):

$$V = \varkappa GA(w' + \psi)\,. \tag{4.25}$$

This is a *constitutive equation for the shear force.* An element of the beam undergoes a shear deformation $w' + \psi$ under the action of a shear force V (Fig. 4.13b). The quantity $\varkappa GA = GA_S$ is called the *shear stiffness* and $A_S = \varkappa A$ is referred to as the "shear area" (compare Sections 4.6.2 and 6.1).

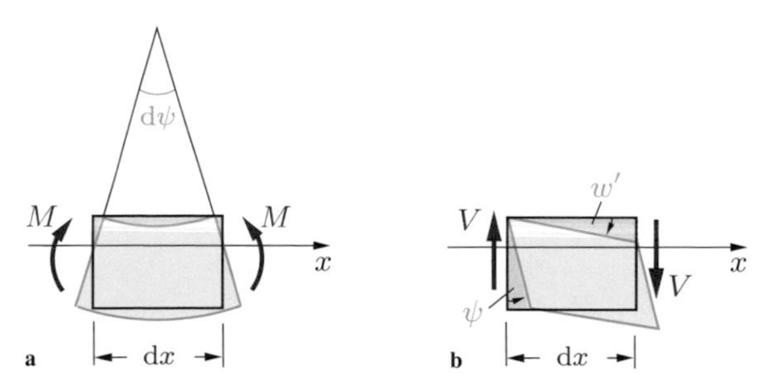

Fig. 4.13

4.4

4.4 Normal Stresses

If $\psi' = M/EI$ according to (4.24) is introduced into (4.23a), the normal stresses in the cross section of a beam are obtained (compare (4.4)):

$$\sigma = \frac{M}{I}\,z\,. \tag{4.26}$$

This equation is called the *bending formula*. It shows that the normal stresses, which are also referred to as the *flexural* or *bending stresses*, are linearly distributed in z-direction as shown in Fig. 4.14. If the bending moment M is positive, the stresses are

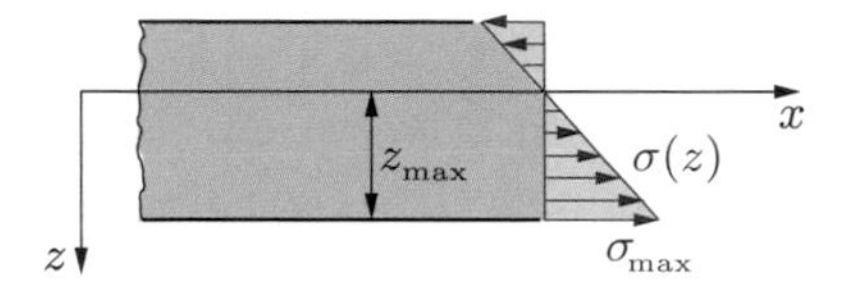

Fig. 4.14

positive (tensile stresses) for $z > 0$ and they are negative (compressive stresses) for $z < 0$. For $z = 0$ (i.e., in the x, y-plane) we have $\sigma = 0$. Since $\varepsilon = \sigma/E$, the strain ε is also zero in the x, y-plane: the fibers in this plane do not undergo any elongation or contraction. Therefore, this plane is called the *neutral surface* of the beam. The intersection of a cross section of the beam with the neutral surface (i.e., the y-axis) is called the *neutral axis*. The bending stresses (tensile or compressive) attain their maximum values at the extreme fibers. With the notation z_{max} for the maximum value of z (often also denoted by c) and

$$W = \frac{I}{|z|_{max}} \tag{4.27}$$

we obtain the maximum tensile or compressive stress, respectively:

$$\sigma_{max} = \frac{M}{W}\,. \tag{4.28}$$

The quantity W (often also denoted by S) is called the *section modulus*.

If the state of stress in a beam is investigated, it often suffices to determine only the normal stresses since the shear stresses are usually negligibly small (slender beams!). There are several different types of problems arising in this context. If, for example, the bending moment M, the section modulus W and the allowable stress σ_{allow} are known, one has to verify that the maximum stress $\sigma_{\max}$ satisfies the requirement

$$\sigma_{\max} \leq \sigma_{\text{allow}} \qquad \rightarrow \qquad \frac{M}{W} \leq \sigma_{\text{allow}} \,.$$

This is called a *stress check*.

On the other hand, if M and σ_{allow} are given, the required section modulus can be calculated from

$$W_{\text{req}} = \frac{M}{\sigma_{\text{allow}}} \,.$$

This is referred to as the *design* of a beam.

Finally, if W and σ_{allow} are given, the allowable load can be calculated from the condition that the maximum bending moment $M_{\max}$ must not exceed the allowable moment $M_{\text{allow}} = W\,\sigma_{\text{allow}}$:

$$M_{\max} \leq W\,\sigma_{\text{allow}} \,.$$

Example 4.4 The cross section of a cantilever beam ($l = 3$ m) consists of a circular ring ($R_i = 4$ cm, $R_a = 5$ cm), see Fig. 4.15. The allowable stress is given by $\sigma_{\text{allow}} = 150$ MPa. **E4.4**

Determine the allowable value of the load F.

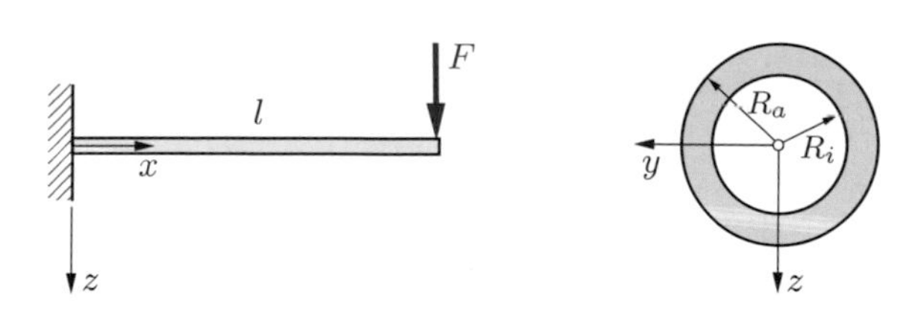

Fig. 4.15

Solution The maximum bending moment is located at the clamping; it has the magnitude

$$M_{\max} = l\,F\,.$$

The maximum stress is then determined by

$$\sigma_{\max} = \frac{M_{\max}}{W} = \frac{l\,F}{W}\,.$$

To obtain the allowable force F, we use the condition $\sigma_{\max} \le \sigma_{\text{allow}}$:

$$\frac{l\,F}{W} \le \sigma_{\text{allow}} \qquad \rightarrow \qquad F \le \frac{W\sigma_{\text{allow}}}{l}\,.$$

According to (4.11), the moment of inertia of a circular ring is given by $I = I_y = \pi(R_a^4 - R_i^4)/4$. This yields the section modulus ($z_{\max} = R_a$):

$$W = \frac{I}{z_{\max}} = \frac{\pi(R_a^4 - R_i^4)}{4\,R_a} = 58\,\text{cm}^3\,.$$

With σ_{allow} =150 MPa and $l = 3$ m we obtain the allowable force:

$$\underline{\underline{F \le 2.9\,\text{kN}}}\,.$$

E4.5

Example 4.5 The simply supported beam (length $l = 10$ m) carries the force $F = 200$ kN (Fig. 4.16).

Find the required side length c of the thin-walled quadratic cross section such that the allowable stress $\sigma_{\text{allow}} = 200$ MPa is not exceeded. The thickness $t = 15$ mm of the profile is given.

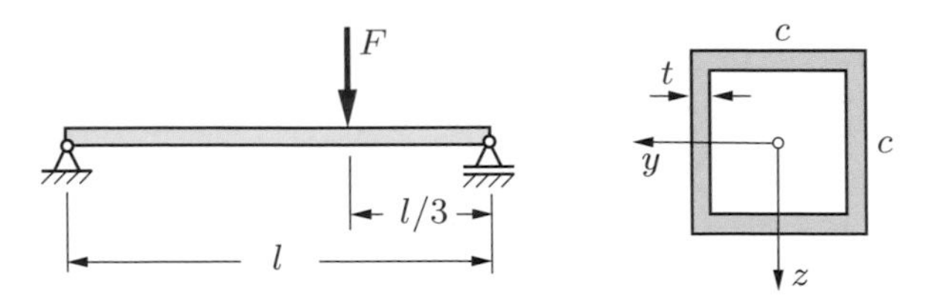

Fig. 4.16

Solution The side length has to be chosen such that the condition

$$W \geq \frac{M}{\sigma_{\text{allow}}} \tag{a}$$

is satisfied. The maximum bending moment (located at the point of application of F) is given by

$$M = \frac{2}{9} l F . \tag{b}$$

The moment of inertia of the cross section ($t \ll c$) is obtained as

$$I \approx 2\left[\frac{t\,c^3}{12} + \left(\frac{c}{2}\right)^2 c\,t\right] = \frac{2}{3} t\,c^3 .$$

This yields the section modulus

$$W = \frac{I}{z_{\max}} \approx \frac{I}{c/2} = \frac{4}{3} t\,c^2 . \tag{c}$$

Introducing (b) and (c) into (a) leads to

$$\frac{4}{3} t\,c^2 \geq \frac{2\,l\,F}{9\,\sigma_{\text{allow}}} \qquad \rightarrow \qquad c \geq \sqrt{\frac{l\,F}{6\,t\,\sigma_{\text{allow}}}}$$

or

$$\underline{\underline{c \geq 333\,\text{mm}}} .$$

4.5 Deflection Curve

4.5

4.5.1 Differential Equation of the Deflection Curve

The Equations (4.18), (4.24) and (4.25) represent four differential equations for the stress resultants V, M and the kinematic quantities ψ, w. They can be further simplified if we assume that the shear rigidity is very large. Assuming $\varkappa GA \rightarrow \infty$ (and the shear force to be finite), Equation (4.25) reduces to

$$w' + \psi = 0 . \tag{4.29}$$

In this case, an element of the beam does not undergo a shear deformation due to the shear force: $\gamma = 0$. The beam is then referred to as being *rigid with respect to shear*. Equation (4.29) allows a simple geometrical interpretation: a cross section which is perpendicular to the undeformed axis of the beam (the x-axis) remains perpendicular to the deformed beam axis during the bending (Fig. 4.17). This statement and the assumption that plane cross sections remain plane, see assumption b) in Section 4.3, are called the *Bernoulli assumptions* (Jakob Bernoulli, 1655–1705). They yield sufficiently accurate results for slender beams; in the special case of pure bending ($V = 0$) they are exact. A beam whose behaviour is investigated according to this theory is referred to as a *Bernoulli beam*.

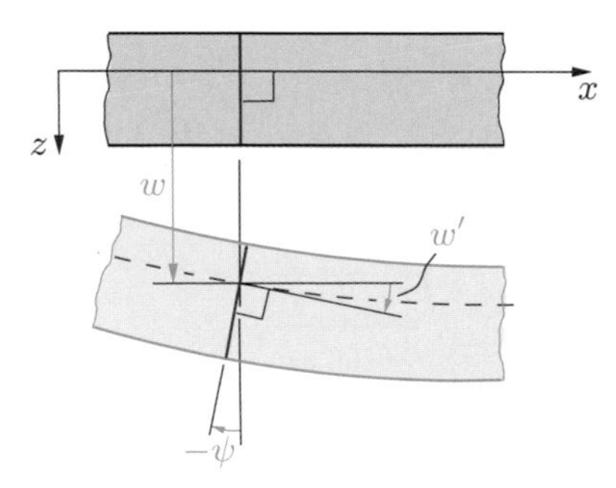

Fig. 4.17

Given the load $q(x)$, the Equations (4.18), (4.24) and (4.29) are four differential equations of first order for the four unknown functions V, M, ψ and w:

$$V' = -q, \qquad M' = V, \qquad \psi' = \frac{M}{EI}, \qquad w' = -\psi. \qquad (4.30)$$

If we eliminate ψ from the third and the fourth equation, we obtain the *differential equation of the deflection curve*:

$$w'' = -\frac{M}{EI}. \qquad (4.31)$$

Provided the bending moment M and the flexural rigidity EI are known, integration of (4.31) yields the slope $w'(x)$ and the deflection $w(x)$, also called the *elastic line.*

The *curvature* $\varkappa_B$ of the deformed axis of the beam is given by

$$\varkappa_B = \frac{w''}{(1+w'^2)^{3/2}} . \tag{4.32a}$$

In the case of a small slope ($w'^2 \ll 1$) the relation (4.32a) reduces to

$$\varkappa_B \approx w'' . \tag{4.32b}$$

The curvature of the deformed axis of the beam is therefore proportional to the bending moment (see (4.31)). It is negative for $M > 0$ and positive for $M < 0$ (Fig. 4.18).

Fig. 4.18

If we use the first and the second equation of (4.30), we obtain a different form of the differential equation of the elastic line. We first differentiate $M = -EIw''$ and insert the result into $V = M'$. This yields

$$V = -(EI\,w'')' . \tag{4.33}$$

Then we differentiate once more and use the relation $V' = -q$ to obtain a differential equation of fourth order:

$$(EI\,w'')'' = q . \tag{4.34a}$$

In the special case of $EI = \text{const}$, (4.34a) reduces to

$$EI\,w^{IV} = q . \tag{4.34b}$$

Table 4.2. Boundary conditions

Support	w	w'	M	V
pin	0	$\neq 0$	0	$\neq 0$
parallel motion	$\neq 0$	0	$\neq 0$	0
fixed end	0	0	$\neq 0$	$\neq 0$
free end	$\neq 0$	$\neq 0$	0	0

If the load q and the flexural rigidity EI are given, the deflection w can be determined through integration of the above differential equation.

The constants of integration are determined from boundary conditions (compare Volume 1, Section 7.2.3). We distinguish between geometrical and statical boundary conditions. *Geometrical boundary conditions* are statements concerning the geometrical (kinematic) quantities w or w', respectively. *Statical boundary conditions* are statements referring to the stress resultants V or M. Consider, for example, a beam that is supported by a pin at one end. Then the deflection w and the bending moment M are zero at this point. No statement concerning the slope w' or the shear force V, respectively, can be made. At a clamping, the deflection and the slope are zero; the shear force and the bending moment are unknown. Note that two boundary conditions can be formulated at each end point of a beam. Table 4.2 displays the boundary conditions for various supports (cf. Volume 1, Equation (7.11)).

Equation (4.31) can be applied to statically determinate problems only, since only in this case the bending moment can be calculated from equilibrium conditions in advance. The two constants of integration which appear through the integration of (4.31) are determined from geometrical boundary conditions; the statical boundary conditions are satisfied automatically. If the problem is statically indeterminate, the differential equation (4.34) has to be used since the bending moment is unknown in this case. The four constants of integration are calculated from geometrical and/or statical boundary conditions.

4.5.2 Beams with one Region of Integration

We will now show with the aid of several examples how the differential equations (4.31) or (4.34) can be used to obtain the deflection curve. In this section we restrict ourselves to beams where the integration can be performed in *one* region, i.e., we assume that each of the quantities $q(x)$, $V(x)$, $M(x)$, $w'(x)$ and $w(x)$ is given by *one* function for the entire length of the beam.

Let us first consider a cantilever beam (flexural rigidity EI) subjected to a concentrated force F (Fig. 4.19a). Since the system is statically determinate, the bending moment can be calculated from the equilibrium conditions (compare Volume 1, Section 7.2). With the coordinate system as shown in Fig. 4.19b, we obtain $M = -F(l - x)$. Introducing into (4.31) and integrating yields

$$EI\,w'' = F\,(-\,x + l)\,,$$

$$EI\,w' = F\Big(-\,\frac{x^2}{2} + l\,x\Big) + C_1\,,$$

$$EI\,w = F\Big(-\,\frac{x^3}{6} + \frac{l\,x^2}{2}\Big) + C_1\,x + C_2\,.$$

The geometrical boundary conditions

$$w'(0) = 0, \qquad w(0) = 0$$

lead to the constants of integration:

$$C_1 = 0, \qquad C_2 = 0\,.$$

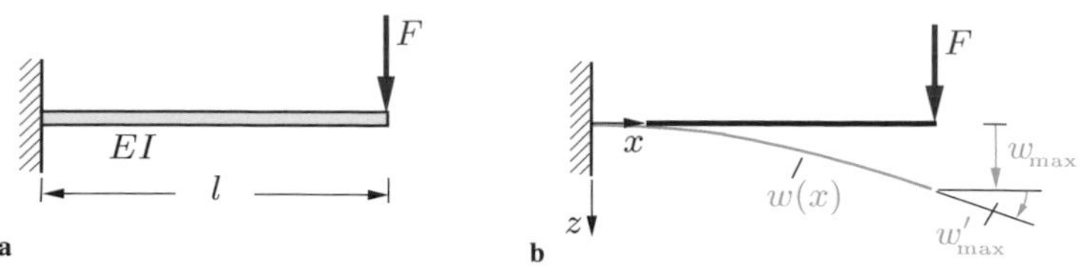

Fig. 4.19

Hence, the slope and the deflection are obtained as

$$w'(x) = \frac{F\,l^2}{2\,EI}\left(-\,\frac{x^2}{l^2} + 2\,\frac{x}{l}\right),$$

$$w(x) = \frac{F\,l^3}{6\,EI}\left(-\,\frac{x^3}{l^3} + 3\,\frac{x^2}{l^2}\right).$$

The maximum slope and the maximum deflection (at $x = l$, see Fig. 4.19b) are

$$w'_{\max} = \frac{F\,l^2}{2\,EI}, \qquad w_{\max} = \frac{F\,l^3}{3\,EI}.$$

Let us now consider three beams (bending stiffness EI) subjected to a constant line load q_0 (Figs. 4.20a-c). The supports in the three cases are different; the systems in the Figs. 4.20a and b are statically determinate, the system in Fig. 4.20c is statically indeterminate. Since in the latter case the bending moment can not be calculated from equilibrium conditions, we will use the differential equation (4.34b) in all three cases. We introduce a coordinate system and integrate (4.34b):

$$\begin{aligned}
EI\,w^{IV} &= q = q_0\,,\\
EI\,w''' &= -\,V = q_0\,x + C_1\,,\\
EI\,w'' &= -\,M = \frac{1}{2}q_0\,x^2 + C_1\,x + C_2\,,\\
EI\,w' &= \frac{1}{6}q_0\,x^3 + \frac{1}{2}C_1\,x^2 + C_2\,x + C_3\,,\\
EI\,w &= \frac{1}{24}q_0\,x^4 + \frac{1}{6}C_1\,x^3 + \frac{1}{2}C_2\,x^2 + C_3\,x + C_4\,.
\end{aligned}$$

These equations are independent of the supports and therefore are valid for all three cases. Different boundary conditions lead to

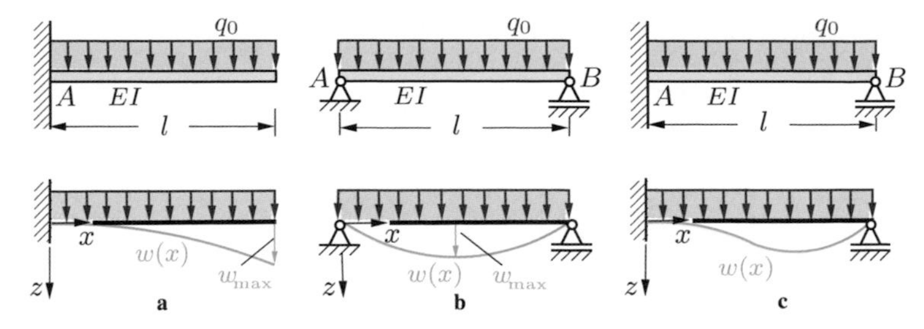

Fig. 4.20

different constants of integration:

a) $$\begin{aligned} w'(0) &= 0 \quad \rightarrow \quad C_3 = 0\,, \\ w(0) &= 0 \quad \rightarrow \quad C_4 = 0\,, \\ V(l) &= 0 \quad \rightarrow \quad q_0\, l + C_1 = 0 \quad \rightarrow \quad C_1 = -\, q_0\, l\,, \\ M(l) &= 0 \quad \rightarrow \quad \frac{1}{2} q_0\, l^2 + C_1\, l + C_2 = 0 \quad \rightarrow \quad C_2 = \frac{1}{2} q_0\, l^2\,, \end{aligned}$$

b) $$\begin{aligned} M(0) &= 0 \quad \rightarrow \quad C_2 = 0\,, \\ M(l) &= 0 \quad \rightarrow \quad \frac{1}{2} q_0\, l^2 + C_1\, l = 0 \quad \rightarrow \quad C_1 = -\, \frac{1}{2} q_0\, l\,, \\ w(0) &= 0 \quad \rightarrow \quad C_4 = 0\,, \\ w(l) &= 0 \quad \rightarrow \quad \frac{1}{24} q_0\, l^4 + \frac{1}{6} C_1\, l^3 + C_3\, l = 0 \quad \rightarrow \quad C_3 = \frac{1}{24} q_0\, l^3\,, \end{aligned}$$

c) $$\begin{aligned} w'(0) &= 0 \quad \rightarrow \quad C_3 = 0\,, \\ w(0) &= 0 \quad \rightarrow \quad C_4 = 0\,, \\ M(l) &= 0 \quad \rightarrow \quad \frac{1}{2} q_0\, l^2 + C_1\, l + C_2 = 0\,, \\ w(l) &= 0 \quad \rightarrow \quad \frac{1}{24} q_0\, l^4 + \frac{1}{6} C_1\, l^3 + \frac{1}{2} C_2\, l^2 = 0 \\ & \qquad \rightarrow \quad C_1 = -\, \frac{5}{8} q_0\, l\,, \quad C_2 = \frac{1}{8} q_0\, l^2\,. \end{aligned}$$

Hence, we obtain the elastic lines (Figs. 4.20a–c)

$$\text{a)} \qquad w(x) = \frac{q_0\, l^4}{24\, EI} \left[\left(\frac{x}{l}\right)^4 - 4\left(\frac{x}{l}\right)^3 + 6\left(\frac{x}{l}\right)^2 \right],$$

$$\text{b)} \qquad w(x) = \frac{q_0\, l^4}{24\, EI} \left[\left(\frac{x}{l}\right)^4 - 2\left(\frac{x}{l}\right)^3 + \left(\frac{x}{l}\right) \right],$$

$$\text{c)} \qquad w(x) = \frac{q_0\, l^4}{24\, EI} \left[\left(\frac{x}{l}\right)^4 - \frac{5}{2}\left(\frac{x}{l}\right)^3 + \frac{3}{2}\left(\frac{x}{l}\right)^2 \right].$$

The maximum deflection in the case a) is given by

$$w_{\max} = w(l) = \frac{q_0\, l^4}{8\, EI}$$

and in case b) it is

$$w_{\max} = w\left(\frac{l}{2}\right) = \frac{5}{384}\,\frac{q_0\, l^4}{EI}.$$

After having determined the constants of integration we also know the slope w', the bending moment $M(x)$ and the shear force $V(x)$. For example,

$$V(x) = -\,\frac{q_0\, l}{8}\left[8\left(\frac{x}{l}\right) - 5\right],$$

$$M(x) = -\,\frac{q_0\, l^2}{8}\left[4\left(\frac{x}{l}\right)^2 - 5\left(\frac{x}{l}\right) + 1\right]$$

for the statically indeterminate beam in case c). The support reactions can be taken from the stress resultants:

$$A = V(0) = \frac{5\, q_0\, l}{8}, \qquad B = -\,V(l) = \frac{3\, q_0\, l}{8},$$

$$M_A = M(0) = -\,\frac{q_0\, l^2}{8}.$$

As a check we verify that the equilibrium conditions

$$\uparrow:\quad A + B - q_0\, l = 0, \qquad \overset{\curvearrowleft}{A}:\quad -\,M_A + l\,B - \frac{l}{2} q_0\, l = 0$$

are satisfied.

E4.6

Example 4.6 A simply supported beam (bending stiffness EI) is loaded by a moment M_0 (Fig. 4.21a).

Determine the location and magnitude of the maximum deflection.

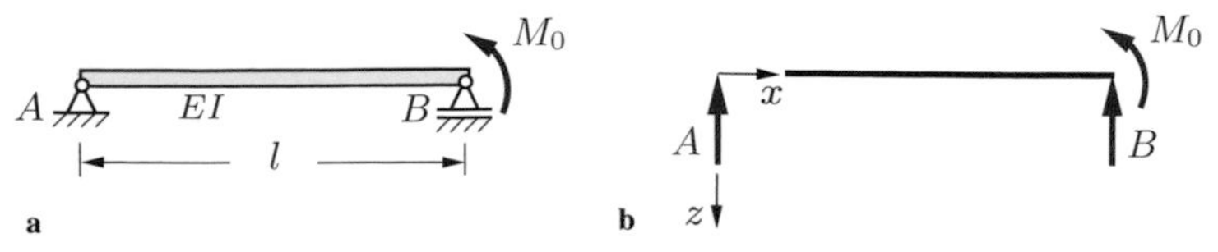

Fig. 4.21

Solution The beam is statically determinate. Therefore we can determine the bending moment from the equilibrium conditions. Using the coordinate system in Fig. 4.21b and the support reactions $A = -B = M_0/l$ we find

$$M(x) = x\,A = M_0\,\frac{x}{l}\,.$$

Substitution into (4.31) followed by two integrations yield

$$EI\,w'' = -\frac{M_0}{l}\,x\,,$$

$$EI\,w' = -\frac{M_0}{2\,l}\,x^2 + C_1\,,$$

$$EI\,w = -\frac{M_0}{6\,l}\,x^3 + C_1\,x + C_2\,.$$

The constants of integration can be calculated from the geometrical boundary conditions:

$$w(0) = 0 \quad \rightarrow \quad C_2 = 0\,,$$

$$w(l) = 0 \quad \rightarrow \quad -\frac{M_0}{6\,l}\,l^3 + C_1\,l = 0 \quad \rightarrow \quad C_1 = \frac{M_0\,l}{6}\,.$$

Thus, we obtain the elastic line

$$w(x) = \frac{1}{EI}\left[-\frac{M_0}{6\,l}x^3 + \frac{M_0\,l}{6}x\right] = \frac{M_0\,l^2}{6\,EI}\left[-\left(\frac{x}{l}\right)^3 + \left(\frac{x}{l}\right)\right].$$

The maximum deflection is characterized by a vanishing slope:

$$w' = 0 \quad \rightarrow \quad -\frac{M_0}{2\,l}x^2 + \frac{M_0\,l}{6} = 0 \quad \rightarrow \quad \underline{\underline{x^* = \frac{1}{\sqrt{3}}\,l}}\,.$$

Hence,

$$\underline{\underline{w_{\max}}} = w(x^*) = \frac{M_0\,l^2}{6\,EI}\left[-\frac{1}{3\sqrt{3}} + \frac{1}{\sqrt{3}}\right] = \underline{\underline{\frac{\sqrt{3}\,M_0\,l^2}{27\,EI}}}\,.$$

E4.7

Example 4.7 The beam in Fig. 4.22 is subjected to a concentrated force F.

Find the deflection at A and the moment M_B at the clamped end B.

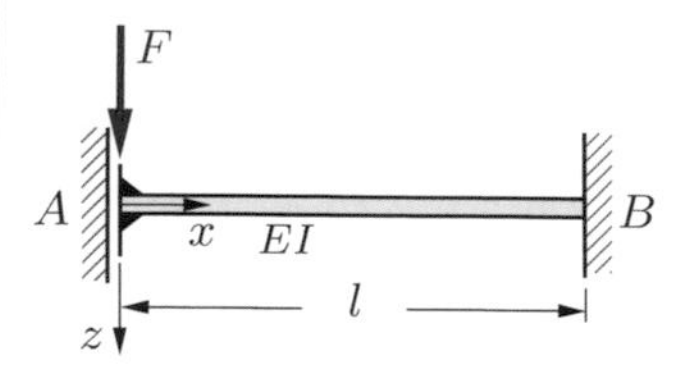

Fig. 4.22

Solution The system is statically indeterminate. Therefore we have to use Equation (4.34b). Integration with $q(x) = 0$ yields

$$\begin{aligned}
EI\,w^{IV} &= 0\,,\\
EI\,w''' &= -\,V = C_1\,,\\
EI\,w'' &= -\,M = C_1\,x + C_2\,,\\
EI\,w' &= \frac{1}{2}C_1\,x^2 + C_2\,x + C_3\,,\\
EI\,w &= \frac{1}{6}C_1\,x^3 + \frac{1}{2}C_2\,x^2 + C_3\,x + C_4\,.
\end{aligned}$$

The constants of integration are determined from the boundary conditions:

$$
\begin{aligned}
V(0) &= -F \quad \rightarrow \quad C_1 = F\,,\\
w'(0) &= 0 \quad \rightarrow \quad C_3 = 0\,,\\
w'(l) &= 0 \quad \rightarrow \quad \frac{1}{2}C_1\,l^2 + C_2\,l = 0 \;\rightarrow\; C_2 = -\frac{1}{2}F\,l\,,\\
w(l) &= 0 \quad \rightarrow \quad \frac{1}{6}C_1\,l^3 + \frac{1}{2}C_2\,l^2 + C_4 = 0 \;\rightarrow\; C_4 = \frac{1}{12}F\,l^3\,.
\end{aligned}
$$

The deflection curve and the bending moment can now be evaluated as

$$
\begin{aligned}
w(x) &= \frac{F\,l^3}{12\,EI}\left[2\left(\frac{x}{l}\right)^3 - 3\left(\frac{x}{l}\right)^2 + 1\right],\\
M(x) &= -\frac{F\,l}{2}\left[2\left(\frac{x}{l}\right) - 1\right].
\end{aligned}
$$

The deflection at A and the moment M_B are

$$
\underline{\underline{w_A = w(0) = \frac{F\,l^3}{12\,EI}}}\,, \qquad \underline{\underline{M_B = M(l) = -\frac{F\,l}{2}}}\,.
$$

Example 4.8 The beam in Fig. 4.23a carries a linearly varying line load. **E4.8**

Determine the shear force and the bending moment.

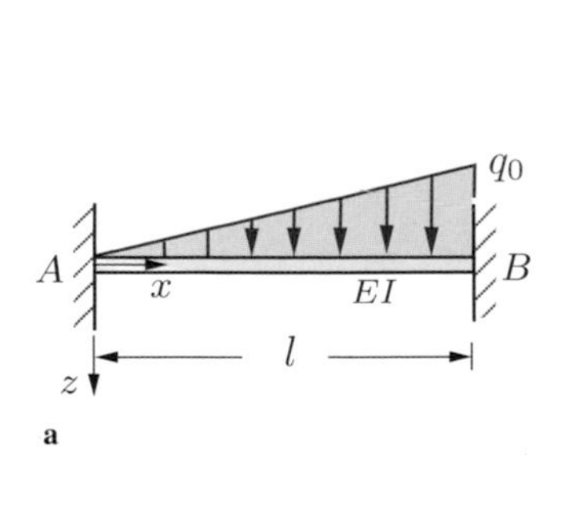

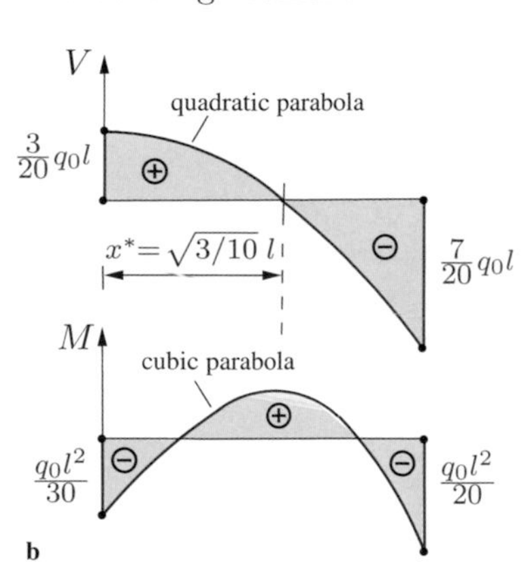

Fig. 4.23

Solution The beam is statically indeterminate. If we use the coordinate system in Fig. 4.23a, we can write $q(x) = q_0\, x/l$. Integration of (4.34b) yields

$$\begin{aligned}
EI\,w^{IV} &= \frac{q_0}{l}\,x\,,\\
EI\,w''' &= -\,V = \frac{1}{2}\frac{q_0}{l}\,x^2 + C_1\,,\\
EI\,w'' &= -\,M = \frac{1}{6}\frac{q_0}{l}\,x^3 + C_1\,x + C_2\,,\\
EI\,w' &= \frac{1}{24}\frac{q_0}{l}\,x^4 + \frac{1}{2}C_1\,x^2 + C_2\,x + C_3\,,\\
EI\,w &= \frac{1}{120}\frac{q_0}{l}\,x^5 + \frac{1}{6}C_1\,x^3 + \frac{1}{2}C_2\,x^2 + C_3\,x + C_4\,.
\end{aligned}$$

The constants of integration are determined from the boundary conditions:

$$\begin{aligned}
w'(0) &= 0 \quad\to\quad C_3 = 0\,,\\
w(0) &= 0 \quad\to\quad C_4 = 0\,,\\
w'(l) &= 0 \quad\to\quad \frac{1}{24}q_0\,l^3 + \frac{1}{2}C_1\,l^2 + C_2\,l = 0\\
w(l) &= 0 \quad\to\quad \frac{1}{120}q_0\,l^4 + \frac{1}{6}C_1\,l^3 + \frac{1}{2}C_2\,l^2 = 0\\
&\qquad\;\;\to\quad C_1 = -\,\frac{3}{20}q_0\,l\,,\qquad C_2 = \frac{1}{30}q_0\,l^2\,.
\end{aligned}$$

Thus, the shear force and bending moment distributions are given by (Fig. 4.23b)

$$\underline{\underline{V(x) = \frac{q_0\,l}{20}\left[-10\left(\frac{x}{l}\right)^2 + 3\right]}}\,,$$

$$\underline{\underline{M(x) = \frac{q_0\,l^2}{60}\left[-10\left(\frac{x}{l}\right)^3 + 9\left(\frac{x}{l}\right) - 2\right]}}\,.$$

The maximum bending moment is characterized by the condition of a vanishing shear force. It is therefore located at $x^* = \sqrt{3/10}\,l$. The support reactions can be taken from the diagrams (Fig. 4.23b):

$$A = V(0) = \frac{3}{20} q_0 \, l \,, \qquad B = -V(l) = \frac{7}{20} q_0 \, l \,,$$

$$M_A = M(0) = -\frac{q_0 \, l^2}{30} \,, \qquad M_B = M(l) = -\frac{q_0 \, l^2}{20} \,.$$

E4.9

Example 4.9 The cantilever beam (modulus of elasticity E) in Fig. 4.24 has a rectangular cross section. It is loaded by a constant line load q_0.

Given the constant width b, determine the height $h(x)$ such that the maximum bending stress at each cross section has the same value σ_0. Find the deflection $w(0)$ at the free end in this case.

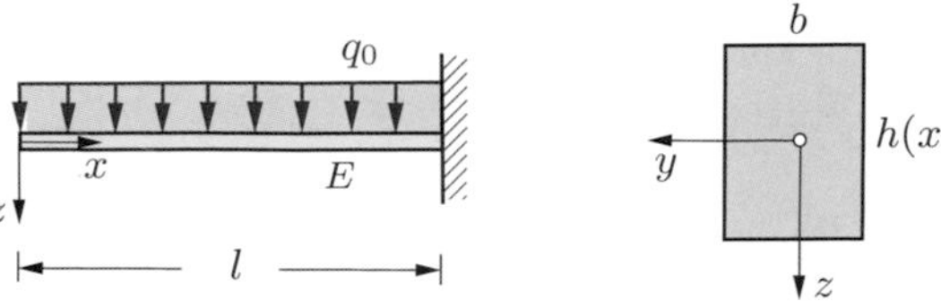

Fig. 4.24

Solution The maximum bending stress at each cross section is equal to σ_0 if, according to (4.28),

$$\sigma_0 = \frac{|M|}{W} \,.$$

The bending moment is given by

$$M(x) = -\frac{1}{2} q_0 \, x^2 \tag{a}$$

and the section modulus for a rectangular cross section is

$$W = \frac{I}{h/2} = \frac{b \, h^3 \, 2}{12 \, h} = \frac{b \, h^2}{6} \,.$$

This yields the required height distribution:

$$\sigma_0 = \frac{q_0 \, x^2 \, 6}{2 \, b \, h^2} \qquad \rightarrow \qquad \underline{\underline{h(x) = \sqrt{\frac{3 \, q_0}{b \, \sigma_0}} \, x}} \,.$$

The moment of inertia follows as

$$I(x) = \frac{b\,h^3}{12} = \frac{b}{12}\left(\frac{3\,q_0}{b\,\sigma_0}\right)^{3/2} x^3 = I_0\,\frac{x^3}{l^3}\,, \qquad \text{(b)}$$

where $I_0 = b\,h^3(l)/12$ is the reference value at $x = l$, that is, $I_0 = I(l)$. If we introduce (a) and (b) into the differential equation (4.31) and integrate twice, we obtain

$$\begin{aligned}
w'' &= -\frac{M}{EI} = \frac{q_0\,l^3}{2\,EI_0}\,\frac{1}{x}\,,\\
w' &= \frac{q_0\,l^3}{2\,EI_0}\ln\frac{x}{C_1}\,,\\
w &= \frac{q_0\,l^3}{2\,EI_0}\left[x\ln\frac{x}{C_1} - x + C_2\right].
\end{aligned}$$

The constants of integration are determined from the boundary conditions:

$$\begin{aligned}
w'(l) &= 0 \quad \rightarrow \quad \ln\frac{l}{C_1} = 0 \quad \rightarrow \quad C_1 = l\,,\\
w(l) &= 0 \quad \rightarrow \quad l\,\ln 1 - l + C_2 = 0 \quad \rightarrow \quad C_2 = l\,.
\end{aligned}$$

Thus, introducing the dimensionless coordinate $\xi = x/l$, the deflection curve is given by

$$w(\xi) = \frac{q_0\,l^4}{2\,EI_0}\left[\xi\,\ln\,\xi - \xi + 1\right].$$

The deflection at the free end ($\xi = 0$) can be calculated with the help of the limit expression $\lim\limits_{\xi\to 0}\xi\,\ln\,\xi = 0$ to give

$$\underline{\underline{w(0) = \frac{q_0\,l^4}{2\,EI_0}}}\,.$$

Note that it is four times the deflection of a cantilever with a constant moment of inertia I_0.

4.5.3 Beams with several Regions of Integration

Frequently, one or several of the quantities q, V, M, w', w or the flexural rigidity EI are given through different functions of x in different portions of the beam. In this case the beam must be divided into several regions and the integration has to be performed separately in each of these regions. The constants of integration can be calculated from both, boundary conditions and *matching conditions*, also called *continuity conditions* (compare Volume 1, Section 7.2.4). The treatment of such problems will be illustrated by means of the following example.

E4.10

Example 4.10 A simply supported beam is subjected to a concentrated force F at $x = a$ (Fig. 4.25).

Determine the deflection w at the location $x = a$.

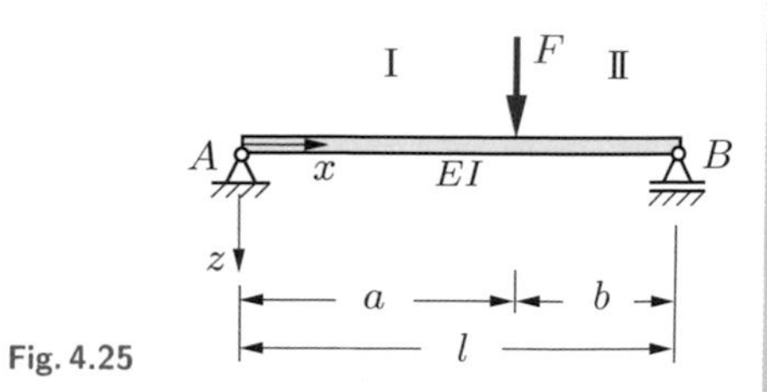

Fig. 4.25

Solution In this example, the shear force V has a jump (discontinuity) at $x = a$ and the bending moment is given by

$$M(x) = \begin{cases} F\dfrac{b}{l}\,x & \text{for } 0 \le x \le a\,, \\[2ex] F\dfrac{a}{l}(l-x) & \text{for } a \le x \le l\,. \end{cases}$$

Introduction into (4.31) and integration in the regions I ($0 \le x \le a$) and II ($a \le x \le l$), respectively, yields

$$\text{I}: \qquad EI\,w_{\text{I}}'' = -\,F\frac{b}{l}\,x,$$

$$EI\,w_{\text{I}}' = -\,F\frac{b}{l}\frac{x^2}{2} + C_1,$$

$$EI\,w_{\text{I}} = -\,F\frac{b}{l}\frac{x^3}{6} + C_1\,x + C_2,$$

$$\text{II}: \qquad EI\,w_{\text{II}}'' = -\,F\frac{a}{l}(l-x),$$

$$EI\,w_{II}' = F\frac{a}{l}\frac{(l-x)^2}{2} + C_3,$$

$$EI\,w_{II} = -\,F\frac{a}{l}\frac{(l-x)^3}{6} - C_3(l-x) + C_4.$$

It is practical to use the distance $(l-x)$ from support B as the variable in region II.

We now have to determine the four constants of integration. There are two boundary conditions:

$$w_I(0) = 0 \quad \rightarrow \quad C_2 = 0,$$

$$w_{II}(l) = 0 \quad \rightarrow \quad C_4 = 0\,.$$

In addition, we can use two matching conditions. Since there are no jumps in the deflection or the slope, these quantities are continuous at $x = a$:

$$w_I(a) = w_{II}(a) \;\rightarrow\; -\,F\frac{b}{l}\frac{a^3}{6} + C_1\,a = -\,F\frac{a}{l}\frac{b^3}{6} - C_3\,b,$$

$$w_I'(a) = w_{II}'(a) \;\rightarrow\; -\,F\frac{b}{l}\frac{a^2}{2} + C_1 = F\frac{a}{l}\frac{b^2}{2} + C_3,$$

$$\rightarrow C_1 = \frac{F\,a\,b(a+2\,b)}{6\,l}, \quad C_3 = -\,\frac{F\,a\,b(b+2\,a)}{6\,l}.$$

This yields the elastic line

$$w(x) = \begin{cases} \dfrac{F\,b\,l^2}{6\,EI}\dfrac{x}{l}\left[1-\left(\dfrac{b}{l}\right)^2-\left(\dfrac{x}{l}\right)^2\right] & \text{for } 0 \le x \le a, \\[2ex] \dfrac{F\,a\,l^2}{6\,EI}\dfrac{(l-x)}{l}\left[1-\left(\dfrac{a}{l}\right)^2-\left(\dfrac{l-x}{l}\right)^2\right] & \text{for } a \le x \le l. \end{cases}$$

The deflection at $x = a$ is found to be

$$\underline{\underline{w(a) = \frac{F\,a^2\,b^2}{3\,EI\,l}}}\,.$$

4.5.4 Method of Superposition

The differential equation of the elastic line is linear, see Equations (4.31) or (4.34), and the deflection depends linearly on the load. Therefore, it is possible to superimpose solutions for different load cases. Let us, for example, consider the beam in Fig. 4.26 which is subjected to a line load q_1 and a force F_2. The deflection w is obtained through a superposition of the deflections w_1 (due to the load q_1) and w_2 (due to the force F_2): $w = w_1 + w_2$. Analogously, we find the slope $w' = w_1' + w_2'$, the bending moment $M = M_1 + M_2$, and the shear force $V = V_1 + V_2$.

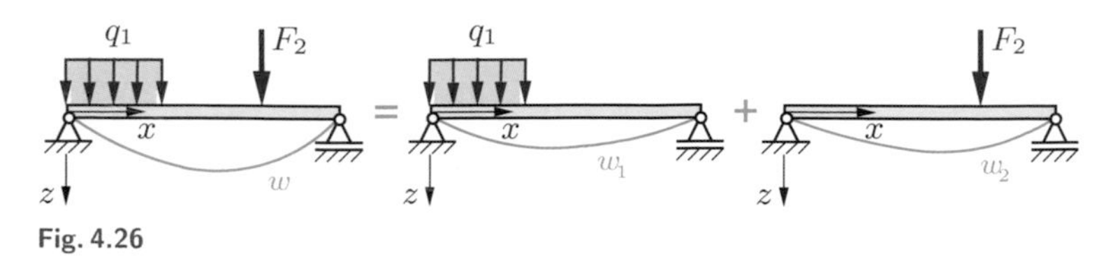

Fig. 4.26

Table 4.3 displays several solutions for statically determinate beams with a constant flexural rigidity EI. They can be used to find the solutions for combined loadings without having to integrate the differential equation of the elastic line.

It is often useful to apply the method of superposition to statically indeterminate systems. In order to illustrate the procedure let us consider the beam in Fig. 4.27a. It is statically indeterminate to the first degree. If we remove support B, we obtain a statically determinate system. This system subjected to the given load only is called " 0 "-system or *primary system*. The deflection $w^{(0)}$ of the beam in the " 0 "-system can be taken from Table 4.3. In particular, the deflection at the free end (at point B) is given by

$$w_B^{(0)} = \frac{q_0 \, l^4}{8 \, EI} .$$

Let us now consider the statically determinate system subjected only to an initially unknown force X ("1"-system). This force acts at the point where the support was removed (here point B). It corresponds to the support reaction B in the given system and is as yet unknown. The deflection at the free end of the beam can

Table 4.3. Elastic lines (see explanations at the end of the table)

Nr.	load	$EI\,w'_A$	$EI\,w'_B$
1	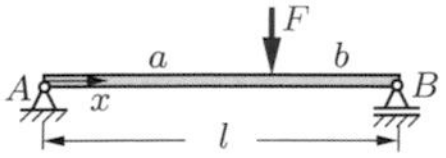	$\dfrac{F\,l^2}{6}(\beta-\beta^3)$	$-\dfrac{F\,l^2}{6}(\alpha-\alpha^3)$
2	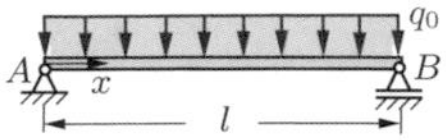	$\dfrac{q_0\,l^3}{24}$	$-\dfrac{q_0\,l^3}{24}$
3	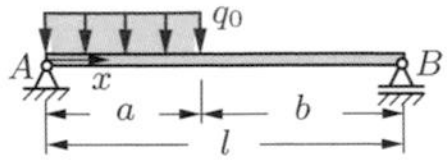	$\dfrac{q_0\,l^3}{24}(1-\beta^2)^2$	$\dfrac{q_0\,l^3}{24}[\,4(1-\beta^3) -6(1-\beta^2) +(1-\beta^2)^2]$
4	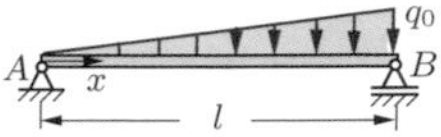	$\dfrac{7\,q_0\,l^3}{360}$	$-\dfrac{q_0\,l^3}{45}$
5	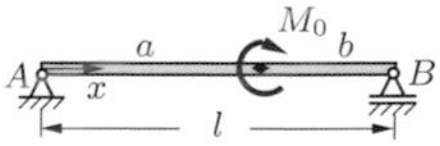	$\dfrac{M_0\,l}{6}(3\beta^2-1)$ $-\dfrac{M_0\,l}{6}$ for $b=0$	$\dfrac{M_0\,l}{6}(3\alpha^2-1)$ $\dfrac{M_0\,l}{3}$ for $b=0$
6	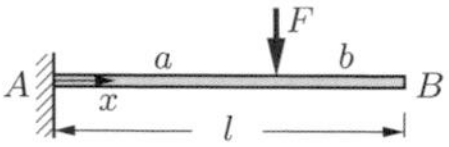	0	$\dfrac{F\,a^2}{2}$
7	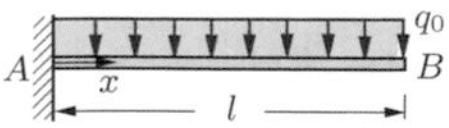	0	$\dfrac{q_0\,l^3}{6}$

$EI\,w(x)$	$EI\,w_{\max}$
$\frac{F\,l^3}{6}[\beta\xi(1-\beta^2-\xi^2)+\langle\xi-\alpha\rangle^3]$	$\frac{F\,l^3}{48}$ for $a=b=l/2$
$\frac{q_0\,l^4}{24}(\xi-2\xi^3+\xi^4)$	$\frac{5\,q_0\,l^4}{384}$
$\frac{q_0\,l^4}{24}[\xi^4-\langle\xi-\alpha\rangle^4-2(1-\beta^2)\xi^3+(1-\beta^2)^2\xi]$	
$\frac{q_0\,l^4}{360}(7\xi-10\,\xi^3+3\,\xi^5)$	
$\frac{M_0\,l^2}{6}[\xi(3\,\beta^2-1)+\xi^3-3\langle\xi-\alpha\rangle^2]$	$\frac{\sqrt{3}\,M_0\,l^2}{27}$ for $a=0$
$\frac{F\,l^3}{6}[3\,\xi^2\alpha-\xi^3+\langle\xi-\alpha\rangle^3]$	$\frac{F\,l^3}{3}$ for $a=l$
$\frac{q_0\,l^4}{24}(6\,\xi^2-4\,\xi^3+\xi^4)$	$\frac{q_0\,l^4}{8}$

Table 4.3. (continued)

Nr.	load	$EI\,w'_A$	$EI\,w'_B$
8	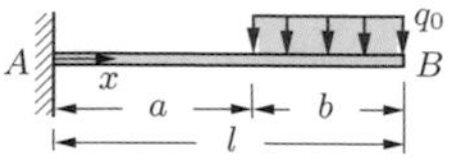	0	$\dfrac{q_0\,l^3}{6}\,\beta(\beta^2-3\,\beta+3)$
9	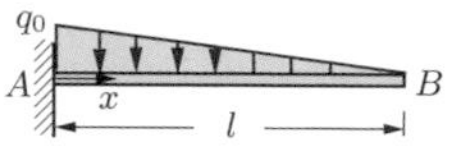	0	$\dfrac{q_0\,l^3}{24}$
10	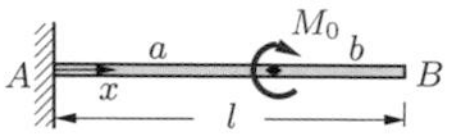	0	$M_0\,a$

Explanations: $\xi = \frac{x}{l}$; $\alpha = \frac{a}{l}$; $\beta = \frac{b}{l}$; $EI = \text{const}$; $w' = \frac{\mathrm{d}w}{\mathrm{d}x}$;

be taken from Table 4.3:

$$w_B^{(1)} = -\frac{X\,l^3}{3\,EI}.$$

The original system is obtained as a superposition of the systems "0" and "1". Since there is a support at B, the deflection at this point has to be zero:

$$w_B = w_B^{(0)} + w_B^{(1)} = 0.$$

This condition of compatibility yields

$$\frac{q_0\,l^4}{8\,EI} - \frac{X\,l^3}{3\,EI} = 0 \qquad \rightarrow \qquad X = B = \frac{3}{8}\,q_0\,l.$$

Now the elastic line $w = w^{(0)} + w^{(1)}$, the slope $w' = w'^{(0)} + w'^{(1)}$, the shear force $V = V^{(0)} + V^{(1)}$ and the bending moment $M = M^{(0)} + M^{(1)}$ are also known. For example, the bending moment is obtained by combining $M^{(0)} = -\frac{1}{2}q_0(l-x)^2$, $M^{(1)} = X(l-x)$

$EI\,w(x)$	$EI\,w_{\max}$
$\dfrac{q_0\,l^4}{24}[\langle\xi-\alpha\rangle^4 - 4\,\beta\,\xi^3 + 6\,\beta(2-\beta)\,\xi^2]$	
$\dfrac{q_0\,l^4}{120}(10\,\xi^2 - 10\,\xi^3 + 5\,\xi^4 - \xi^5)$	$\dfrac{q_0\,l^4}{30}$
$\dfrac{M_0\,l^2}{2}(\xi^2 - \langle\xi-\alpha\rangle^2)$	$\dfrac{M_0\,l^2}{2}$ for $a=l$

$$\langle\xi-\alpha\rangle^n = \begin{cases} (\xi-\alpha)^n & \text{for } \xi > \alpha\,, \\ 0 & \text{for } \xi < \alpha\,. \end{cases}$$

with the known value of X to give (compare Section 4.5.2)

$$M = -\,\frac{1}{2}q_0(l-x)^2 + X(l-x) = -\,\frac{q_0\,l^2}{8}\left[4\left(\frac{x}{l}\right)^2 - 5\left(\frac{x}{l}\right) + 1\right].$$

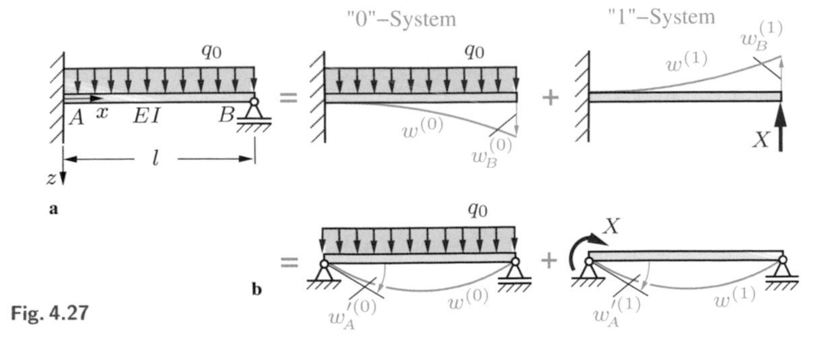

Fig. 4.27

In particular, the moment at the clamped end A of the beam is

$$M_A = M(0) = -\frac{q_0\, l^2}{8}.$$

A statically indeterminate system to the first degree has one excess support reaction. This additional reaction is called *redundant.*

There are various possible ways to choose a statically determinate "0"-system, i.e., to choose the redundant support reaction. In the present example we may replace the clamping A by a hinged support as shown in Fig. 4.27b instead of removing the support B. In order to replace the constraint that was removed, the beam is subjected to a moment X in the "1"-system. Since the superposition has to yield the original system, the condition of compatibility requires a vanishing slope at the clamped end:

$$w'_A = w'^{(0)}_A + w'^{(1)}_A = 0.$$

We take the slopes $w'^{(0)}_A = q_0\, l^3/24\, EI$ and $w'^{(1)}_A = X\, l/3\, EI$ from Table 4.3 and obtain again the moment at the clamped end as

$$X = M_A = -\frac{q_0\, l^2}{8}.$$

The problem can be treated with essentially the same procedure if the right-hand end of the beam is supported by a strut (axial rigidity EA, length a) instead of a simple support (Fig. 4.28). If we remove the strut, we obtain a "0"-system which corresponds to the one as shown in Fig. 4.27a. Thereby the strut is unloaded; its change of length is zero: $\Delta a^{(0)} = 0$. The "1"-system consists of the beam and the strut, both subjected to the unknown force $X = S$. Note that this force acts in opposite directions at the beam and the strut (actio = reactio). Now the strut experiences a shortening $\Delta a^{(1)} = Xa/EA$. Since the beam and the strut are connected in the original system, their displacements at B have to coincide. Therefore the condition of compatibility is given by

$$w_B = \Delta a \qquad \rightarrow \qquad w^{(0)}_B + w^{(1)}_B = \Delta a^{(1)}.$$

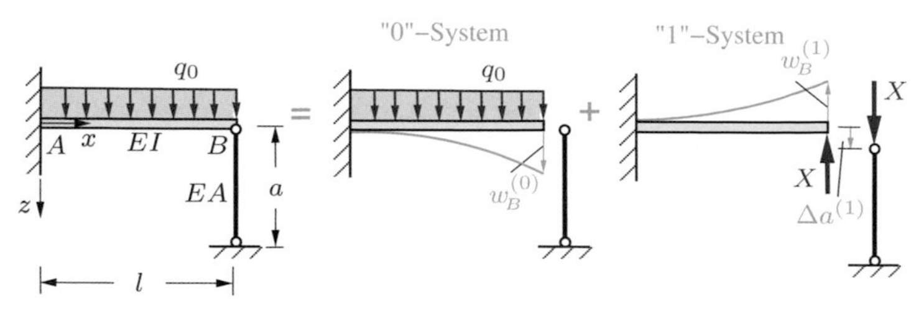

Fig. 4.28

With $w_B^{(0)} = q_0\, l^4/8\, EI$ and $w_B^{(1)} = -X\, l^3/3\, EI$ (see Table 4.3) we obtain

$$\frac{q_0\, l^4}{8\; EI} - \frac{X\, l^3}{3\; EI} = \frac{X\, a}{EI} \qquad \rightarrow \qquad X = S = \frac{3}{8} q_0\, l \frac{1}{1 + \dfrac{3\, EI\, a}{EA\, l^3}}.$$

Once we have determined the force in the strut, the elastic line of the beam and the stress resultants can easily be calculated. In the special case of $3\, EIa/EA\, l^3 \ll 1$, this fraction may be neglected. Hence, if $EA \rightarrow \infty$, the force in the strut becomes $X = 3\, q_0\, l/8$. This result coincides with the result for support B in the system shown in Fig. 4.27a.

In the case of a statically indeterminate system to the first degree, we remove one constraint to obtain a statically determinate "0"-system. The auxiliary "1"-system is then subjected to a load according to the constraint that was removed. If the system is statically indeterminate of degree n, it has n redundant support reactions. Therefore, we have to remove n constraints in order to obtain the "0"-system. In addition, we have to consider n auxiliary systems which are subjected to a force or a moment, according to the constraints that were removed. These n forces/moments can be determined from n conditions of compatibility (compare Section 1.6).

As an example let us consider the beam shown in Fig. 4.29. It is statically indeterminate to the second degree. To obtain the statically determinate "0"-system, we remove the supports B and C. The auxiliary systems "1" and "2" are subjected to the forces $X_1 = B$ and $X_2 = C$, respectively. Since the deflections are zero

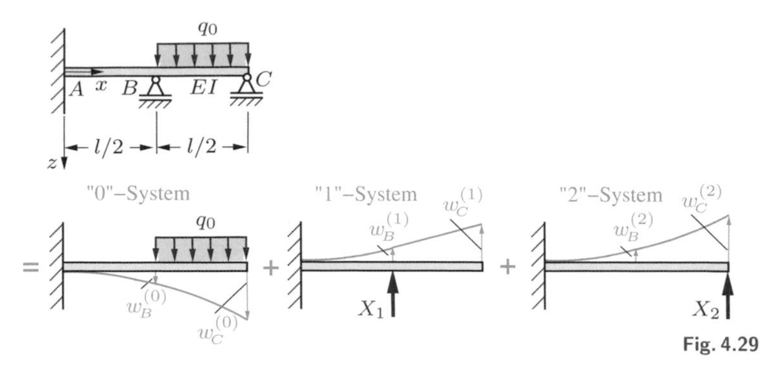

Fig. 4.29

at B and C in the original system, the compatibility conditions are

$$w_B = 0 \quad \rightarrow \quad w_B^{(0)} + w_B^{(1)} + w_B^{(2)} = 0,$$

$$w_C = 0 \quad \rightarrow \quad w_C^{(0)} + w_C^{(1)} + w_C^{(2)} = 0.$$

From Table 4.3 we take

$$w_B^{(0)} = \frac{14\, q_0\, l^4}{384\, EI}, \qquad w_B^{(1)} = -\frac{2X_1\, l^3}{48\, EI}, \qquad w_B^{(2)} = -\frac{5X_2\, l^3}{48\, EI},$$

$$w_C^{(0)} = \frac{41 q_0\, l^4}{384\, EI}, \qquad w_C^{(1)} = -\frac{5X_1\, l^3}{48\, EI}, \qquad w_C^{(2)} = -\frac{16X_2\, l^3}{48\, EI}.$$

Solving for the unknowns yields

$$X_1 = B = \frac{19}{56}\, q_0\, l, \qquad X_2 = C = \frac{12}{56}\, q_0\, l.$$

The method of superposition may also be applied to investigate the deformations of frames by combining the deformations of the individual beams making up the frame. Note, however, that the deformation of a beam has an effect on the displacement of the adjacent one.

As an example we determine the vertical displacement w_C of the angled member at point C (Fig. 4.30a). First, we consider only the deformation of part ① and its effect on the displacement of part ② (which is assumed to be rigid in the first step). Part ① is subjected to the internal moment $M = b\,F$ and the axial load F at its end B (Fig. 4.30b). The moment M causes a deflection

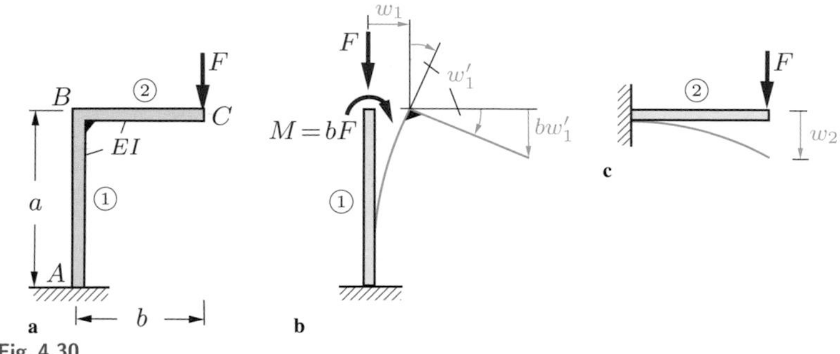

Fig. 4.30

w_1 and a slope w_1' at B. If we assume the axial rigidity to be infinite ($EA \to \infty$) the force F does not cause any deformation. This leads to a vertical displacement $b\,w_1'$ at point C of part ② (small angles!). In a second step we consider part ① to be rigid and part ② to be elastic. This is equivalent to a clamping at B and yields the deflection w_2 of a cantilever (Fig. 4.30c). Taking the individual terms from Table 4.3, the total deflection at C is given by

$$w_C = b\,w_1' + w_2 = b\frac{(b\,F)\,a}{EI} + \frac{F\,b^3}{3\,EI} = \frac{F\,b^2}{3\,EI}(3\,a + b).$$

If the axial rigidity EA is finite, part ① of the structure will be shortened by an amount Fa/EA. Then the total deflection of point C is

$$w_C = \frac{F\,b^2}{3\,EI}(3\,a + b) + \frac{F\,a}{EA}.$$

Usually, the second term is small as compared with the term resulting from the bending.

Example 4.11 Determine the support reactions and the deflection at point D for the beam in Fig. 4.31a.

E4.11

Solution The system is statically indeterminate to the first degree. In order to obtain a simple "0"-system we cut the beam and introduce a pin at the support B (Fig. 4.31b). According to the removed constraint, both parts of the original beam are subjected

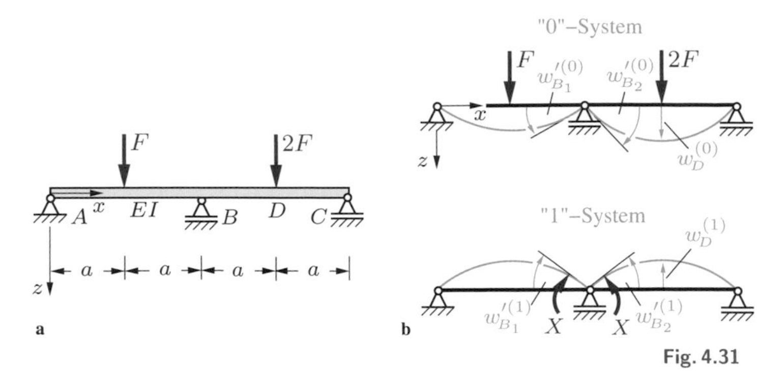

Fig. 4.31

to a moment X in the “1“-system.

There is no pin at B in the original system. Therefore the slopes in both parts of the system have to coincide at B. This leads to the compatibility condition

$$w'_{B_1} = w'_{B_2} \qquad \rightarrow \qquad w'^{(0)}_{B_1} + w'^{(1)}_{B_1} = w'^{(0)}_{B_2} + w'^{(1)}_{B_2} .$$

The support reactions (counted as positive upwards), the deflection at D and the slope at B are taken from Table 4.3. They are given by

$$A^{(0)} = \frac{F}{2}, \qquad B^{(0)} = \frac{F}{2} + F = \frac{3}{2}F, \qquad C^{(0)} = F,$$
$$w^{(0)}_D = \frac{F\,a^3}{3\,EI}, \qquad w'^{(0)}_{B_1} = -\,\frac{F\,a^2}{4\,EI}, \qquad w'^{(0)}_{B_2} = \frac{2\,F\,a^2}{4\,EI}$$

for the “0“-system and

$$A^{(1)} = -\,\frac{X}{2\,a}, \qquad B^{(1)} = \frac{X}{2\,a} + \frac{X}{2\,a} = \frac{X}{a}, \qquad C^{(1)} = -\,\frac{X}{2\,a},$$
$$w^{(1)}_D = -\,\frac{X\,a^2}{4\,EI}, \qquad w'^{(1)}_{B_1} = \frac{2\,X\,a}{3\,EI}, \qquad w'^{(1)}_{B_2} = -\,\frac{2\,X\,a}{3\,EI}.$$

for the “1“-system. Substitution into the compatibility condition yields

$$-\,\frac{F\,a^2}{4\,EI} + \frac{2\,X\,a}{3\,EI} = \frac{2\,F\,a^2}{4\,EI} - \frac{2\,X\,a}{3\,EI} \qquad \rightarrow \qquad X = \frac{9}{16}\,a\,F.$$

Thus, we obtain the support reactions and the deflection at D for the given system:

$$\underline{\underline{A}} = A^{(0)} + A^{(1)} = \underline{\underline{\frac{7}{32}F}}, \qquad \underline{\underline{B}} = B^{(0)} + B^{(1)} = \underline{\underline{\frac{66}{32}F}},$$

$$\underline{\underline{C}} = C^{(0)} + C^{(1)} = \underline{\underline{\frac{23}{32}F}}, \qquad \underline{\underline{w_D}} = w_D^{(0)} + w_D^{(1)} = \underline{\underline{\frac{37\,F\,a^3}{192\,EI}}}.$$

4.6 Influence of Shear

4.6

4.6.1 Shear Stresses

In Section 4.3 it was shown that the assumptions concerning the displacements lead to shear stresses that are constant in a cross section (recall (4.23b)). This constant distribution of the shear stresses is only a first approximation. A better result can be obtained with the aid of the normal stresses (4.26) and the equilibrium conditions. At first, we restrict our attention to prismatic beams with solid cross sections, where the y-axis and the z-axis are assumed to be principal axes. In addition, we make the following assumptions:

a) Only the component in z-direction of the shear stress τ is relevant (see Fig. 4.32a).
b) The shear stress τ is independent of y, i.e., $\tau = \tau(z)$, analogously to the normal stress $\sigma = \sigma(z)$.

Both assumptions are not exactly satisfied in reality. First, the shear stress always has the direction of the tangent to the boundary of an arbitrarily shaped cross section (Fig. 4.32b), and secondly, the shear stress depends to a certain degree on y. Therefore, the shear stress which is obtained using the above assumptions is only an *average shear stress* over the width $b(z)$.

In order to determine the shear stresses, we separate an element of length $\mathrm{d}x$ from the rest of the beam. Then we isolate a part of this element by an additional cut perpendicular to the z-axis

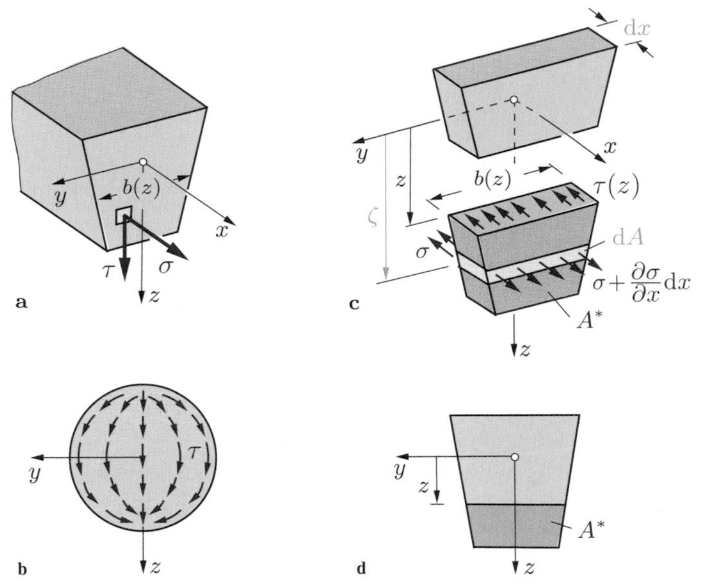

Fig. 4.32

at an arbitrary position z (Fig. 4.32c). Let us now consider the forces acting on this subelement. The unknown shear stresses $\tau(z)$ act in z-direction at its front face. According to Section 2.1, these shear stresses are accompanied by shear stresses of equal magnitude acting in areas perpendicular to the front face (complementary shear stresses). Hence, the top face of the subelement is also acted upon by $\tau(z)$, see Fig. 4.32c. Since only the forces in x-direction are needed in the following derivation, only the corresponding stresses are shown in Fig. 4.32c. The resultant of the shear stresses at the top surface is given by $\tau(z)\, b(z)\mathrm{d}x$; it points into the direction of the negative x-axis. The two areas which are perpendicular to the x-axis (front and back surface) are subjected to the resultant forces $\int_{A^*} \sigma \mathrm{d}A$ and $\int_{A^*} (\sigma + (\partial\sigma/\partial x)\mathrm{d}x)\mathrm{d}A$. Here, the area A^* is the area of the front surface of the subelement, i.e., it is that portion of the cross sectional area A which lies beyond the level z at which the shear stress is being evaluated, see Fig. 4.32d. The bottom face of the element (outer surface of the beam) is not subjected to a load. Hence, the equilibrium condition in

x-direction yields

$$-\tau(z)\,b(z)\mathrm{d}x - \int\limits_{A^*} \sigma\,\mathrm{d}A + \int\limits_{A^*} \left(\sigma + \frac{\partial\sigma}{\partial x}\mathrm{d}x\right)\mathrm{d}A = 0$$

or

$$\tau(z)\,b(z) = \int\limits_{A^*} \frac{\partial\sigma}{\partial x}\,\mathrm{d}A.$$

We denote the distance of the area element $\mathrm{d}A$ from the y-axis by ζ (Fig. 4.32c). Then the flexural stress is given by $\sigma = (M/I)\zeta$, see (4.26). The bending moment M is independent of y and z. Therefore, $\partial M/\partial x = \mathrm{d}M/\mathrm{d}x$. With $\mathrm{d}M/\mathrm{d}x = V$, we get

$$\frac{\partial\sigma}{\partial x} = \frac{V}{I}\,\zeta \tag{4.35}$$

and therefore

$$\tau(z)\,b(z) = \frac{V}{I}\int\limits_{A^*} \zeta\,\mathrm{d}A.$$

The integral on the right-hand side is the first moment S of the area A^* with respect to the y-axis:

$$S(z) = \int\limits_{A^*} \zeta\,\mathrm{d}A. \tag{4.36}$$

Hence, the shear stress distribution over a cross section is found to be

$$\tau(z) = \frac{V\,S(z)}{I\,b(z)}\,. \tag{4.37}$$

This equation is called the *shear formula*. It can be used to calculate the shear stress at any point of a beam with a solid cross section.

As an example we determine the distribution of the shear stress due to a shear force in a rectangular cross section (Fig. 4.33a). The width b, the height h, the cross sectional area $A = bh$ and

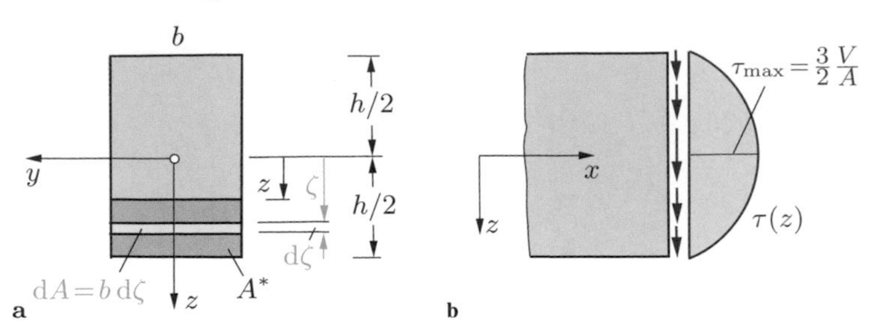

Fig. 4.33

the moment of inertia $I = b\,h^3/12$ of the entire cross section (see (4.8a)) are known. The first moment of the area A^* with respect to the y-axis is obtained as

$$S(z) = \int\limits_{z}^{h/2} \zeta(b\,\mathrm{d}\zeta) = \frac{b}{2}\zeta^2 \Big|_{z}^{h/2} = \frac{b\,h^2}{8}\left(1 - \frac{4\,z^2}{h^2}\right)$$

(see Volume 1, Section 4.3). This yields the distribution of the shear stress in the cross section:

$$\tau(z) = \frac{V}{8}\frac{b\,h^2\,12}{b\,h^3\,b}\left(1 - \frac{4\,z^2}{h^2}\right) = \frac{3}{2}\frac{V}{A}\left(1 - \frac{4\,z^2}{h^2}\right). \tag{4.38}$$

It has the shape of a parabola as shown in Fig. 4.33b. The maximum shear stress $\tau_{\max} = \frac{3}{2}\,V/A$ is located at $z = 0$; it is 50 % larger than the average shear stress $\tilde{\tau} = V/A$. The shear stress is zero at the extreme fibers ($z = \pm h/2$). This is due to the fact that there are no stresses acting at the outer surfaces (top and bottom) of the beam. Therefore, the shear stress in the extreme fibers also has to be zero (complementary shear stresses!).

Since the shear strain is given by $\gamma = \tau/G$, its distribution over the cross section is also parabolic. This implies that the cross sections do not remain plane during the bending: they will *warp* (Fig. 4.34). Therefore the hypothesis of Bernoulli that the cross sections remain plane is only a first approximation and the shear strain $w' + \psi$ (see (4.25)) is an *average shear strain* $\tilde{\gamma}$.

It should be noted again that in addition to the vertical shear stresses in the cross sections there are horizontal shear stresses

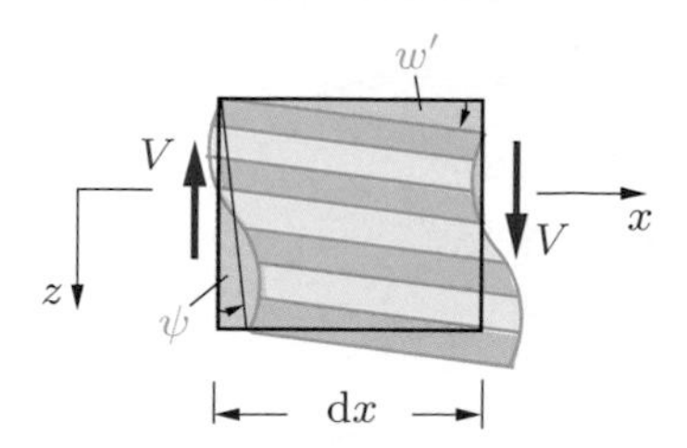

Fig. 4.34

that act between horizontal layers of the beam (complementary stresses). This fact can easily be demonstrated with the aid of two smooth beams ① and ② lying on top of each other (Fig. 4.35). During bending, the beams move relative to each other in the area of contact (smooth surfaces!). This relative sliding may be prevented if the beams are bonded together by welding, gluing or riveting. This will generate shear stresses in the contact area which have to be supported by the bonding (e.g. the weld seam).

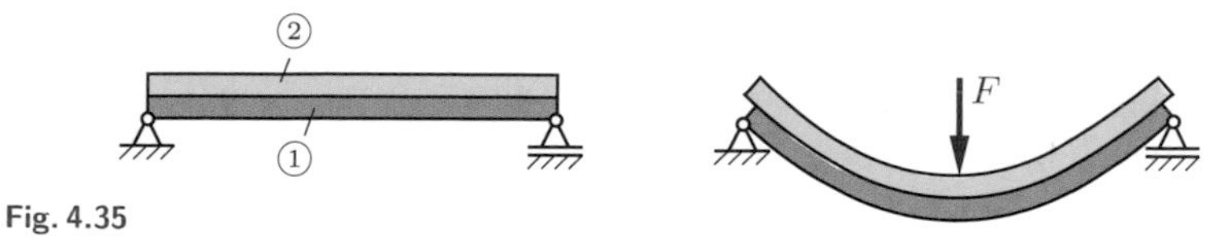

Fig. 4.35

We will now determine the shear stresses in beams with thin-walled cross sections, restricting ourselves to open cross sections. We assume that the shear stresses τ at a position s of the cross section are constant across the thickness t and that they are parallel to the boundary (Fig. 4.36a). The magnitude and the direction of the shear stresses and the thickness may depend on the coordinate (arc length) s. As in the case of a solid cross section we apply the equilibrium condition to an isolated element of the beam (Fig. 4.36b):

$$\tau(s)\, t(s)\, \mathrm{d}x = \int\limits_{A^*} \frac{\partial \sigma}{\partial x}\, \mathrm{d}x\, \mathrm{d}A.$$

With (4.35) we obtain

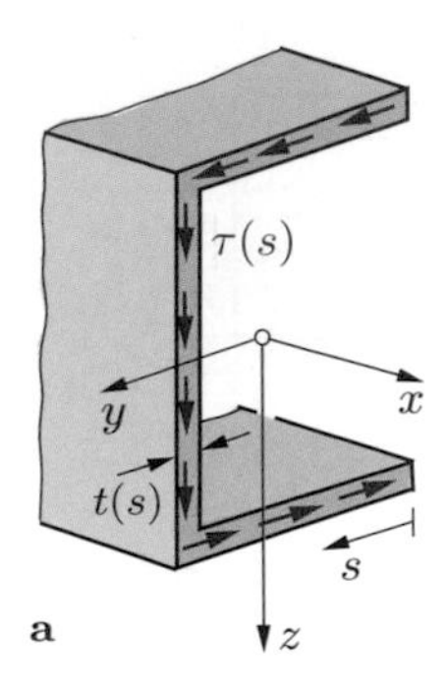

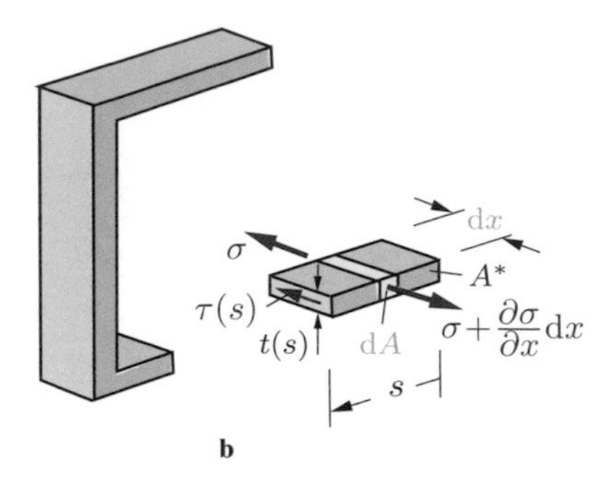

Fig. 4.36

$$\tau(s) = \frac{V\,S(s)}{I\,t(s)}\,. \tag{4.39}$$

Here, $S(s) = \int_{A^*} \zeta\,\mathrm{d}A$ is the first moment of the area A^* with respect to the y-axis.

As an example we determine the shear stresses due to a shear force V in the thin-walled cross section shown in Fig. 4.37a. The moment of inertia of the entire cross section (note that $t \ll a$) is obtained as

$$I = \frac{t(2\,a)^3}{12} + 2[a^2(a\,t)] = \frac{8}{3}\,t\,a^3\,.$$

The first moments of the areas A^* (green areas in the Figs. 4.37b, c) are found to be

$$S(s) = z_s^* A^* = a(t\,s) = a\,t\,s$$

for the bottom flange and

$$S(z) = a(t\,a) + \frac{a+z}{2}[(a-z)t] = \frac{t}{2}(3\,a^2 - z^2)$$

for the web. Thus, (4.39) yields the shear stresses

$$\tau(s) = \frac{3\,V\,a\,t\,s}{8\,t\,a^3\,t} = \frac{3\,V}{8\,t\,a}\frac{s}{a}$$

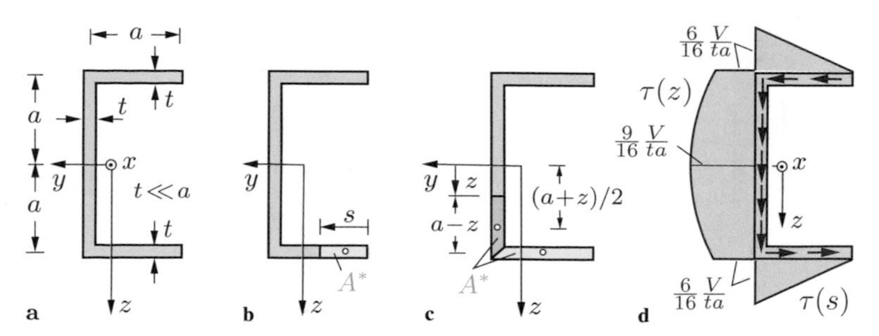

Fig. 4.37

in the bottom flange and

$$\tau(z) = \frac{3\,V\,t\,(3\,a^2 - z^2)}{8\,t\,a^3\,t\,2} = \frac{3\,V}{16\,t\,a}\left(3 - \frac{z^2}{a^2}\right)$$

in the web. The shear stresses in the top flange are distributed as those in the bottom flange; however, they act in the opposite direction. Fig. 4.37d shows the distribution of the shear stresses: linear in the flanges and quadratic in the web.

As can be seen by inspection of Fig. 4.37d, the shear stresses cause a resultant moment about the x-axis. This is due to the fact that the cross section is not symmetrical with respect to the z-axis. In order that the shear force V is statically equivalent to the shear stresses, i.e., that it has the same moment, its line of action has to be located to the left of the z-axis. In the present example, the resultant of the shear stresses in the upper flange is given by $P_u = \frac{1}{2}(\frac{6}{16}V/t\,a)a\,t = \frac{3}{16}V$; it points to the left (Fig. 4.38a). The resultant $P_l = \frac{3}{16}V$ in the lower flange has the same magnitude; it points to the right. Finally, the resultant force in the web is $P_w = V$. The condition of the equivalence of the moments (the lever arms can be taken from Fig. 4.38a) yields

$$y_O\,V = a\frac{3}{16}V + \frac{a}{4}V + a\frac{3}{16}V \qquad \rightarrow \qquad y_O = \frac{5}{8}a.$$

The point O on the y-axis is called the *shear center*. In order to prevent a torsion of the beam (see Chapter 5) the applied forces F have to act in a plane that has the distance y_O from the x,z-plane (Fig. 4.38b). Only then the bending moment and the shear

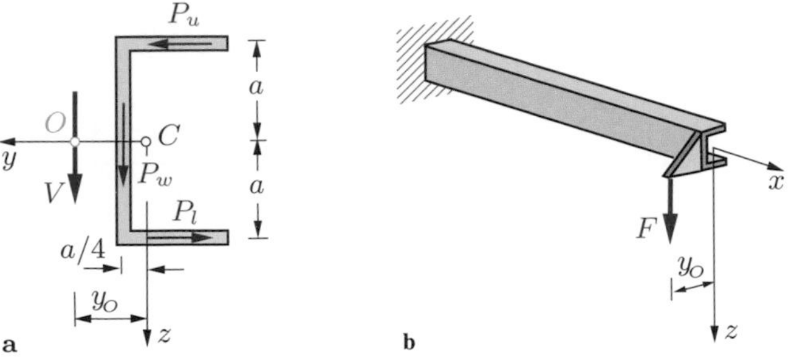

Fig. 4.38

force are in equilibrium with the applied loads and a torque is not caused by the loads.

E4.12

Example 4.12 Determine the distribution of the shear stresses due to a shear force in a solid circular cross section with radius r (Fig. 4.39).

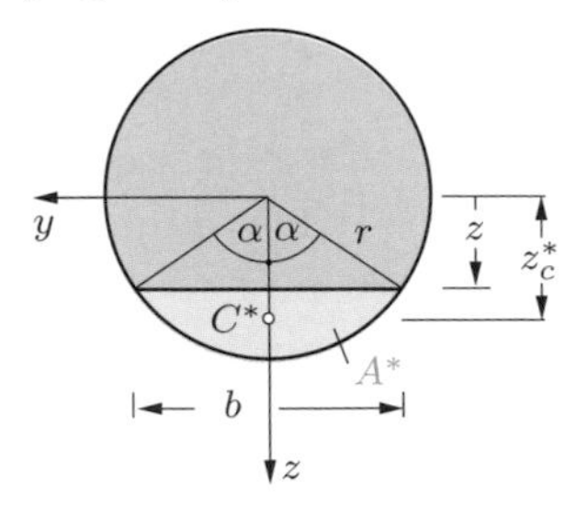

Fig. 4.39

Solution The first moment S of the circular segment A^* (shown in green) is obtained as the product of the area A^* and the distance z_c^* of its centroid C^* from the y-axis (Fig. 4.39). Introducing the auxiliary angle α we have

$$A^* = \frac{r^2}{2}(2\,\alpha - \sin 2\,\alpha), \qquad z_c^* = \frac{4\,r}{3}\frac{\sin^3\alpha}{2\,\alpha - \sin 2\,\alpha},$$

(see Volume 1, Table 4.1). Thus,

$$S = z_c^*\,A^* = \frac{2}{3}\,r^3\,\sin^3\alpha.$$

With $I = \pi\,r^4/4$ (see Table 4.1), $b = 2\,r\sin\alpha$, $A = \pi\,r^2$, and

$z = r\cos\alpha$ we obtain from (4.37)

$$\underline{\underline{\tau}} = \frac{V\,S}{I\,b} = \frac{4}{3}\frac{V}{\pi\,r^2}\sin^2\alpha = \underline{\underline{\frac{4}{3}\frac{V}{A}\left(1 - \frac{z^2}{r^2}\right)}}.$$

The maximum shear stress $\tau_{\max} = \frac{4}{3}V/A$ is located at $z = 0$.

Example 4.13 E4.13 Fig. 4.40a shows a welded thin-walled wide-flange cantilever beam ($t \ll h, b$) subjected to the force F.

Determine the shear stress in the weld seam at the bottom of the web.

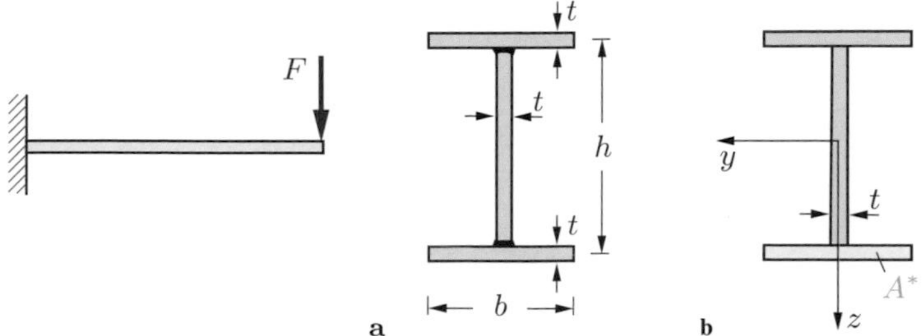

Fig. 4.40

Solution The shear force is constant in the beam, $V = F$. The first moment of the area A^* of the lower flange (Fig. 4.40b) is given by

$$S(z = \frac{h}{2}) = \frac{h}{2}\,t\,b\,.$$

The moment of inertia of the entire cross section is taken from Example 4.2:

$$I = \frac{t\,h^3}{12} + 2\left[\left(\frac{h}{2}\right)^2 t\,b\right] = \frac{t\,h^2}{12}(h + 6\,b)\,.$$

With the width t at the weld seam (lower end of the web) we obtain from (4.39)

$$\underline{\underline{\tau}} = \frac{V\,S}{I\,t} = \frac{12\,F\,h\,t\,b}{2\,t\,h^2\,(h + 6\,b)t} = \underline{\underline{\frac{6\,F\,b}{t\,h\,(h + 6\,b)}}}.$$

E4.14

Example 4.14 Determine the shear stresses in a thin-walled open circular ring (Fig. 4.41a) due to a shear force V. Locate the shear center.

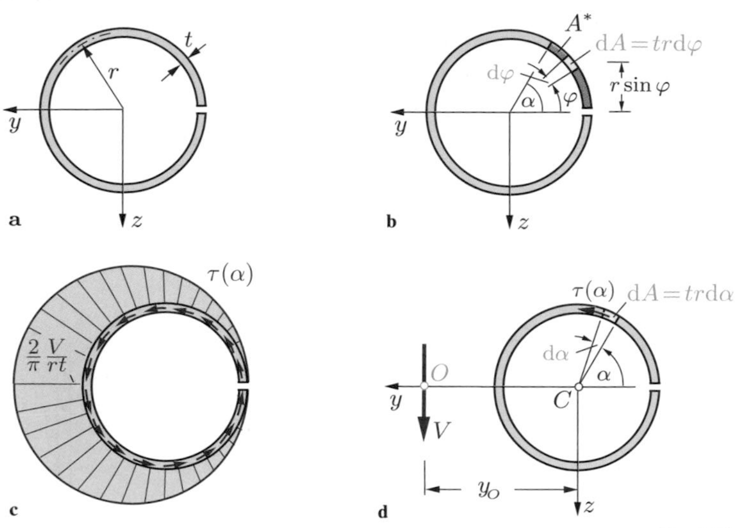

Fig. 4.41

Solution The shear stresses follow from (4.39). The first moment of the area A^* is obtained as (see Fig. 4.41b)

$$S = \int_0^\alpha (r \sin \varphi)(t\, r\, \mathrm{d}\varphi) = r^2\, t(1 - \cos \alpha).$$

The moment of inertia $I = \pi\, r^3\, t$ is taken from Table 4.1. This yields the distribution of the shear stresses τ as a function of α (Fig. 4.41c):

$$\underline{\underline{\tau}} = \frac{V\,S}{I\,t} = \underline{\underline{\frac{V(1-\cos \alpha)}{\pi\, r\, t}}}.$$

To locate the shear center O we use the condition that the moment of the shear force with respect to O has to coincide with the moment of the shear stresses (Fig. 4.41d):

$$y_O\, V = \int r\, \tau\, \mathrm{d}A = \int_0^{2\pi} r\, \frac{V(1-\cos \alpha)}{\pi\, r\, t}(t\, r\, \mathrm{d}\alpha) \quad \rightarrow \quad \underline{\underline{y_O = 2\, r}}.$$

4.6.2 Deflection due to Shear

The average shear strain $\tilde{\gamma} = w' + \psi$ is proportional to the shear force V (see (4.25)):

$$w' + \psi = \frac{V}{GA_S}. \tag{4.40}$$

In Section 4.5, the deflections of beams were determined with the assumption that the beams are rigid with respect to shear, i.e., the right-hand side of (4.40) was set equal to zero (Bernoulli beam). Now we want to verify that this assumption is justified for slender beams by investigating the effect of the shear strain (caused by the shear force) on the deflection. If we introduce the notations $w'_B = -\psi$ (see (4.29)) and

$$w'_S = \frac{V}{GA_S} \tag{4.41}$$

we can write (4.40) in the form

$$w' = w'_B + w'_S\,. \tag{4.42}$$

Hence, the slope w' of the beam is the sum of the slopes w'_B of the Bernoulli beam and w'_S due to shear. Similarly, the deflection w is written as

$$w = w_B + w_S\,. \tag{4.43}$$

In order to obtain an estimate of the deflection w_S due to shear we consider a cantilever beam with a rectangular cross section subjected to a force F (Fig. 4.42a). Integration of (4.41) with

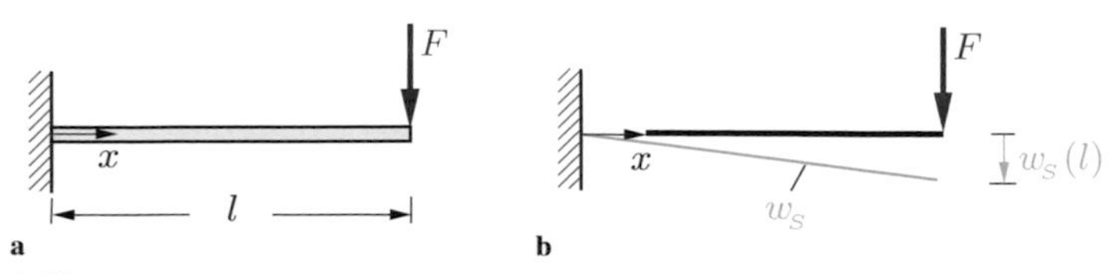

Fig. 4.42

$V = F$ yields

$$w_S = \frac{F}{GA_S}\, x + C.$$

The boundary condition $w_S(0) = 0$ leads to $C = 0$ and to (see Fig. 4.42b)

$$w_S = \frac{F}{GA_S}\, x. \tag{4.44}$$

The total deflection w is the sum of the deflections $w_B = (F\,l^3/6\,EI)[-x^3/l^3 + 3\,x^2/l^2]$ (see Section 4.5.2) and the deflection w_S, given by (4.44). The deflection at the free end is therefore found to be

$$w(l) = w_B(l) + w_S(l) = \frac{F\,l^3}{3\,EI} + \frac{F\,l}{GA_S} = \frac{F\,l^3}{3\,EI}\left(1 + \frac{3\,EI}{GA_S\,l^2}\right).$$

If we introduce the relation $G = E/2(1+\nu)$, the moment of inertia $I = r_g^2\,A$ (r_g = radius of gyration) and $A_S = \varkappa\,A$ ($\varkappa$ = shear correction factor), we obtain

$$w(l) = \frac{F\,l^3}{3\,EI}\left[1 + \frac{6(1+\nu)}{\varkappa}\frac{r_g^2}{l^2}\right] = \frac{F\,l^3}{3\,EI}\left[1 + \frac{c}{\lambda^2}\right].$$

The second term in the brackets represents the influence of the shear. It depends on the *slenderness ratio* $\lambda = l/r_g$ and on the constant $c = 6(1+\nu)/\varkappa$. In Section 6.1 it will be shown that the shear correction factor for a rectangular cross section has the value $\varkappa = 5/6$. With the radius of gyration $r_g = h/2\sqrt{3}$ (see (4.8e)) and Poisson's ratio $\nu = 1/3$ we obtain $c/\lambda^2 = 4\,h^2/5\,l^2$ for the rectangular cross section. As a numerical example, if we choose the ratio $l/h = 5$, we get $c/\lambda^2 \approx 0.03$. In this example, the deflection due to shear is only about 3 % of the deflection due to the bending moment.

The influence of the shear decreases if the slenderness λ of the beam increases. In general, the deflection due to shear may be neglected if the length of the beam is larger than five times the height of the cross section.

4.7 Unsymmetric Bending

Frequently a beam undergoes a deflection w in z-direction as well as a deflection v in y-direction. In this case the bending is referred to as *unsymmetric bending*. This type of bending occurs if the beam carries loads in z- as well as in y-direction or if the cross section is not symmetrical. These cases give rise to the shear forces V_y, V_z and the bending moments M_y, M_z (see Volume 1, Section 7.4). We will restrict ourselves to Bernoulli beams (i.e., we neglect the influence of shear) and we will distinguish between the following two cases.

Case 1: We assume that y and z are the principal axes of the cross section. Then we can use the results of the ordinary bending (Sections 4.4 and 4.5) if we consider the loads in z- and y-directions separately. The load in z-direction causes normal stresses σ and deflections w. They are given by (see (4.26) and (4.31))

$$\sigma = \frac{M_y}{I_y}\, z, \qquad w'' = -\frac{M_y}{EI_y}\,.$$

Analogously, the load in y-direction leads to

$$\sigma = -\frac{M_z}{I_z}\, y, \qquad v'' = \frac{M_z}{EI_z}\,.$$

The different algebraic signs are due to the sign convention (see Volume 1, Section 7.4): positive moments at a positive face point into the direction of the positive coordinate axes (Fig. 4.43). Superposition yields the total normal stresses:

$$\sigma = \frac{M_y}{I_y}\, z - \frac{M_z}{I_z}\, y\,. \tag{4.45}$$

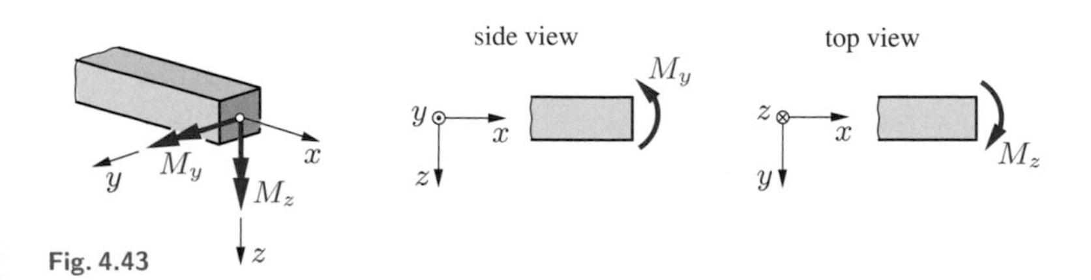

Fig. 4.43

The deflections w and v are independent of each other. They are obtained through integration of

$$w'' = -\frac{M_y}{EI_y}, \qquad v'' = \frac{M_z}{EI_z}. \tag{4.46}$$

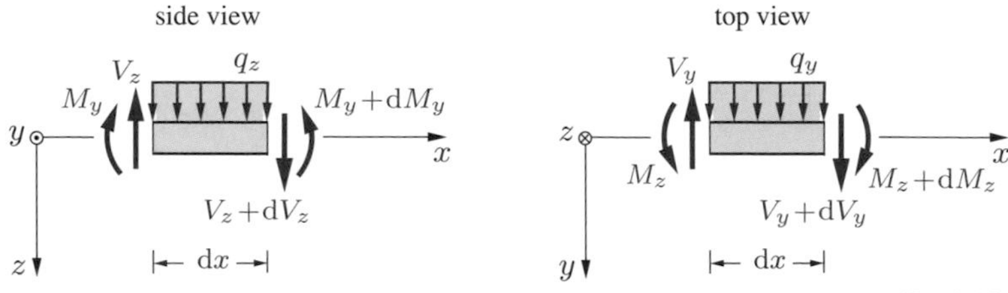

Fig. 4.44

Case 2: Now we assume that y and z are not principal axes of the cross section. To derive the relevant equations we proceed as in the case of ordinary bending. Let us consider the forces and moments that act on an element (length $\mathrm{d}x$) of the beam. They are shown in Fig. 4.44 (note the sign convention of the stress resultants). The conditions of equilibrium are

$$\begin{aligned} \frac{\mathrm{d}V_z}{\mathrm{d}x} &= -\, q_z, & \frac{\mathrm{d}V_y}{\mathrm{d}x} &= -\, q_y, \\ \frac{\mathrm{d}M_y}{\mathrm{d}x} &= V_z, & \frac{\mathrm{d}M_z}{\mathrm{d}x} &= -\, V_y. \end{aligned} \tag{4.47}$$

As in case 1, we assume that the deflections v and w are independent of y and z: $v = v(x)$, $w = w(x)$. In addition, we apply the hypotheses of Bernoulli (see Section 4.5.1): the cross sections remain plane and stay perpendicular to the deformed axis of the beam. Now we introduce the angles of rotation ψ_y and ψ_z of the cross section about the y-axis and the z-axis, respectively (positive sense of rotation according to the cork-screw rule). In the following, we first determine the displacement u in axial direction of an arbitrary point P of the cross section with the coordinates y, z (Fig. 4.45). Due to a *small* rotation ψ_y only, this point is displaced by an amount $z\,\psi_y$ in the positive x-direction. Similarly, a small rotation ψ_z leads to the displacement $-y\,\psi_z$. Therefore, the

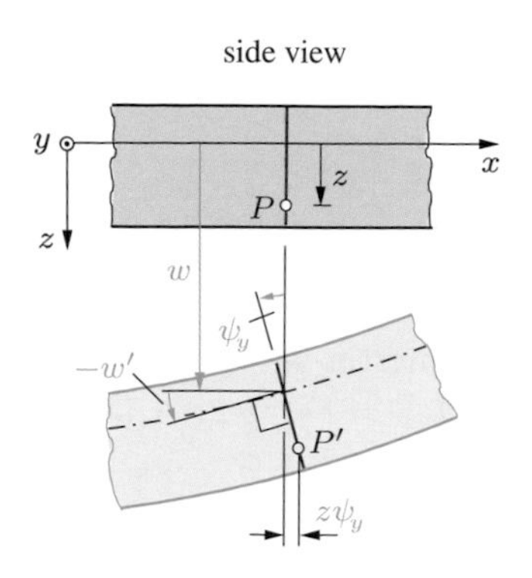

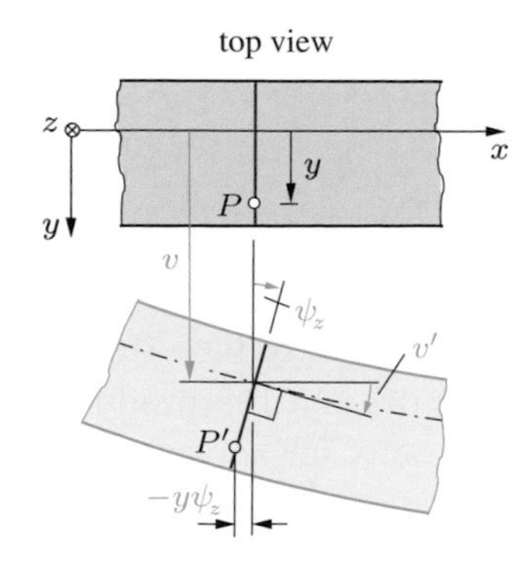

Fig. 4.45

total displacement is obtained as

$$u = z\,\psi_y - y\,\psi_z .$$

With the relations

$$\psi_y = -\,w', \qquad \psi_z = +\,v'$$

(see Fig. 4.45 and note that the cross section is perpendicular to the deformed axis of the beam) we get

$$u = -\,(w'\,z + v'\,y).$$

The strain $\varepsilon = \partial u/\partial x$ is therefore given by

$$\varepsilon = -\,(w''\,z + v''\,y), \tag{4.48}$$

and Hooke's law $\sigma = E\,\varepsilon$ finally yields

$$\sigma = -\,E(w''\,z + v''\,y). \tag{4.49}$$

The bending moments M_y and M_z are the resultant moments of the normal stresses σ in the cross section (note the senses of rotation):

$$M_y = \int z\,\sigma\,\mathrm{d}A, \qquad M_z = -\int y\,\sigma\,\mathrm{d}A. \tag{4.50}$$

With (4.49) we obtain

$$M_y = -\,E\left[w''\int z^2\,\mathrm{d}A + v''\int y\,z\,\mathrm{d}A\right],$$

$$M_z = E\left[w''\int y\,z\,\mathrm{d}A + v''\int y^2\,\mathrm{d}A\right].$$

We now introduce the moments of inertia $I_y = \int z^2\,\mathrm{d}A, I_z = \int y^2\,\mathrm{d}A$ and the product of inertia $I_{yz} = -\int y\,z\,\mathrm{d}A$ (see (4.6)) and solve for w'' and v'':

$$\begin{aligned} E\,w'' &= \frac{1}{\Delta}[-\,M_y\,I_z + M_z\,I_{yz}],\\ E\,v'' &= \frac{1}{\Delta}[M_z\,I_y - M_y\,I_{yz}]\,. \end{aligned} \tag{4.51}$$

Here, $\Delta = I_y\,I_z - I_{yz}^2$. The deflections w and v can be determined from (4.51) through integration if the bending moments are known.

If we insert (4.51) into (4.49) we obtain the normal stress:

$$\sigma = \frac{1}{\Delta}[(M_y\,I_z - M_z\,I_{yz})\,z - (M_z\,I_y - M_y\,I_{yz})\,y]\,. \tag{4.52}$$

Thus, the distribution of the normal stress is linear in y and z ((4.52) is the equation of a plane). The normal stress is zero for

$$\frac{z}{y} = \frac{M_z\,I_y - M_y\,I_{yz}}{M_y\,I_z - M_z\,I_{yz}}\,. \tag{4.53a}$$

This equation defines the *neutral axis* in the cross section. The maximum normal stress $\sigma_{\max}$ is located at the point that has the maximum distance from the neutral axis.

In the special case that y and z are the principal axes of the cross section, the product of inertia vanishes: $I_{yz} = 0$. Then (4.51) and (4.52) reduce to (4.46) and (4.45), respectively, and the neutral axis is given by

$$\frac{z}{y} = \frac{M_z\,I_y}{M_y\,I_z}. \tag{4.53b}$$

We may use the equations of either case 1 or those of case 2 to treat unsymmetric bending. In the first case we have to determine the principal axes of the cross section and then resolve the applied loads and the bending moments in the corresponding coordinate system. The normal stress and the displacements are given by (4.45) and (4.46) with respect to the principal axes. In case 2, the stress and the displacements are determined by (4.52) and (4.51) in an arbitrary coordinate system.

Example 4.15 The beam in Fig. 4.46a is supported by ball-and-socket joints and subjected to a force F that acts at an angle $\alpha = 30°$ to the vertical. It has a rectangular cross section (width b, height $h = 2b$).

E4.15

Determine the normal stresses and the deflections at the center of the beam.

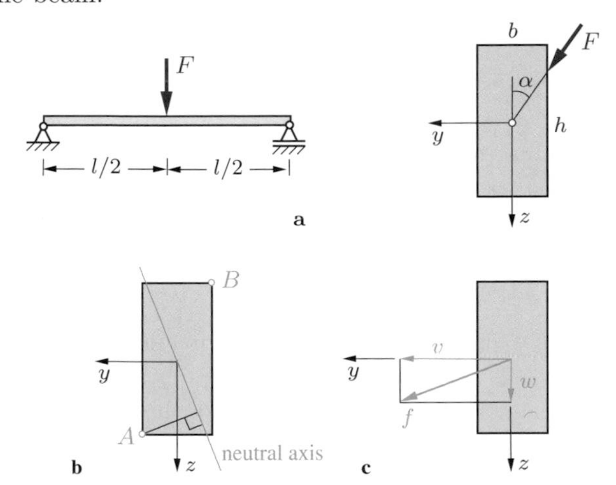

Fig. 4.46

Solution The normal stresses can be determined from (4.45) since y and z are the principal axes of the cross section. We resolve the force F into its components in y- and z-direction:

$$F_y = F \sin\alpha = \frac{F}{2}, \qquad F_z = F \cos\alpha = \frac{\sqrt{3}}{2} F\,.$$

This yields the bending moments

$$M_y = \frac{l}{2}\frac{F_z}{2} = \frac{\sqrt{3}\, l\, F}{8}, \qquad M_z = -\frac{l}{2}\frac{F_y}{2} = -\frac{l\, F}{8} \tag{a}$$

at the center of the beam (note the algebraic signs). The moments of inertia are given by (compare Table 4.1)

$$I_y = \frac{b\, h^3}{12} = \frac{2}{3}\, b^4, \qquad I_z = \frac{h\, b^3}{12} = \frac{1}{6}\, b^4. \tag{b}$$

Inserting (a) and (b) into (4.45) we obtain

$$\underline{\underline{\sigma}} = \frac{\sqrt{3}\, l\, F\, 3}{8 \cdot 2\, b^4}\, z + \frac{l\, F\, 6}{8\, b^4}\, y = \underline{\underline{\frac{3\, l\, F}{4\, b^4}\left(\frac{\sqrt{3}}{4}\, z + y\right)}}.$$

The neutral axis follows from the condition $\sigma = 0$ as

$$y = -\frac{\sqrt{3}}{4}\, z.$$

This axis is shown in Fig. 4.46b. As can be seen by inspection, the points A and B are at maximum distance from the neutral axis. With the coordinates $y_A = b/2, z_A = b$ of point A we obtain the maximum normal stress (tension)

$$\sigma_{\max} = \frac{3\, l\, F}{4\, b^4}\left(\frac{\sqrt{3}}{4}\, b + \frac{b}{2}\right) = \frac{3\, l\, F}{8\, b^3}\left(\frac{\sqrt{3}}{2} + 1\right).$$

The normal stress at point B has the same magnitude but it is a compressive stress.

The displacements w and v at the center of the beam are taken from Table 4.3. The force component F_z causes the displacement

$$\underline{\underline{w}} = \frac{F_z\, l^3}{48\, E I_y} = \underline{\underline{\frac{\sqrt{3}\, F\, l^3}{64\, E\, b^4}}}.$$

Analogously, the force component F_y leads to the displacement

$$\underline{\underline{v}} = \frac{F_y\, l^3}{48\, E I_z} = \underline{\underline{\frac{4\, F\, l^3}{64\, E\, b^4}}}.$$

The total displacement f (see Fig. 4.46c) is obtained as

$$f = \sqrt{w^2 + v^2} = \frac{\sqrt{19}\, F\, l^3}{64\, E\, b^4}\,.$$

Example 4.16 A cantilever beam with a thin-walled cross section ($t \ll a$) carries a force F that acts in z-direction (Fig. 4.47a).

E4.16

Determine the deflection at the free end B. Locate and determine the maximum bending stress.

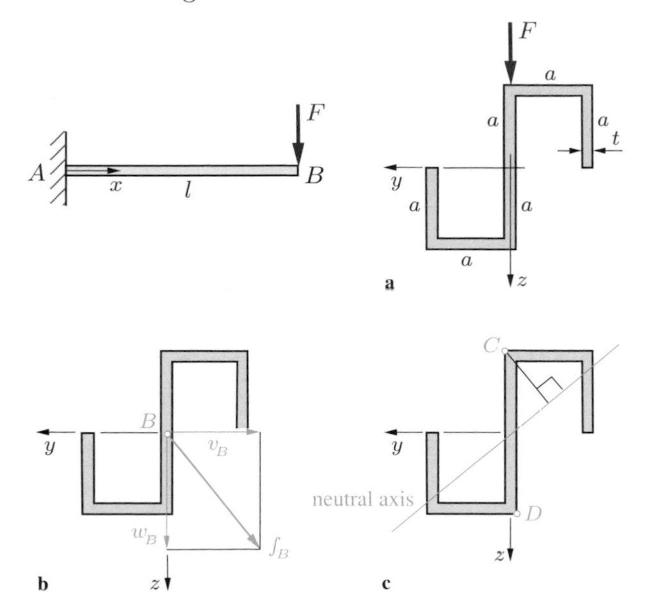

Fig. 4.47

Solution The bending moments are given by

$$M_y = -\,F(l - x), \qquad M_z = 0. \tag{a}$$

The moments of inertia of the cross section are taken from Example 4.3:

$$I_y = \frac{10}{3}\,t\,a^3, \qquad I_z = \frac{8}{3}\,t\,a^3, \qquad I_{yz} = -\,\frac{6}{3}\,t\,a^3. \tag{b}$$

Since $I_{yz} \neq 0$, the axes y and z are not the principal axes of the cross section. Therefore, we determine the displacements w and v with the aid of (4.51). With (a), (b) and $\Delta = I_y\,I_z - I_{yz}^2 = \frac{44}{9}\,t^2\,a^6$, the differential equations for the deflections are

$$E\,w'' = -\frac{M_y\,I_z}{\Delta} = \frac{6\,F}{11\,t\,a^3}(l-x),$$

$$E\,v'' = -\frac{M_y\,I_{yz}}{\Delta} = -\frac{9\,F}{22\,t\,a^3}(l-x).$$

We integrate twice, use the boundary conditions $w(0) = 0$, $w'(0) = 0$, $v(0) = 0$, $v'(0) = 0$ and obtain the deflections (recall Section 4.5.2)

$$w(x) = \frac{F\,l^3}{11\,E\,t\,a^3}\left[3\left(\frac{x}{l}\right)^2 - \left(\frac{x}{l}\right)^3\right],$$

$$v(x) = -\frac{3\,F\,l^3}{44\,E\,t\,a^3}\left[3\left(\frac{x}{l}\right)^2 - \left(\frac{x}{l}\right)^3\right].$$

This yields

$$w_B = \frac{2\,F\,l^3}{11\,E\,t\,a^3}, \qquad v_B = -\frac{3\,F\,l^3}{22\,E\,t\,a^3}$$

at the free end $B\,(x = l)$, and the total deflection f_B is found to be (Fig. 4.47b)

$$\underline{\underline{f_B}} = \sqrt{w_B^2 + v_B^2} = \underline{\underline{\frac{5\,F\,l^3}{22\,E\,t\,a^3}}}.$$

If we insert (a), (b) and Δ into (4.52) we obtain the normal stress

$$\sigma = \frac{M_y}{\Delta}(I_z\,z + I_{yz}\,y) = -\frac{3\,F(l-x)}{22\,t\,a^3}(4\,z - 3\,y).$$

The maximum stress is located at the clamping ($x = 0$) at those points of the cross section that have the maximum distance from the neutral axis. The neutral axis is determined from the condition $\sigma = 0$ which yields $y = \frac{4}{3}\,z$. Fig. 4.47c shows that the points C and D are at the maximum distance from the neutral axis. With the coordinates $y_C = 0, z_C = -a$ of point C we get

$$\underline{\underline{\sigma_{\max} = \frac{6\,F\,l}{11\,t\,a^2}}}\,.$$

The normal stress at D has the same magnitude but is a compressive stress.

4.8 Bending and Tension/Compression

The bending moments M_y and M_z in a system of *principal axes* lead to the bending stresses

$$\sigma = \frac{M_y}{I_y}\,z - \frac{M_z}{I_z}\,y$$

(see (4.45)). If a member is subjected to a normal force N only, the normal stress is constant in the cross section (see (1.1)):

$$\sigma = \frac{N}{A}\,.$$

If bending moments as well as a normal force exist, the resulting stress is obtained through superposition:

$$\sigma = \frac{N}{A} + \frac{M_y}{I_y}\,z - \frac{M_z}{I_z}\,y\,. \tag{4.54a}$$

In the case of ordinary bending about the y-axis ($M_z = 0$), Equation (4.54a) reduces to ($M_y = M, I_y = I$)

$$\sigma = \frac{N}{A} + \frac{M}{I}\,z\,. \tag{4.54b}$$

The superposition of the terms resulting from bending and from tension/compression is also valid for the deformation. The bending moments cause the deflections $w(x)$ and $v(x)$, whereas the normal force causes only a displacement $u(x)$ in the direction of the beam axis. These deformations may be superimposed.

Note that the deformation due to the normal force is usually much smaller than the deformation caused by the bending moments. Consequently the change in length of the beam may be neglected and the beam may be considered to be rigid with respect to tension/compression.

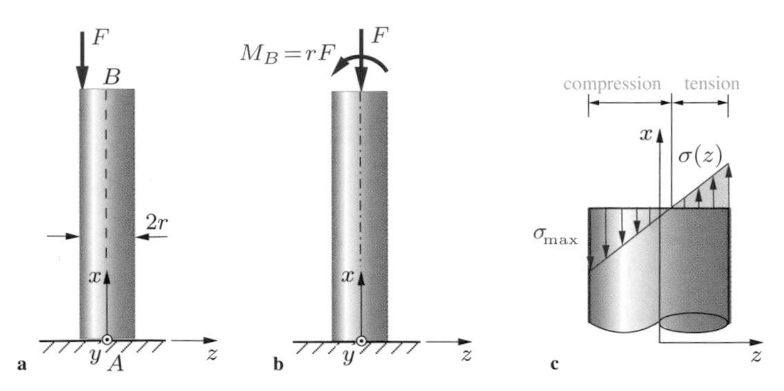

Fig. 4.48

As an example let us consider a column with a circular cross section (radius r) that is subjected to an eccentrically acting force F (Fig. 4.48a). This is statically equivalent to the system of the same force F with the action line equal to the x-axis and the moment $M_B = r\,F$ (Fig. 4.48b). If we neglect the weight of the column, then the stress resultants N and $M_y = M_B$ are independent of x:

$$N = -\,F, \qquad M = M_B = r\,F.$$

The bending moment about the z-axis is equal to zero: $M_z = 0$ (ordinary bending). With $A = \pi\,r^2$ and $I = \pi\,r^4/4$ (see Table 4.1), the normal stress follows from (4.54b):

$$\sigma = -\,\frac{F}{\pi\,r^2} + \frac{r\,F\,4}{\pi\,r^4}\,z = \frac{F}{\pi\,r^2}\left[-\,1 + 4\frac{z}{r}\right].$$

It is shown in Fig. 4.48c. The maximum stress is found at $z = -r$:

$$|\sigma|_{max} = \frac{5\,F}{\pi\,r^2}.$$

According to (1.18), the change of length Δl of the beam due to the compressive force F is given by

$$\Delta l = -\frac{F\,l}{EA}.$$

The deflection of point B due to the moment $M_0 = r\,F$ can be taken from Table 4.3:

$$w_B = -\frac{M_0\,l^2}{2\,EI} = -\frac{r\,F\,l^2}{2\,EI}.$$

Inserting A and I, the ratio between Δl and w_B is found to be

$$\frac{\Delta l}{w_B} = \frac{r}{2\,l}.$$

As a numerical example, we choose $l/r = 20$ which yields $\Delta l/w_B = 1/40$. Hence, the shortening of the beam is small as compared with the deflection.

E4.17

Example 4.17 A simply supported beam with a thin-walled cross section ($t \ll a$) carries a constant line load q_0 that acts in the x, z-plane (Fig. 4.49a).

Determine the maximum normal stress.

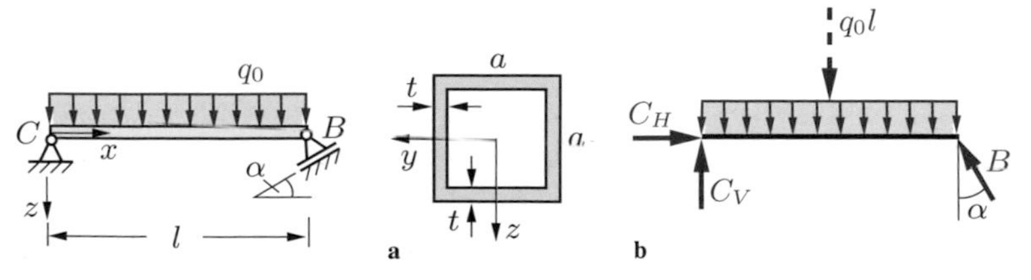

Fig. 4.49

Solution In a first step we calculate the support reactions (Fig. 4.49b)

$$\overset{\frown}{C}: \quad -\frac{l}{2}\,q_0\,l + l\,B\cos\alpha = 0 \to B = \frac{q_0\,l}{2\cos\alpha},$$

$$\overset{\frown}{B}: \quad -l\,C_V + \frac{l}{2}\,q_0\,l = 0 \to C_V = \frac{q_0\,l}{2},$$

$$\to: \quad C_H - B\sin\alpha = 0 \to C_H = B\sin\alpha = \frac{q_0\,l}{2}\tan\alpha.$$

The normal force is constant: $N = -C_H = -\frac{1}{2}q_0\, l \tan\alpha$; the maximum bending moment is located at $x = l/2$: $M_{\max} = q_0\, l^2/8$. With

$$A = 4\,t\,a, \qquad I = 2\left[\frac{t\,a^3}{12} + \left(\frac{a}{2}\right)^2 a\,t\right] = \frac{2}{3}\,t\,a^3$$

we obtain the distribution of the normal stress at $x = l/2$ from (4.54b):

$$\begin{aligned}\sigma &= -\frac{q_0\, l \tan\alpha}{2 \cdot 4\,t\,a} + \frac{q_0\, l^2\, 3}{8 \cdot 2\,t\,a^3}\, z \\ &= \frac{q_0\, l}{8\,t\,a}\left(-\tan\alpha + \frac{3\,l\,z}{2\,a^2}\right).\end{aligned}$$

For $\tan\alpha > 0$ (i.e., $0 < \alpha < \pi/2$) the maximum stress (compressive stress) is located at $z = -a/2$:

$$\underline{\underline{|\sigma|_{\max} = \frac{q_0\, l}{8\,t\,a}\left(\tan\alpha + \frac{3\,l}{4\,a}\right).}}$$

4.9

4.9 Core of the Cross Section

Let us consider a rod or a column that is loaded by an eccentrically acting compressive force F whose point of application in the cross section is given by y_F, z_F (Fig. 4.50a, b). With $N = -F$, $M_y = -z_F\, F$, $M_z = y_F\, F$ and the radii of gyration $r_{gy}^2 = I_y/A$, $r_{gz}^2 = I_z/A$, the normal stress in the cross section is obtained as (see (4.54a))

$$\sigma = -\frac{F}{A}\left[\frac{z_F}{r_{gy}^2}\, z + \frac{y_F}{r_{gz}^2}\, y + 1\right].$$

The neutral axis ($\sigma = 0$) is therefore given by the straight line

$$\frac{z_F}{r_{gy}^2}\, z + \frac{y_F}{r_{gz}^2}\, y + 1 = 0 \qquad \text{or} \qquad \frac{z}{z_0} + \frac{y}{y_0} = 1 \tag{4.55}$$

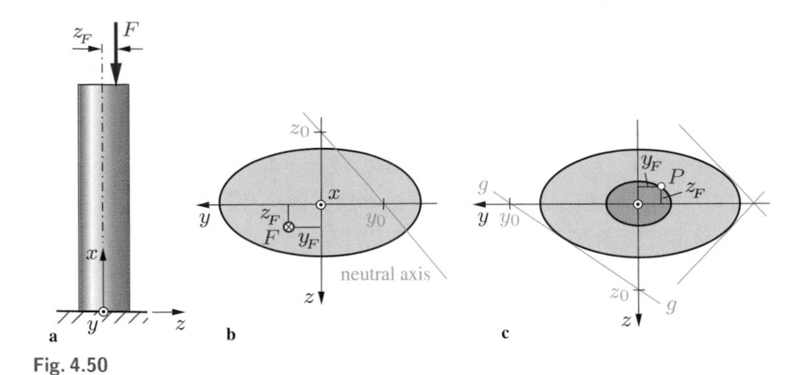

Fig. 4.50

where the quantities

$$y_0 = -\frac{r_{gz}^2}{y_F}, \qquad z_0 = -\frac{r_{gy}^2}{z_F} \tag{4.56}$$

are determined by the intersections of the neutral axis with the coordinate axes (Fig. 4.50b).

The neutral axis may lie outside the cross section. In this case the stresses throughout the entire cross section are compressive stresses. This is an important consideration if the material of the column (e.g., concrete) is very weak in tension. We now want to determine the allowable region of the points of application y_F, z_F of the force so that only compressive stresses are caused in the cross section. This region is called the *core* of the cross section. To find the core of the cross section we assume that the neutral axis $g - g$ is a tangent to the cross section (Fig. 4.50c). Then the quantities y_0, z_0 correspond to a point P on the boundary of the core. According to (4.56) the coordinates of this point are

$$y_F = -\frac{r_{gz}^2}{y_0}, \qquad z_F = -\frac{r_{gy}^2}{z_0}. \tag{4.57}$$

The tangents to the cross section determine the boundary of the core. If the point of application of the force F lies inside the core, the neutral axis lies outside the cross section and the column is subjected to compression only.

As an example let us consider a rectangular cross section (Fig. 4.51a). Here, $r_{gy}^2 = h^2/12$ and $r_{gz}^2 = b^2/12$. First we choose the straight line g_1 to be the neutral axis. Then $y_0 \to \infty$, $z_0 = h/2$. Equation (4.57) yields the point P_1 (coordinates $y_F = 0$, $z_F = -h/6$) on the boundary of the core (Fig. 4.51b). Similarly, the neutral axes $g_2 \ldots g_4$ lead to the points $P_2 \ldots P_4$ on the y-axis or the z-axis, respectively, that have the distances $b/6$ or $h/6$ from the x-axis. In addition, it can be shown that the neutral axes g_i at the corners of the cross section correspond to points of application of the force along the straight lines between P_1 and P_2, between P_2 and P_3, etc. Therefore, the core of the rectangular cross section is a rhombus.

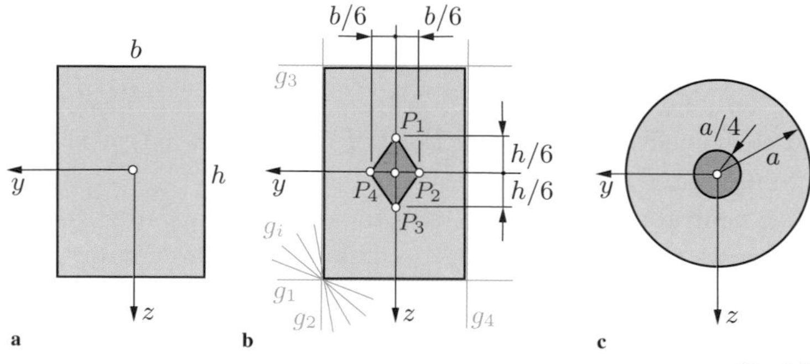

Fig. 4.51

As a second example consider a circular cross section with radius a (Fig. 4.51c). With $r_{gy}^2 = a^2/4$, $z_0 = a$ and with symmetry considerations the core is obtained to be an inner circle of radius $a/4$.

4.10

4.10 Thermal Bending

A change in temperature causes a material to expand or to contract. If the change of the temperature is constant throughout the cross section of a beam it changes the length of the beam (if this is not prevented by constraints, see Section 1.4). However, if the change of the temperature is not evenly distributed, it will also cause a *thermal bending*. In the following we will determine the stresses and deformations caused by a “ thermal load ”, restricting ourselves to ordinary bending of a Bernoulli beam.

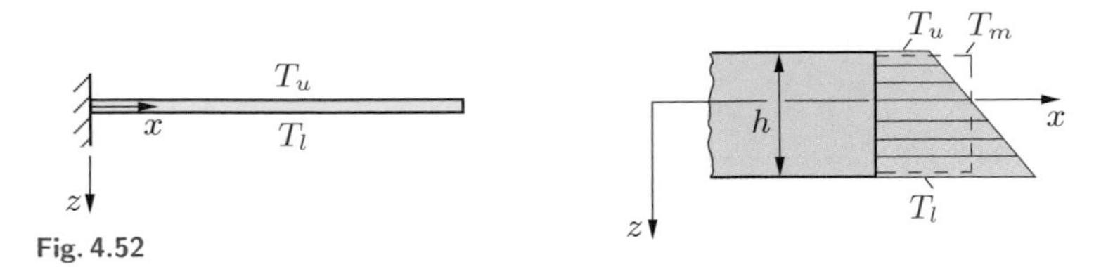

Fig. 4.52

Let us consider a straight cantilever beam which undergoes a rise of its originally constant temperature. For the sake of simplicity we assume that the distribution $\Delta T(z)$ of this change over the height h is *linear* in z-direction (Fig. 4.52):

$$\Delta T = T_m + (T_l - T_u)\frac{z}{h}. \tag{4.58}$$

Here, T_l and T_u are the temperature changes at the extreme lower fiber and the extreme upper fiber, respectively. The constant term $T_m = \frac{1}{A}\int \Delta T \mathrm{d}A$ (average change in the cross section) only causes a change of the length of the beam. Therefore, we investigate only the effect of the linear term

$$\Delta T^* = (T_l - T_u)\frac{z}{h}. \tag{4.59}$$

Hooke's law is given by (1.12):

$$\sigma = E\,\varepsilon - E\,\alpha_T \Delta T = E\,\varepsilon - E\,\alpha_T\,(T_l - T_u)\frac{z}{h}. \tag{4.60}$$

With $\varepsilon = \partial u/\partial x$ and the assumptions of Bernoulli $u = z\,\psi, \psi = -w'$ (recall (4.22b), (4.29)) we obtain

$$\sigma = -\,E\,w''\,z - E\,\alpha_T\,(T_l - T_u)\,\frac{z}{h}. \tag{4.61}$$

The bending moment M is the resultant of the normal stresses:

$$M = \int z\,\sigma\,\mathrm{d}A.$$

If we insert (4.61) and $I = \int z^2 \, dA$, we find

$$M = -\,EI\,w'' - EI\alpha_T\,\frac{T_l - T_u}{h}$$

or

$$w'' = -\,\frac{M}{EI} - \alpha_T\,\frac{T_l - T_u}{h}\,. \tag{4.62}$$

This is the differential equation of the deflection curve.

One can see that the difference $T_l - T_u$ of the temperature causes a curvature of the beam axis, just as the bending moment M does. Therefore, we introduce the notion of a "temperature moment"

$$M_{\Delta T} = EI\,\alpha_T\,\frac{T_l - T_u}{h}\,. \tag{4.63}$$

Then (4.62) can be written in the form

$$w'' = -\,\frac{M + M_{\Delta T}}{EI}\,. \tag{4.64}$$

In the special case $M_{\Delta T} = 0$, (4.64) reduces to (4.31).

If we eliminate w'' from (4.61) and (4.62), we find the stress distribution in the cross section (recall (4.26)):

$$\sigma = \frac{M}{I}\,z.$$

As an illustrative example let us consider the clamped beam in Fig. 4.53a. We will assume that the temperature difference $T_l - T_u$ is independent of x. Then the temperature moment $M_{\Delta T}$ is constant. Since there is no applied force, the bending moment M and the normal stress σ are zero. The elastic curve is determined

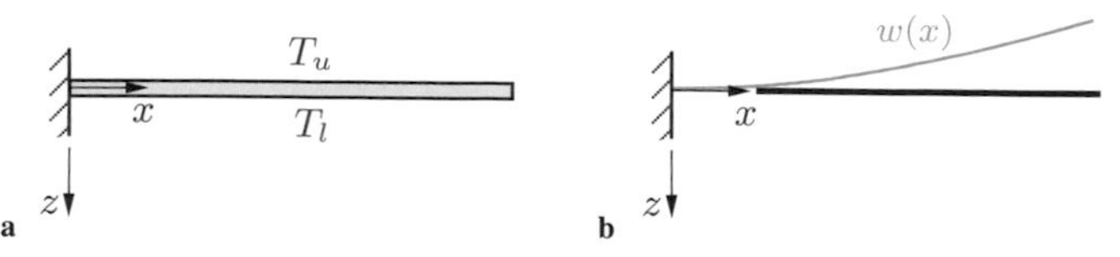

Fig. 4.53

through integration of (4.64):

$$w'' = -\frac{M_{\Delta T}}{EI} = \text{const},$$

$$w' = -\frac{M_{\Delta T}}{EI}\,x + C_1,$$

$$w = -\frac{M_{\Delta T}}{EI}\frac{x^2}{2} + C_1\,x + C_2.$$

The boundary conditions $w'(0) = 0$ and $w(0) = 0$ lead to $C_1 = 0$ and $C_2 = 0$. Hence, the elastic curve is given by (Fig. 4.53b)

$$w = -\frac{M_{\Delta T}}{2\,EI}\,x^2 = -\frac{\alpha_T\,(T_l - T_u)}{2\,h}\,x^2. \tag{4.65}$$

Let us now consider the beam in Fig. 4.54. Since there is a redundant support at B, the system is statically indeterminate. We can obtain the elastic curve with the method of superposition. The deflection $w^{(0)}$ of the "0"-system (redundant support B removed) is given by (4.65). In particular, at point B we have

$$w_B^{(0)} = -\frac{\alpha_T\,(T_l - T_u)}{2\,h}\,l^2.$$

The elastic curve $w^{(1)}$ of the "1"-system (cantilever subjected to the redundant force $X = B$) can be taken from Table 4.3. In particular,

$$w_B^{(1)} = -\frac{X\,l^3}{3\,EI}.$$

Substitution into the compatibility condition $w_B^{(0)} + w_B^{(1)} = 0$ yields

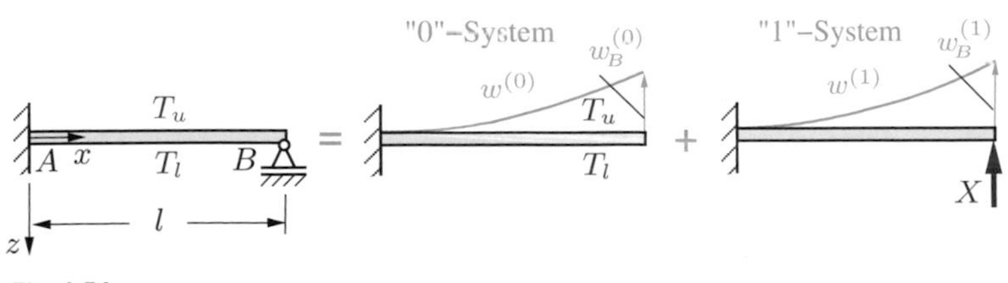

Fig. 4.54

$$X = B = -\frac{3\,EI\,\alpha_T\,(T_l - T_u)}{2\,h\,l}.$$

This determines the elastic curve $w = w^{(0)} + w^{(1)}$. The bending moment M is obtained by combining $M^{(0)} = 0$, $M^{(1)} = X(l - x)$ and the known value of X to give

$$M = M^{(0)} + M^{(1)} = -\frac{3\,EI\,\alpha_T\,(T_l - T_u)}{2\,h\,l}(l - x).$$

4.11

4.11 Supplementary Examples

Detailed solutions to the following examples are given in (**A**) D. Gross et al. *Formeln und Aufgaben zur Technischen Mechanik 2*, Springer, Berlin 2010 or (**B**) W. Hauger et al. *Aufgaben zur Technischen Mechanik 1-3*, Springer, Berlin 2008.

E4.18

Example 4.18 Calculate the principal moments of inertia for the thin-walled cross section ($t \ll a$) about its centroidal axes (Fig. 4.55). Find the principal axes.

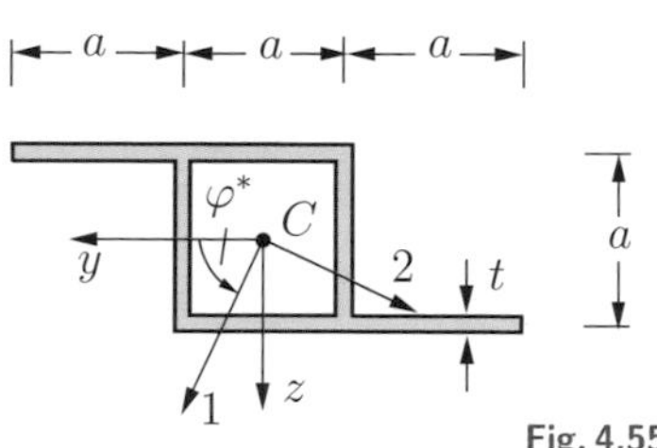

Fig. 4.55

Results: see (**B**) $I_1 = 3.3\,a^3 t\,, \quad I_2 = 0.7\,a^3 t\,, \quad \varphi^* = 65°.$

E4.19

Example 4.19 A quarter-circular area (radius a) and two sets of coordinate axes (centroidal axes y, z) are given in Fig. 4.56.

Calculate

a) $I_{\bar{y}}$, $I_{\bar{z}}$, $I_{\bar{y}\bar{z}}$,

b) I_y, I_z, I_{yz} and the principal axes and principal moments of inertia.

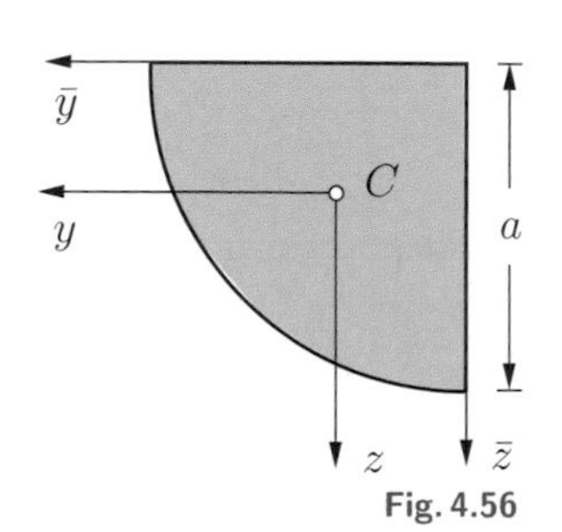

Fig. 4.56

Results: see (**A**)

$$I_{\bar{y}} = I_{\bar{z}} = \frac{\pi\, a^4}{16}\,, \quad I_{\bar{y}\bar{z}} = -\frac{a^4}{8}\,,$$
$$I_y = I_z = \Big(\frac{\pi}{16} - \frac{4}{9\pi}\Big)a^4, \quad I_{yz} = \Big(-\frac{1}{8} - \frac{4}{9\pi}\Big)a^4\,,$$
$$\varphi_1^* = \pi/4\,, \quad \varphi_2^* = 3\,\pi/4\,,$$
$$I_1 = \Big(\frac{\pi}{16} - \frac{1}{8}\Big)a^4\,, \quad I_2 = \Big(\frac{\pi}{16} - \frac{8}{9\pi} + \frac{1}{8}\Big)a^4\,.$$

E4.20

Example 4.20 Calculate the moments of inertia I_y, I_z and the product of inertia I_{yz} for the unsymmetrical, thin-walled Z-section ($t \ll b, h$) as shown in Fig 4.57.

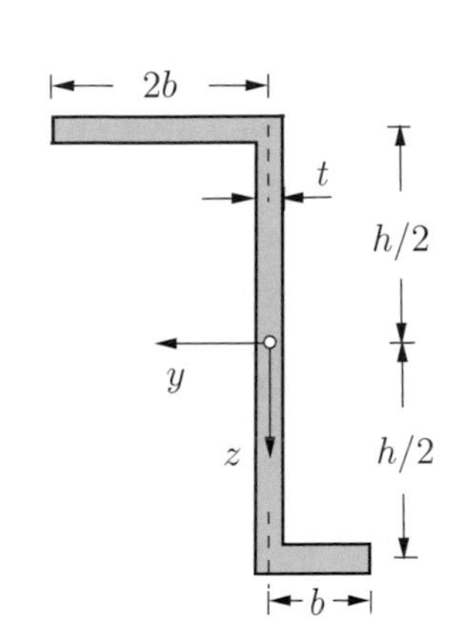

Fig. 4.57

Results: see (**A**)

$$I_y = b\,t h^2\Big(\frac{3}{4} + \frac{1}{12}\,\frac{h}{b}\Big)\,, \quad I_z = 3\,t\,b^3\,, \quad I_{yz} = \frac{5}{4}\,t\,b^2 h\,.$$

E4.21

Example 4.21 A cantilever beam with a thin-walled cross section ($t \ll b$) is subjected to a uniform line load q_0 (Fig. 4.58). The allowable stress σ_{allow} is given.

Determine the allowable length l_{allow} of the beam.

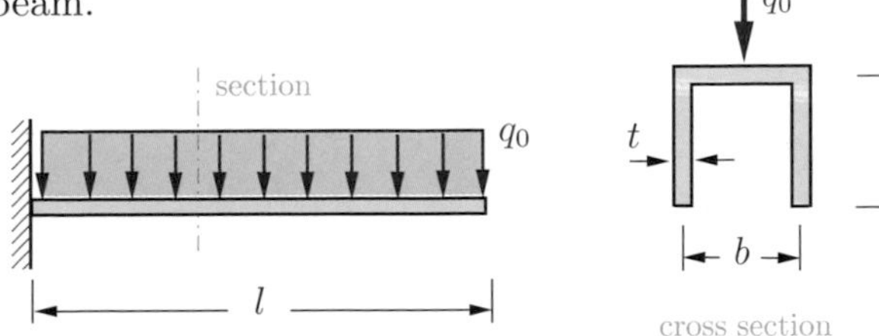

Fig. 4.58

Result: see (**B**) $l_{\text{allow}} \le b\sqrt{\dfrac{t\,\sigma_{\text{allow}}}{q_0}}$.

E4.22

Example 4.22 A compound beam consists of a steel I-beam which is bonded to a concrete beam (width b, height h), see Fig. 4.59. The system is subjected to a bending moment M.

a) Determine the width b so that only compressive stresses act in the concrete part (Con) and only tensile stresses in the steel part (St).

b) How large are then the stresses at the extreme fibers?

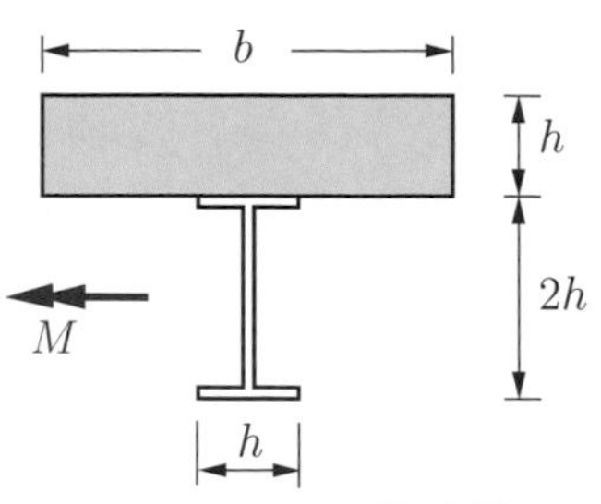

Fig. 4.59

Given:
$M = 2000$ kNm,
$E_{\text{Con}} = 3.5 \cdot 10^4$ MPa,
$E_{\text{St}} = 2.1 \cdot 10^5$ MPa,
$h = 500$ mm,
$A_{\text{St}} = h^2/6$, $I_{\text{St}} = h^4/18$.

Results: see (**A**)

$$\text{a)}\quad b = 2h = 1000\,\text{mm}\,,$$

$$\text{b)}\quad \sigma_{\text{Con}} = -8\,\text{MPa}\,, \qquad \sigma_{\text{St}} = 96\,\text{MPa}\,.$$

E4.23

Example 4.23 The column in Fig. 4.60 consists of three layers with different Young's moduli. It is subjected to a vertical force F_1 and a horizontal force F_2.

Determine the distribution of the normal stress at the clamped cross section.

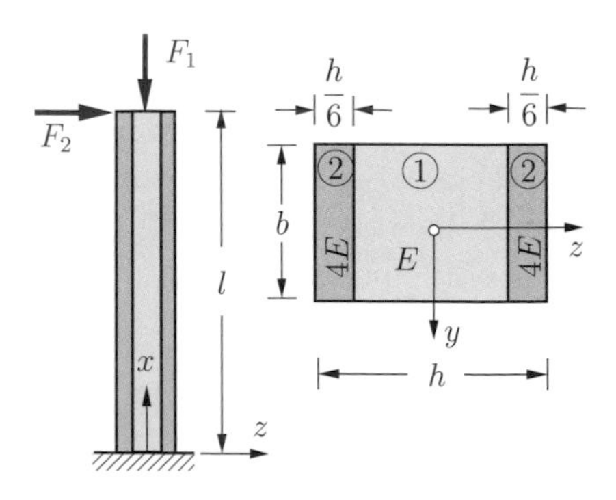

Fig. 4.60

Results: see (**A**) Selected values:

$$\text{Material ①}: \quad \sigma_1\left(\frac{h}{3}\right) = -\frac{F_1}{2bh} - \frac{9F_2 l}{7bh^2},$$

$$\text{Material ②}: \quad \sigma_2\left(\frac{h}{2}\right) = -\frac{2F_1}{bh} - \frac{54F_2 l}{7bh^2}.$$

E4.24

Example 4.24 A beam (flexural rigidity EI) is clamped at its left end and supported by a linear torsion spring (spring constant k_T) at its right end as shown in Fig. 4.61.

Determine the deflection curve due to the moment M_0 applied at the right end.

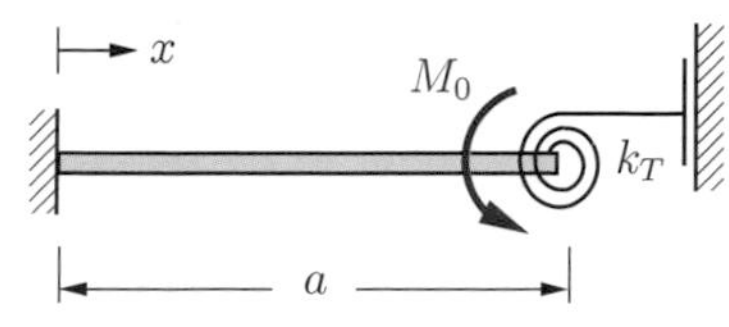

Fig. 4.61

Result: see (**B**) $w(x) = \left(\frac{k_T a}{EI + k_T a} - 1\right)\frac{M_0}{2EI}\,x^2\,.$

E4.25

Example 4.25 A beam (flexural rigidity EI) is supported as shown in Fig. 4.62. The stiffnesses of the torsion spring at A ($k_A = EI/a$) and of the linear spring at B ($k_B = EI/a^3$) are given. In the unloaded case both springs are unstretched.

Calculate the moment M_A in the spring at A and the force F_B in the spring at B if the beam is loaded at its middle by a force F.

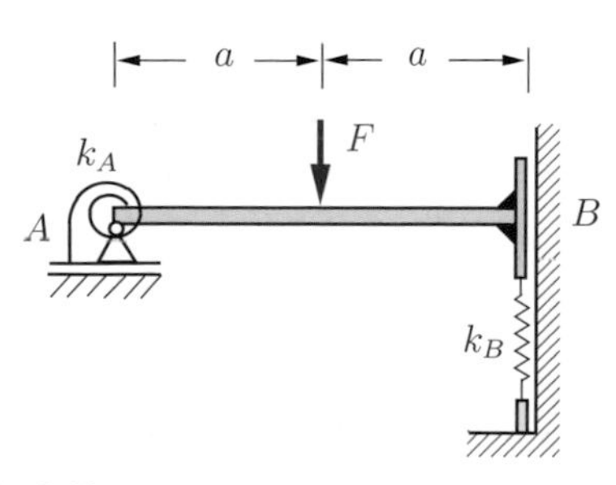

Fig. 4.62

Results: see (**B**) $M_A = -\frac{11\,Fa}{42}\,, \quad F_B = \frac{5\,F}{14}\,.$

E4.26

Example 4.26 Fig. 4.63 shows a leaf spring. The cross section has the constant height t and the variable width $b = b_0\,l/(l + x)$. The spring is loaded by a line load q_0 whose resultant is $F = q_0 b_0/2$.

Calculate the deflection at the free end.

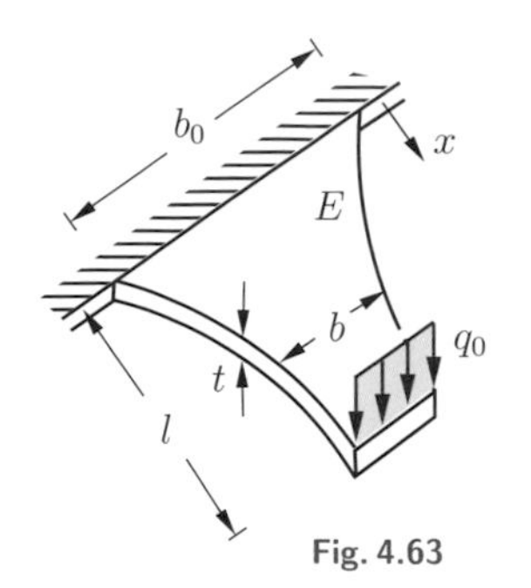

Fig. 4.63

Result: see (**A**) $w(l) = 5\,\dfrac{Fl^3}{Eb_0t^3}$.

E4.27

Example 4.27 Two cantilever beams (flexural rigidity EI) are connected with a bar (axial rigidity EA, coefficient of thermal expansion α_T) as shown in Fig. 4.64.

Determine the internal force S in the bar which is caused by a change ΔT of its temperature and by the force F applied at point C.

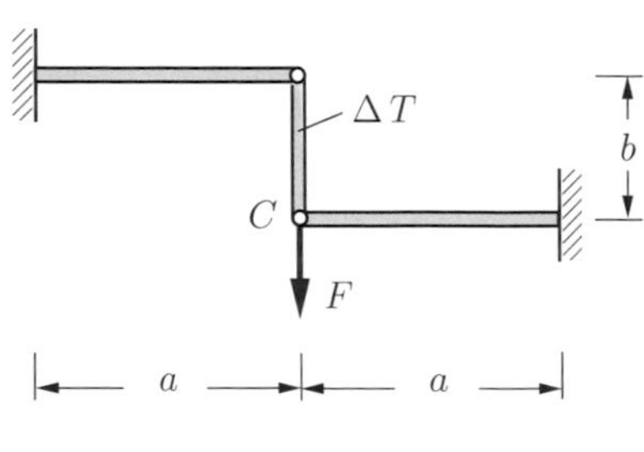

Fig. 4.64

Result: see (**B**) $S = \dfrac{\dfrac{Fa^3}{3EI} - \alpha_T \Delta T\, b}{\dfrac{2a^3}{3EI} + \dfrac{b}{EA}}$.

E4.28

Example 4.28 Two parallel beams (flexural rigidity EI, length a) are clamped to a wall at a distance l (Fig. 4.65). A bar (axial rigidity EA, length $l+\delta$ with $\delta \ll l$) is squeezed between the cantilever beams at the distance $a/2$ from the wall.

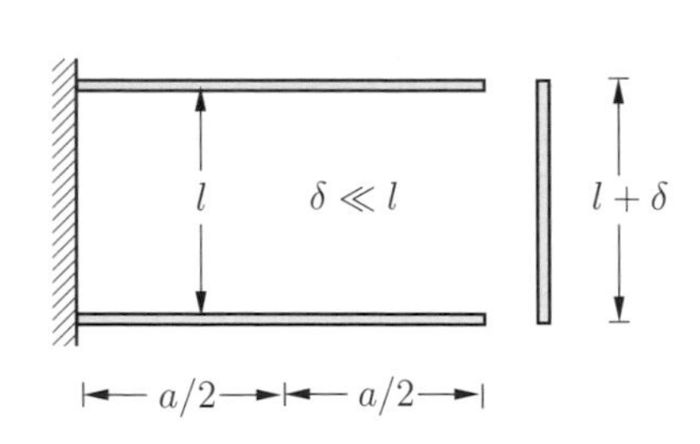

Fig. 4.65

a) Determine the force S in the bar.
b) Calculate the change e of the distance l of the two free ends.

Result: see (**A**) Selected value: $e = \frac{5}{24}\frac{a^3 EA}{l\,EI}\frac{\delta}{1+\frac{a^3 EA}{12\,l\,EI}}$.

E4.29

Example 4.29 The temporary bridge shown in Fig. 4.66 is pin supported at both ends and loaded with a uniform line load q_0. In addition, it rests on a pontoon which has the shape of a rectangular block (horizontal cross section A).

Investigate how far (depth f) the pontoon becomes immersed in the water due to the load q_0. Assume that the upper surface of the pontoon stays above the surface of the water. Note that the water exerts a buoyant force F_b on the pontoon. This force is equal to the weight of the displaced water: $F_b = \rho g f A$.
Given: $EI/Al^3\rho g = 1/12$.

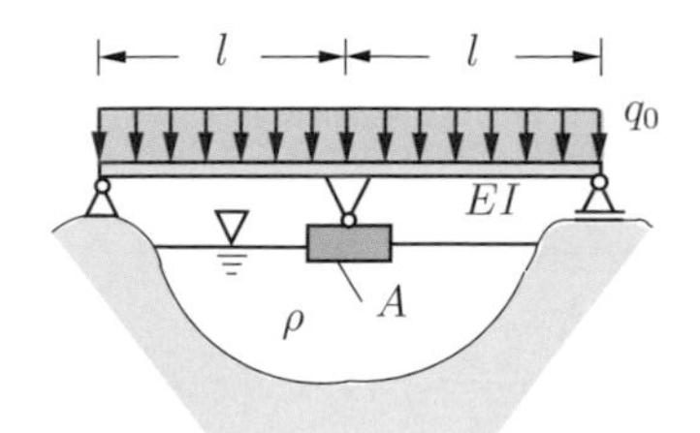

Result: see (**A**) $f = \frac{5}{72}\frac{q_0 l^4}{EI}$.

Fig. 4.66

E4.30

Example 4.30 A column with a thin-walled cross section ($t \ll a$) is clamped at its base and subjected to a horizontal force F at its free end as shown in Fig. 4.67.

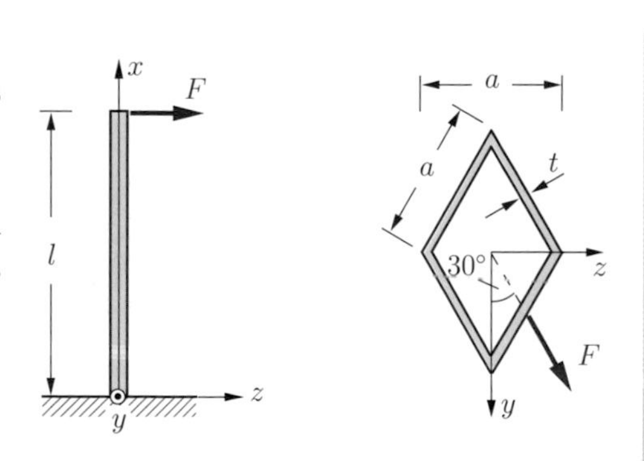

Fig. 4.67

Determine the maximum normal stress and its location. Calculate the displacement of the free end of the column.

Results: see (**B**) $\sigma_{\max} = \dfrac{3}{4}\dfrac{F\,l}{a^2 t}\,,\quad v = \dfrac{\sqrt{3}}{6}\dfrac{F\,l^3}{Eta^3}\,,\quad w = \dfrac{1}{2}\dfrac{F\,l^3}{Eta^3}\,.$

$\sigma_{\max}$ at the clamping in the parts of the profile in the first and the third quadrant.

E4.31

Example 4.31 A cantilever beam with a thin-walled cross section ($t \ll b$) is subjected to a uniform line load q_0 (Fig. 4.68, recall Example 4.21).

Determine the distribution of the shear stress in the cross section. Calculate the ratio $\tau_{\max}/\sigma_{\max}$ between the maximum shear stress and the maximum normal stress.

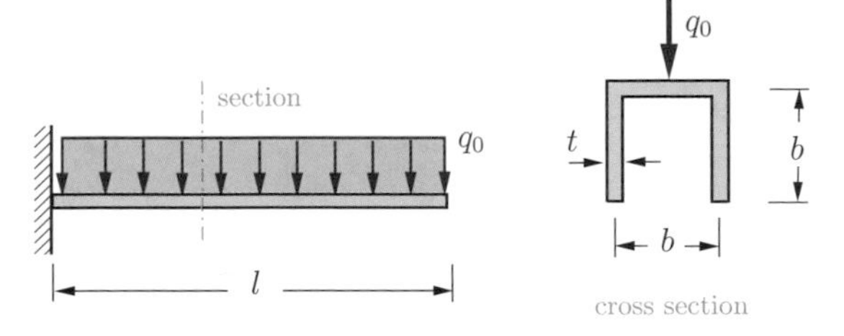

Fig. 4.68

Result: see (**B**) Selected value: $\tau_{\max}/\sigma_{\max} = 2b/3l$.

E4.32

Example 4.32 Locate the shear center O of the thin-walled cross section ($t \ll r$) as shown in Fig. 4.69.

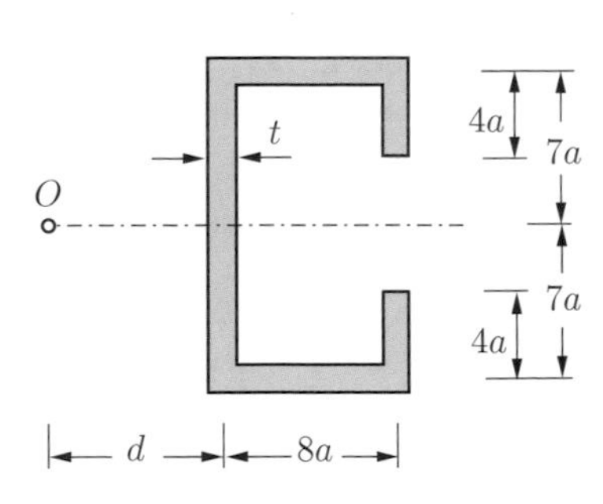

Fig. 4.69

Result: see (**B**) $d = 4.85\,a$.

4.12 Summary

- Moments of inertia:

$$I_y = \int z^2 \mathrm{d}A\,, \quad I_z = \int y^2 \mathrm{d}A\,, \quad I_{yz} = I_{zy} = -\int yz \mathrm{d}A\,.$$

  The transformation relations for a rotation of the coordinate system are analogous to those for the stress tensor. The parallel-axis theorem relates the moments of inertia for a centroidal axis and an axis parallel to it.
- Ordinary bending:

  Normal stress $\sigma(z) = \dfrac{M}{I}\,z\,, \quad \sigma_{\max} = \dfrac{M}{W}\,,$

  Shear stress $\tau(z) = \dfrac{V\,S(z)}{I\,b(z)}$ (solid cross section),

  DE deflection curve $EIw'' = -M$ or $(EIw'')'' = q\,.$

- Integration of the differential equation of the deflection curve leads to constants of integration. They are determined with the aid of boundary conditions and, if applicable, matching (continuity) conditions.
- Frequently, the solution of statically indeterminate problems can be found through a superposition of known solutions (Table 4.3).
- The shear deformation can be neglected for slender beams.
- Unsymmetric bending (y, z principal axes):

  Normal stress $\sigma = \dfrac{M_y}{I_y}\,z - \dfrac{M_z}{I_z}\,y\,,$

  DEs deflection curve $EI_y w'' = -M_y\,, \quad EI_z v'' = M_z\,.$

- If a beam is subjected to bending and tension/compression, the stresses and deformations are obtained through superposition of the solutions for the different loads.
- A nonuniform change of the temperature across the cross section causes a temperature moment leading to a curvature of the beam axis.

Chapter 5

Torsion

5

5 Torsion

5.1	Introduction	**191**
5.2	Circular Shaft	**192**
5.3	Thin-Walled Tubes with Closed Cross Sections	**203**
5.4	Thin-Walled Shafts with Open Cross Sections	**212**
5.5	Supplementary Examples	**220**
5.6	Summary	**228**

Objectives: In this chapter we investigate shafts that are twisted by external torques. As in the previous chapters, we calculate the deformations and stresses that are caused by the loading. For simplicity, we restrict ourselves to shafts with circular cross sections and to thin-walled shafts. By reading this chapter students will become familiar with the basic equations of torsion. They will also learn how to apply the theory to solve statically determinate and indeterminate problems.

5.1 Introduction

In the previous chapters we considered two different types of loading which may be applied to slender straight members. In the case of external forces acting *in* the direction of the longitudinal axis the only internal forces are the normal forces. The corresponding stresses and deformations were already discussed in Chapter 1. If a beam is subjected to forces perpendicular to its longitudinal axis or to moments about axes perpendicular to its longitudinal axis, then shear forces and bending moments act in the beam. In Chapter 4 we derived the formulas needed to calculate the stresses and deformations caused by bending. Now we want to analyse the case of an external moment which acts to twist a member *about the longitudinal axis*. This type of loading is associated with a *torque* acting on the cross section of the member.

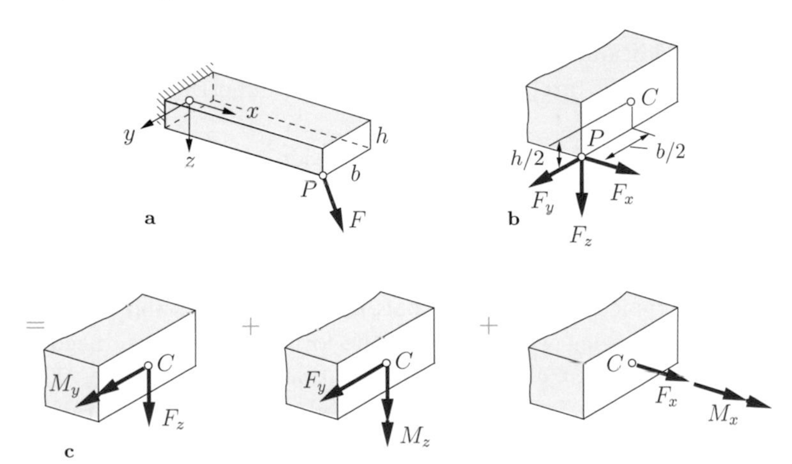

Fig. 5.1

In general, combinations of all these different types of loading will occur. For example, an eccentric load in the longitudinal direction causes a normal force *and* a bending moment (cf. Section 4.8). Let us consider a second example in order to demonstrate the possible coupling between the different types of stressing. For this purpose, we consider a cantilever with a rectangular cross section. At point P of the free end it is subjected to a load F that acts in an arbitrary direction (Fig. 5.1a). First, we decompose the load into its Cartesian components F_x, F_y and F_z (Fig. 5.1b). Then we move the action lines of the components without changing their directions until they pass through the centroid of the cross section at the free end. To avoid changing the effect of the force components on the cantilever, the respective moments of the forces have to be taken into account in addition to the force components (cf. Volume 1, Section 3.1.2). Therefore, the single eccentric force F is equivalent to the three force components acting at the centroid of the cross section and the three moments, see Fig. 5.1c. Here the individual forces and moments have been separated according to their different mechanical meanings:

1) The transverse load F_z and the external moment $M_y = \frac{h}{2}F_x$ lead to symmetric bending (cf. Section 4.3).
2) If, in addition, the transverse load F_y and the external moment $M_z = -\frac{b}{2}F_x$ are acting, we have unsymmetric bending (cf. Section 4.7).
3) The longitudinal load F_x causes tension in the bar (cf. Chapter 1). The external moment $M_x = \frac{b}{2}F_z - \frac{h}{2}F_y$ causes *torsion* of the member.

This example shows how a single force simultaneously can lead to the three typical loadings of a bar: tension, bending and torsion.

In the following we will derive the formulas which are needed to calculate the stresses and deformations due to torsion. The theory of torsion for arbitrarily shaped cross sections is rather complicated, therefore we restrict ourselves to specical cases. As an introductory problem we examine the torsion of a circular shaft in the next section.

5.2

5.2 Circular Shaft

We consider a straight circular shaft with a constant radius R. The shaft is clamped at one end and subjected to an external torque M_x (acting about the longitudinal axis) at its free end (Fig. 5.2a).

To derive the basic equations, we need to combine relations from kinematics, statics and Hooke's law. We start by making the following kinematical assumptions:

a) The cross sections remain unchanged during torsion, i.e. all the points of a cross section undergo the same twist. Points on a straight line within the cross section before twisting remain on a straight line after the deformation: radial lines of a cross section remain straight.

b) Plane cross sections remain plane, i.e. they do not warp. Therefore, we do not observe any deformation perpendicular to the cross sections.

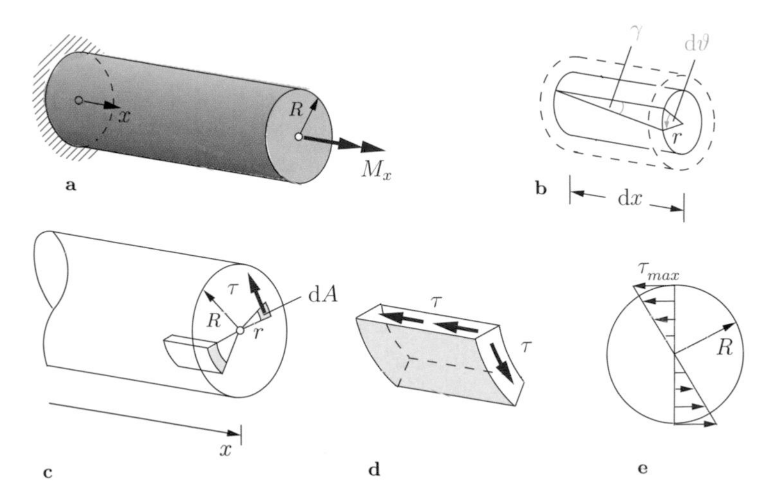

Fig. 5.2

With the aid of the theory of elasticity it can be shown that these assumptions are exactly fulfilled for a circular shaft (see Volume 4, Chapter 2.6.3). Therefore, an infinitesimal cylinder with arbitrary radius r isolated from the circular shaft remains a cylinder after twisting. We solely observe a relative rotation of two adjacent cross sections (distance $\mathrm{d}x$) by an infinitesimal *angle of twist* $\mathrm{d}\vartheta$ (Fig. 5.2b). Thereby, the angle of twist is positive if it rotates according to the *right-hand rule (corkscrew rule)*, i.e. if we look in the direction of the positive x-axis, a positive angle of twist rotates the cross section clockwise. For small deformations the relation between the infinitesimal angle of twist $\mathrm{d}\vartheta$ and the shear strain γ is (see Fig. 5.2b)

$$r\,\mathrm{d}\vartheta = \gamma\,\mathrm{d}x \qquad \rightarrow \qquad \gamma = r\frac{\mathrm{d}\vartheta}{\mathrm{d}x}\,. \tag{5.1}$$

A linear distribution of the shear strain γ corresponds to a linear distribution of the shear stress τ along any radial line of the cross section. At the boundary of the cross section the radial components of the shear stresses vanish since there are no applied forces at the boundary (symmetry of the stress tensor: complementary stresses, cf. (2.3)). Therefore, the shear stresses are tangential to the outer surface and also perpendicular to any radial line on the cross section. The shear stresses acting on an element isolated from the shaft are depicted in Fig. 5.2d. Using Hooke's law $\tau = G\,\gamma$ (cf. (3.10)) and inserting (5.1) yields

$$\tau = G\,r\frac{\mathrm{d}\vartheta}{\mathrm{d}x} = G\,r\,\vartheta'\,, \tag{5.2}$$

where we have used the abbreviation $\vartheta' = \frac{\mathrm{d}\vartheta}{\mathrm{d}x}$. Hence, the shear stress τ varies linearly from zero at $r = 0$ to a maximum value at the outer surface $r = R$ of a circular shaft (Fig. 5.2e).

The torque M_T must be statically equivalent to the moment resulting from the shear stresses shown in Figs. 5.2c, that is:

$$M_T = \int r\,\tau\,\mathrm{d}A\,. \tag{5.3}$$

The torque is positive if it points in the positive direction of the coordinate at a positive face (cf. Volume 1, Section 7.4).

Inserting (5.2) into (5.3) yields

$$M_T = G\,\vartheta' \int r^2\,\mathrm{d}A = G\,\vartheta'\,I_p\,. \tag{5.4}$$

The integral I_p in (5.4) is a purely geometrical quantity and is known as the polar moment of inertia, see (4.6c). In order to ensure that we use a consistent notation when dealing with arbitrary cross sections, we now identify this geometrical quantity as *torsion constant* I_T (cf. Table 5.1). In the case of a circular shaft we have $I_T = I_p$ and (5.4) can be rewritten as

$$GI_T \, \vartheta' = M_T \, . \tag{5.5}$$

The quantity GI_T is known as *torsional rigidity*. Given the torque M_T and the torsional rigidity GI_T, we can calculate the angle of twist ϑ from (5.5). Note that $I_T \neq I_p$ in the case of a non-circular cross section (see Section 5.4).

Let us consider again a circular shaft which is clamped at one end and subjected to a torque M_x at the free end, see Fig. 5.2a. In each arbitrary section perpendicular to the x-axis the stress resultant is a torque M_T, which is constant over the length l of the member and equal to the external moment:

$$M_T = M_x \, . \tag{5.6}$$

The total angle of twist ϑ_l at the free end in the case of constant GI_T is found through integration:

$$\vartheta_l = \int_0^l \vartheta' \mathrm{d}x \qquad \rightarrow \qquad \vartheta_l = \frac{M_T \, l}{GI_T} \, . \tag{5.7}$$

A comparison with (1.18) shows an analogy between the the tensile and the torsional member.

If we eliminate ϑ' from (5.2) with the aid of (5.5) we obtain the *torsion formula* which gives the distribution of the shear stress:

$$\tau = \frac{M_T}{I_T} \, r \, . \tag{5.8}$$

The maximum value appears at the outer boundary, i.e. at $r = R$:

$$\tau_{\max} = \frac{M_T}{I_T} \, R \, ,$$

see Fig. 5.2e. In order to obtain an analogy to bending (cf. (4.28)), we introduce the so-called *section modulus of torsion* W_T:

$$\tau_{\max} = \frac{M_T}{W_T} . \tag{5.9}$$

For the circular shaft $W_T = I_T/R$ holds. Using (4.10a) we obtain

$$I_T = I_p = \frac{\pi}{2} R^4 , \qquad W_T = \frac{\pi}{2} R^3 . \tag{5.10}$$

The formulas (5.1) to (5.9) do not only hold for solid but also for *hollow circular cross sections*. In this case the torsion constant I_T and the section modulus of torsion W_T are

$$I_T = \frac{\pi}{2}(R_a^4 - R_i^4) , \qquad W_T = \frac{\pi}{2} \frac{R_a^4 - R_i^4}{R_a} , \tag{5.11}$$

where R_a and R_i denote the outer and the inner radius, respectively. For a *thin walled* circular tube with wall thickness $t = R_a - R_i$ and mean radius $R_m = (R_a + R_i)/2$ we can derive the following approximations (cf. (4.12))

$$I_T \approx 2\,\pi\, R_m^3\, t , \qquad W_T \approx 2\,\pi\, R_m^2\, t . \tag{5.12}$$

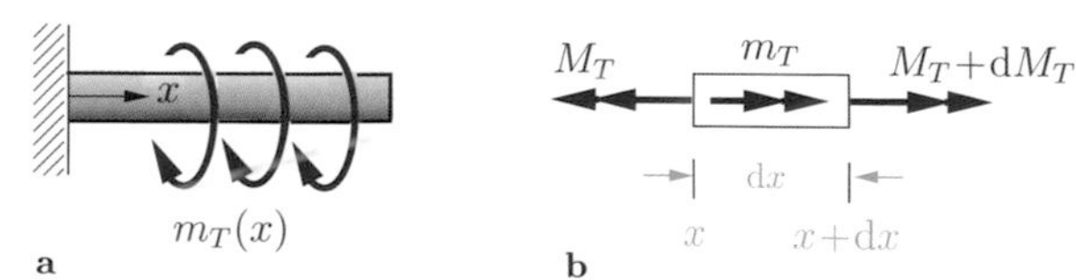

Fig. 5.3

If a distributed torque per unit length $m_T(x)$ acts along the longitudinal axis of a rod (Fig. 5.3a), we obtain

$$\mathrm{d}M_T + m_T\, \mathrm{d}x = 0$$

or

$$\frac{\mathrm{d}M_T}{\mathrm{d}x} = M_T' = -\, m_T . \tag{5.13}$$

This relation is a consequence of the moment equilibrium condition at an infinitesimal rod element (Fig. 5.3b). Hence, for $m_T = 0$ we get $M_T = \text{const.}$

Differentiating (5.5) with respect to x and inserting (5.13) yields the second-order differential equation for the angle of twist:

$$(GI_T\,\vartheta')' = -\,m_T\,. \tag{5.14}$$

Two integration constants appear after the integration over the entire length of the shaft. They can be calculated from the boundary conditions (one for each boundary): at a given boundary we can either prescribe the angle of twist ϑ or the torque $M_T = GI_T\,\vartheta'$. For example, ϑ vanishes at a clamped boundary, or if the free end of a shaft is subjected to an applied torque M_x we have $M_T = M_x$. A comparison of (5.14) with (1.20) shows again the analogy between tension and torsion.

Let us now apply the formulas derived above to calculate the spring constant k of a coil spring. We assume that the spring is tightly wound, i.e. the helix angle (angle of inclination) is approximately zero. Furthermore, we assume that the diameter d of the cross section of the spring wire (Fig. 5.4a) is small compared to the radius a of the spring winding ($d \ll a$).

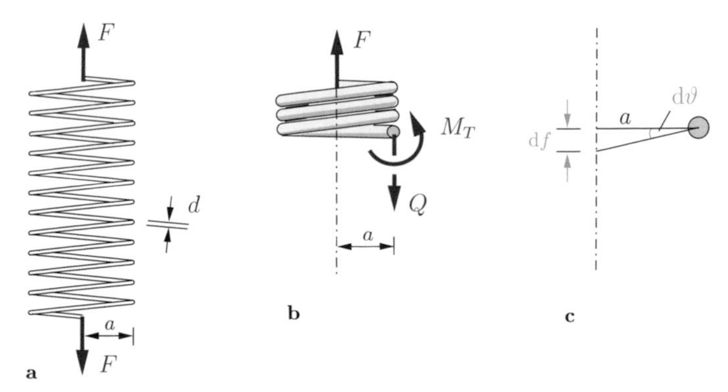

Fig. 5.4

The spring is subjected to a force F acting on its axis. In order to compute the elongation of the spring we take an imaginary cut at an arbitrary position (Fig. 5.4b). Evaluating the equilibrium conditions yields a constant shear force $V = F$ and an internal torque $M_T = a\,F$. Let us now assume that only an element of length $\mathrm{d}s$ of the spring wire is elastic, whereas the remaining part of the spring is rigid. Then the relative angle of twist between the cross sections at the ends of the infinitesimal element is $\mathrm{d}\vartheta$. Therefore the lower part of the helical spring undergoes a displacement $\mathrm{d}f = a\,\mathrm{d}\vartheta$ (Fig. 5.4c). Substituting $\mathrm{d}\vartheta = (M_T/GI_T)\mathrm{d}s$ in the aforementioned equation yields (cf. (5.5))

$$\mathrm{d}f = a\,\frac{M_T}{GI_T}\,\mathrm{d}s = \frac{F\,a^2}{GI_T}\,\mathrm{d}s\,.$$

In the case of a flat spring with n windings the total length of the spring wire is approximately $(2\,\pi\,a)n$. Therefore the total elongation of the spring is obtained through integration over the total length of the spring wire:

$$f = \int \mathrm{d}f = \frac{F\,a^3}{GI_T}\,2\,\pi\,n\,.$$

Inserting the torsion constant as given in (5.10) with $R = d/2$, we obtain the spring constant

$$k = \frac{F}{f} = \frac{GI_T}{2\,\pi\,a^3\,n} \qquad \rightarrow \qquad k = \frac{G\,d^4}{64\,a^3\,n}\,.$$

This result shows that the spring constant decreases with an increasing number of windings n and an increasing radius a, whereas the spring constant increases with an increasing diameter d of the wire.

E5.1

Example 5.1 A homogeneous shaft with a circular cross section (diameter d) is clamped at point A and subjected to two external torques M_0 and M_1 at points B and C (Fig. 5.5a).

a) The external torque M_0 is given. Determine M_1 so that the angle of twist is zero at the free end C.

b) Calculate the maximum shear stress and its location in this case.

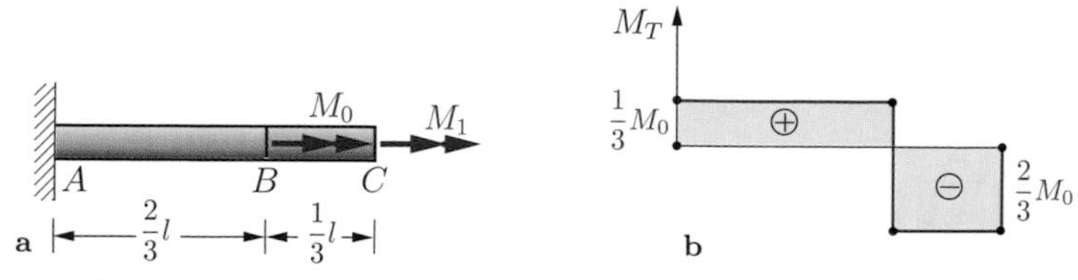

Fig. 5.5

Solution a) The region $\overline{AB}$ is subjected to the torque $M_0 + M_1$, whereas the region $\overline{BC}$ is subjected to M_1. The angle of twist ϑ_C at the free end follows from the superposition of the angles of twist of both parts of the shaft according to (5.7) as

$$\vartheta_C = \vartheta_{AB} + \vartheta_{BC} = \frac{M_0 + M_1}{GI_T}\frac{2}{3}l + \frac{M_1}{GI_T}\frac{l}{3} = \frac{l}{3\,GI_T}(2\,M_0 + 3\,M_1)\,.$$

This angle is zero for

$$\underline{\underline{M_1 = -\frac{2}{3}M_0\,.}}$$

Fig. 5.5b depicts the corresponding moment diagram.

b) The maximum shear stress occurs in the cross sections where the maximum torque appears. According to Fig. 5.5b, the shaft is subjected to the maximum stressing in the region $\overline{BC}$:

$$|M|_{\max} = \frac{2}{3}M_0.$$

Thus, according to (5.8), we find the maximum stress as

$$\tau_{\max} = \frac{M_{\max}}{W_T} = \frac{2}{3}\frac{M_0}{W_T}\,.$$

With the section modulus of torsion $W_T = \pi R^3/2$ for a circular shaft (cf. (5.10)) and with $R = d/2$ we obtain

$$\underline{\underline{\tau_{\max} = \frac{32}{3}\frac{M_0}{\pi d^3}\,.}}$$

E5.2

Example 5.2 A compound shaft (torsional rigidities GI_{T_1} and GI_{T_2}, respectively) is loaded by a torque per unit length m_T (Fig. 5.6a).

Determine the moment diagram.

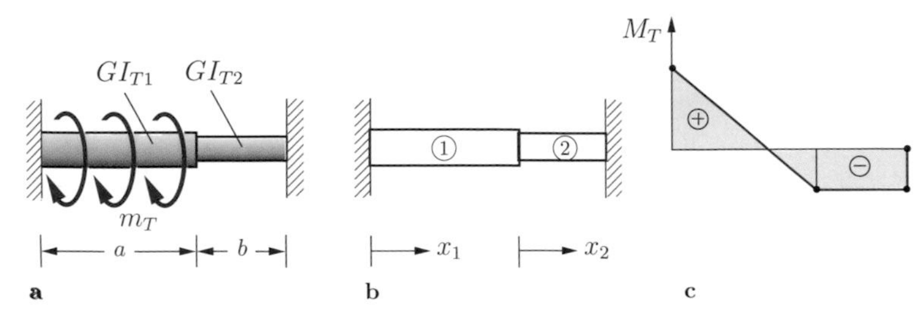

Fig. 5.6

Solution The shaft is clamped at both ends and thus statically indeterminate. We want to solve the problem by integration over each part of the shaft and therefore introduce the coordinates x_1 and x_2 (Fig. 5.6b). We apply (5.14) and obtain for both segments

$$GI_{T_1}\vartheta_1'' = -\,m_T\,, \qquad GI_{T_2}\vartheta_2'' = 0,$$

$$GI_{T_1}\vartheta_1' = -\,m_T x_1 + C_1\,, \qquad GI_{T_2}\vartheta_2' = C_3\,,$$

$$GI_{T_1}\vartheta_1 = -\,m_T\frac{x_1^2}{2} + C_1\,x_1 + C_2, \quad GI_{T_2}\vartheta_2 = C_3\,x_2 + C_4\,.$$

The four constants of integration follow from two boundary and two matching conditions:

$$\vartheta_1(x_1 = 0) = 0 \quad \rightarrow \quad C_2 = 0,$$

$$\vartheta_2(x_2 = b) = 0 \quad \rightarrow \quad C_3\,b + C_4 = 0,$$

$$\vartheta_1(x_1 = a) = \vartheta_2(x_2 = 0) \quad \rightarrow \frac{1}{GI_{T_1}}\left(-m_T\frac{a^2}{2} + C_1\,a\right) = \frac{C_4}{GI_{T_2}}\,,$$

$$M_{T_1}(x_1 = a) = M_{T_2}(x_2 = 0) \rightarrow GI_{T_1}\vartheta_1'(x_1 = a) = GI_{T_2}\vartheta_2'(x_2 = 0)$$

$$\rightarrow\; -\,m_T\,a + C_1 = C_3\,.$$

Solving these equations yields

$$C_1 = \frac{m_T\, a}{2} \frac{a\, GI_{T_2} + 2\, b\, GI_{T_1}}{a\, GI_{T_2} + b\, GI_{T_1}}, \qquad C_3 = -\frac{m_T\, a}{2} \frac{a\, GI_{T_2}}{a\, GI_{T_2} + b\, GI_{T_1}},$$

$$C_4 = \frac{m_T\, a}{2}\, b\, \frac{a\, GI_{T_2}}{a\, GI_{T_2} + b\, GI_{T_1}}.$$

Thus, we obtain the moments in both segments:

$$\underline{\underline{M_{T_1} = -\, m_T \left(x_1 - \frac{a}{2} \frac{a\, GI_{T_2} + 2\, b\, GI_{T_1}}{a\, GI_{T_2} + b\, GI_{T_1}} \right)}},$$

$$\underline{\underline{M_{T_2} = -\, \frac{m_T\, a}{2} \frac{a\, GI_{T_2}}{a\, GI_{T_2} + b\, GI_{T_1}}}}.$$

The diagram of the internal torque is qualitatively given in Fig. 5.6c.

Example 5.3 A circular shaft (length a) is subjected to a force F by means of a lever (length b), see Fig. 5.7a. **E5.3**

Determine the required radius R using the criteria of the maximum distortion energy for the given parameters $a = 3$ m, $b = 1$ m, $F = 5 \cdot 10^3$ N and $\sigma_{\text{allow}} = 180$ MPa.

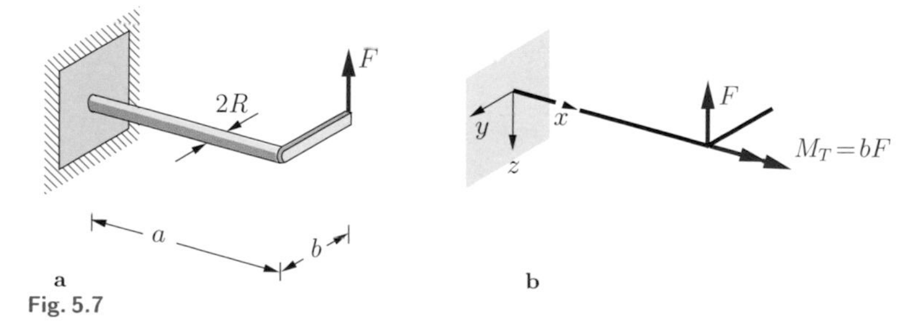

Fig. 5.7

Solution The force F at the free end of the lever is statically equivalent to a force F and a moment $M_T = b\,F$ at the end of the straight circular shaft (Fig. 5.7b). The shaft is subjected to bending (cf. Chapter 4) by the force F in Fig. 5.7b which causes normal stresses σ. The maximum stress is found at the outer radius ($z = \pm R$) at the fixed end ($x = 0$) and has, according to (4.28), the value

$$\sigma_{\max} = \frac{|M|_{\max}}{W}$$

with $|M|_{\max} = a\,F$ and $W = \pi\,R^3/4$. The moment M_T subjects the circular shaft to torsion and causes shear stresses τ. Their maximum value is found at the boundary of the cross section and is constant along the length of the shaft. With the help of (5.9) and (5.10) we obtain

$$\tau_{\max} = \frac{M_T}{W_T} \qquad \text{with} \qquad W_T = \pi\,R^3/2\,.$$

To determine the required radius R_{required} of the shaft we calculate the equivalent stress σ_e according to (3.18) at the locations of the maximum stresses (upper and lower point of the boundary of the cross section at the fixed end). With $\sigma_x = \sigma_{\max}$, $\tau_{xy} = \tau_{\max}$ and $\sigma_y = 0$ we obtain

$$\sigma_{e_{\max}} = \sqrt{\left(\frac{|M|_{\max}}{W}\right)^2 + 3\left(\frac{M_T}{W_T}\right)^2}\,.$$

The condition $\sigma_{e_{\max}} \le \sigma_{\text{allow}}$, see (3.15), leads to

$$\sqrt{\frac{16a^2F^2}{\pi^2R^6} + 3\frac{4\,b^2\,F^2}{\pi^2R^6}} \le \sigma_{\text{allow}}\,.$$

Thus,

$$R^6 \ge \frac{4\,F^2(4\,a^2 + 3\,b^2)}{\pi^2\,\sigma^2_{\text{allow}}}\,.$$

Inserting the given parameters yields the required radius

$$\underline{\underline{R_{\text{required}} = 48\ \text{mm}}}\,.$$

5.3 Thin-Walled Tubes with Closed Cross Sections

The theory of torsion for arbitrary cross sections is rather complicated, as already mentioned in the introduction to this chapter. An exception besides the circular shaft is the thin-walled tube with closed cross sections. For this type of geometries, we can derive useful formulas for approximate solutions based on a few suitable assumptions for the stress distributions. Now we will focus on this type of cross sections, since they are of major importance in practical applications (box girder in bridge constructions, wing constructions in aeronautics, etc.).

We assume that the dimensions of the thin-walled tubes with closed cross sections (= hollow cylinders) do not vary along x, i.e. we consider tubes with constant yet arbitrary thin-walled cross sections. Furthermore, we assume that the cross sections are subjected to a constant torque M_T (Fig. 5.8a). As a coordinate along the center line (also called median line) of the profile (Fig. 5.8c) we introduce the arc length s. The wall thickness of the tube may vary with the arc length: $t = t(s)$.

The applied torque causes shear stresses in the cross sections. No loads are applied at the outer and inner boundaries of the cross section. Therefore we conclude that the shear stresses must be tangential to the boundaries. Since the wall thickness is small, we assume that the shear stresses are uniformly distributed across the thickness of the tube (= average shear stress). Hence, we can express them by a resulting force quantity, namely the *shear flow*

$$T = \tau\, t\,. \tag{5.15}$$

The shear flow T has the dimension force/length and is acting tangential to the median line of the profile (Fig. 5.8c).

Let us now consider a rectangular element of the tube having the infinitesimal length $\mathrm{d}x$ and height $\mathrm{d}s$ as depicted in Fig. 5.8b. The left face (at x) is subjected to the shear flow T and the right face (at $x + \mathrm{d}x$) is subjected to the shear flow $T + (\partial T/\partial x)\mathrm{d}x$. There are no normal stresses in the s-direction. Therefore the equilibrium condition in this direction yields

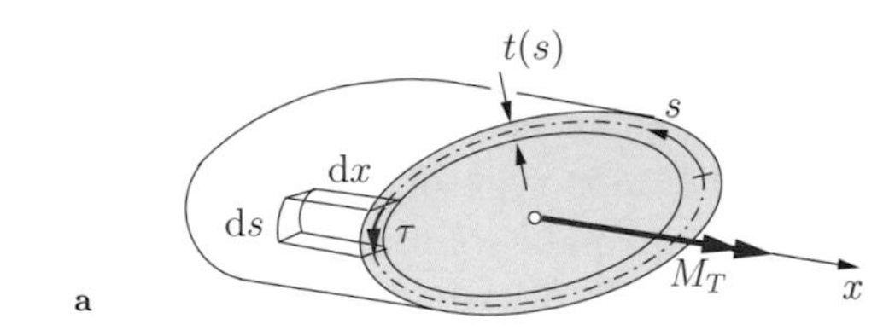

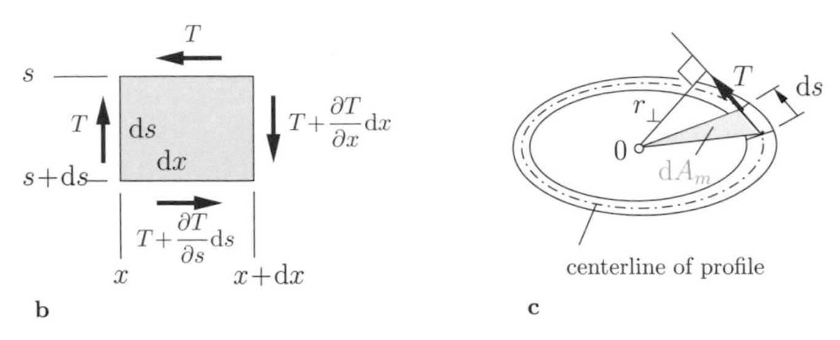

Fig. 5.8

$$\downarrow: \quad \left(T + \frac{\partial T}{\partial x}\mathrm{d}x\right)\mathrm{d}s - T\,\mathrm{d}s = 0 \qquad \rightarrow \qquad \frac{\partial T}{\partial x} = 0\,.$$

Hence, the shear flow is constant in the x-direction. If we further assume that there are also no normal stresses in the x-direction (free warping of the cross section), then we obtain from the equilibrium condition in x-direction:

$$\rightarrow: \quad \left(T + \frac{\partial T}{\partial s}\mathrm{d}s\right)\mathrm{d}x - T\,\mathrm{d}x = 0 \qquad \rightarrow \qquad \frac{\partial T}{\partial s} = 0\,.$$

Thus, the shear flow has the same value at every point s of the cross section, i.e.

$$T = \tau\, t = \text{const}\,. \tag{5.16}$$

We will now derive a relation between the torque M_T and the shear flow T. According to Fig. 5.8c the shear flow generates a force $T\mathrm{d}s$ acting on the centerline of the tube's wall. The infinitesimal moment with respect to an arbitrarily chosen point 0 is given by

$$\mathrm{d}M_T = r_\perp \, T \, \mathrm{d}s \,.$$

Here, $r_\perp$ is the moment arm of the force with respect to point 0. The total moment generated by the shear flow is identical to the applied torque M_T:

$$M_T = \oint \mathrm{d}M_T = T \oint r_\perp \mathrm{d}s \,. \tag{5.17}$$

The *line integral* (circle at the integral sign) means that, starting from an arbitrary point $s = 0$, we have to perform the integration along the arc length s of the whole boundary of the cross section. It can be seen from Fig. 5.8c that $r_\perp \mathrm{d}s$ (= height × base line) is twice the area of the green triangle: $r_\perp \mathrm{d}s = 2\,\mathrm{d}A_m$. Thus the line integral yields

$$\oint r_\perp \, \mathrm{d}s = 2\,A_m \,. \tag{5.18}$$

Here, A_m is the area enclosed within the boundary of the median line of the profile. One must be careful not to confuse this geometrical quantity with the area $A = \oint t\,\mathrm{d}s$ of the cross section. Inserting (5.18) into (5.17) yields

$$M_T = 2\,A_m\,T \,. \tag{5.19}$$

Using (5.16) the shear stresses follow as

$$\tau = \frac{T}{t} = \frac{M_T}{2\,A_m\,t} \,. \tag{5.20}$$

This relation is known as *Bredt's first formula* (Rudolf Bredt, 1842–1900) or as torsion formula for thin-walled tubes.

The largest shear stress appears at the point with the smallest wall thickness $t_{\min}$: $\tau_{\max} = T/t_{\min} = M_T/2\,A_m\,t_{\min}$. If we introduce, in analogy to (5.9), a *torsional section modulus* W_T we obtain

$$\tau_{\max} = \frac{M_T}{W_T} \qquad \text{with} \qquad W_T = 2\,A_m\,t_{\min}\,. \tag{5.21}$$

In the case of a thin-walled circular tube with mean radius R_m we obtain $A_m = \pi R_m^2$ and for a constant wall thickness t we obtain the same value for W_T as in (5.12).

In order to calculate the shear stresses according to (5.20) we have made two statical assumptions:

a) the shear stresses are constant across the wall thickness of the tube,
b) in the sections at $x = \text{const}$ no normal stresses occur.

The second assumption is correlated to kinematic requirements, which have to be additionaly introduced when computing the twist of the shaft. We assume that

c) the form of the cross section does not change during the deformation (as was assumed for the circular shaft),
d) in contrast to the circular shaft we observe longitudinal displacements in x-direction for an arbitrary profile: the cross section *warps (bulges)*. This warping is assumed not to be restricted.

If the warping is restrained by supports (or if M_T varies along x), additional normal stresses occur. The evaluation of these is the subject of the theory of *warping torsion*, which cannot be covered within this introductory book.

The displacements of an arbitrary point P on the centerline of the profile in x- or s-direction are denoted by u and v, respectively. If the cross section rotates about the infinitesimal angle $\mathrm{d}\vartheta$ (the shape of the profile remains unchanged due to the first kinematic assumption), point P is shifted by $r\,\mathrm{d}\vartheta$ to the position P' (Fig. 5.9). The component of this displacement in the direction of the tangent at the centerline of the profile is $\mathrm{d}v = r\,\mathrm{d}\vartheta\cos\alpha$. Here α denotes the angle between the line orthogonal to r and the tangent of the centerline of the profile. The same angle occurs between r and the perpendicular distance $r_\perp$ of the tangent in P (the sides of the angle are pairwise perpendicular). Using $r_\perp = r\cos\alpha$ we obtain

$$dv = r_\perp d\vartheta \,. \tag{5.22}$$

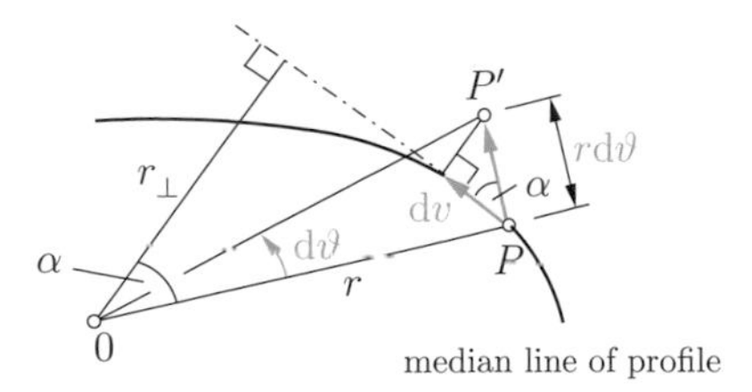

Fig. 5.9

The shear strain γ of an element of the tube wall is given by $\gamma = \partial v/\partial x + \partial u/\partial s$ by analogy to (3.2). Hooke's law $\tau = G\gamma$ (cf. (3.10)) relates the shear stress τ with the shear strain. Eliminating the shear stress in

$$\frac{\tau}{G} = \gamma = \frac{\partial v}{\partial x} + \frac{\partial u}{\partial s}$$

with the shear flow according to (5.20) and taking into account (5.22), we obtain

$$\frac{T}{Gt} = r_\perp \vartheta' + \frac{\partial u}{\partial s} \,. \tag{5.23}$$

This equation still contains the longitudinal displacement u in x-direction, which is as yet unknown. In order to eliminate this displacement we integrate $\partial u/\partial s$ along the arc length s from an initial point A to a final point E:

$$\int_{s_A}^{s_E} \frac{\partial u}{\partial s}\, ds = u_E - u_A \,.$$

If we integrate along the whole perimeter, then the initial point and the final point coincide. In this case, the gap $u_E - u_A$ has to vanish for tubes with closed cross sections, i.e. $\oint (\partial u/\partial s) ds = 0$.

Therefore, we obtain from (5.23)

$$\oint \frac{T}{Gt}\,\mathrm{d}s = \vartheta' \oint r_\perp \mathrm{d}s\,.$$

Solving this equation for ϑ' and using (5.18) and (5.19) yields

$$\vartheta' = \frac{\displaystyle\oint \frac{M_T}{2\,A_m\,G\,t}\,\mathrm{d}s}{2\,A_m} = \frac{M_T \displaystyle\oint \frac{\mathrm{d}s}{t}}{4\,G\,A_m^2}\,.$$

This relation can be written in the form

$$\vartheta' = \frac{M_T}{GI_T} \tag{5.24}$$

with the abbreviation

$$I_T = \frac{(2\,A_m)^2}{\displaystyle\oint \frac{\mathrm{d}s}{t}}\,. \tag{5.25}$$

The relation (5.25) for the torsion constant I_T is also referred to as *Bredt's second formula*. According to (5.24) we can determine the angle of twist ϑ of a thin-walled tube exactly like the one of a circular shaft (cf. (5.5)), if we insert for I_T the value obtained in (5.25). In particular, the relative angle of twist of two cross sections of distance l is given by (cf. (5.7))

$$\vartheta_l = \frac{M_T}{GI_T}\,l\,. \tag{5.26}$$

In the special case $t = \text{const}$ and denoting the circumference of the profile $U = \oint \mathrm{d}s$ in (5.25), we obtain the torsion constant

$$I_T = \frac{(2\,A_m)^2\,t}{U}\,. \tag{5.27}$$

If we apply this formula to a thin-walled tube with a circular cross section of mean radius R_m and constant thickness t, with

$U = 2\pi R_m$ and $A_m = \pi R_m^2$ we obtain the torsion constant

$$I_T = \frac{(2\pi R_m^2)^2 t}{2\pi R_m} = 2\pi R_m^3 t\,,$$

in accordance with (5.12).

Using (5.24) we can compute the angle of twist of a bar under torsion through integration. In order to calculate the *axial displacements* u (= warping) of the points of an arbitrary thin-walled cross section, we start from (5.23):

$$\frac{\partial u}{\partial s} = \frac{T}{Gt} - r_\perp \vartheta'\,.$$

Integration with respect to s (the quantities G, T and ϑ' are independent of s) yields

$$u = \frac{T}{G}\int \frac{\mathrm{d}s}{t} - \vartheta' \int r_\perp \mathrm{d}s + C\,. \tag{5.28}$$

For instance, let us now determine the axial displacements u (the warping function) of a thin-walled tube having a rectangular cross section subjected to a torque M_T as depicted in Fig. 5.10a. To evaluate (5.28) we first compute the shear flow according to (5.19) with $A_m = b\,h$:

$$T = \frac{M_T}{2A_m} = \frac{M_T}{2\,b\,h}\,.$$

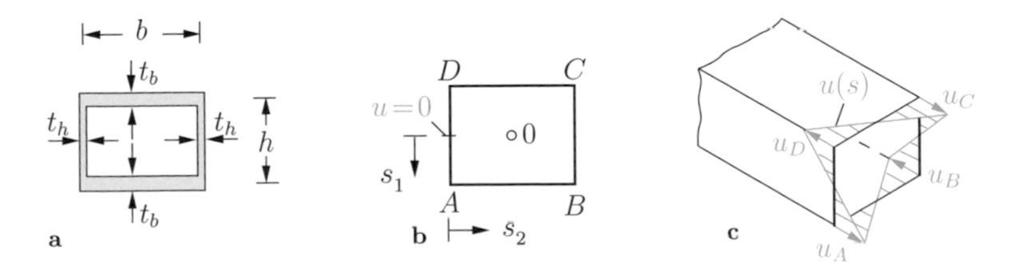

Fig. 5.10

The torsion constant

$$I_T = \frac{(2\,A_m)^2}{\oint \dfrac{\mathrm{d}s}{t}} = \frac{4\,b^2\,h^2}{2\left(\dfrac{b}{t_b} + \dfrac{h}{t_h}\right)}$$

follows from (5.25), and the evaluation of (5.24) yields

$$\vartheta' = \frac{M_T}{G I_T} = \frac{2\,b\,h\,T \cdot 2\left(\dfrac{b}{t_b} + \dfrac{h}{t_h}\right)}{4\,b^2\,h^2\,G} = \frac{T}{G\,b\,h}\left(\frac{b}{t_b} + \frac{h}{t_h}\right).$$

Now we introduce ϑ' in (5.28) and start the integration around the entire profile at the center of the edge $\overline{DA}$ (Fig. 5.10b) where $u = 0$ at this point due to antisymmetry and therefore the constant of integration C is zero. With the reference point 0 at the centroid of the rectangle and with $r_\perp = b/2$ the displacements of the points along the edge $\overline{DA}$ are

$$\begin{aligned} u_1 &= \frac{T}{G}\int\limits_0^{s_1} \frac{\mathrm{d}\bar{s}_1}{t_h} - \frac{T}{G\,b\,h}\left(\frac{b}{t_b} + \frac{h}{t_h}\right)\int\limits_0^{s_1} \frac{b}{2}\,\mathrm{d}\bar{s}_1 \\ &= \frac{T}{G}\left[\frac{1}{t_h} - \frac{1}{2h}\left(\frac{b}{t_b} + \frac{h}{t_h}\right)\right] s_1 . \end{aligned}$$

The displacements vary linearly with s_1. At the corner A ($s_1 = h/2$) we obtain

$$u_A = \frac{T}{4\,G}\left(\frac{h}{t_h} - \frac{b}{t_b}\right).$$

Along the edge $\overline{AB}$ we find (at $s_2 = 0$ we have $u_2 = u_A$)

$$u_2 = u_A + \frac{T}{G}\int\limits_0^{s_2} \frac{\mathrm{d}\bar{s}_2}{t_b} - \vartheta'\int\limits_0^{s_2} \frac{h}{2}\,\mathrm{d}\bar{s}_2$$

and the value at the corner B is

$$u_B = u_2(s_2 = b) = u_A + \frac{T}{G}\frac{b}{t_b} - \vartheta'\,\frac{h}{2}\,b = -\,u_A .$$

Analogous calculations yield

$$u_C = u_A \qquad \text{and} \qquad u_D = u_B \,.$$

The warping of the whole cross section is depicted in Fig. 5.10c.

The warping vanishes for $h/t_h = b/t_b$. Therefore, in the case of a constant wall thickness $t_b = t_h$ we obtain the relation $h = b$: a thin-walled tube having a square cross section and a constant wall thickness exhibits no warping. It should be noted that this fact does *not* hold for thick-walled tubes or solid square cross sections.

E5.4

Example 5.4 An element of a bridge construction is made of a thin-walled tube ($t \ll b$) and subjected to an eccentric force F (Fig. 5.11).

Determine the maximum shear stresses, the angle of twist of the cross section at the free end and the deflection of the point of application of the force.

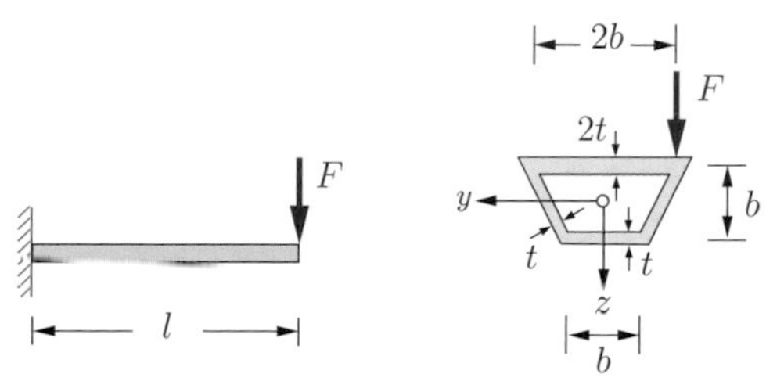

Fig. 5.11

Solution The beam is subjected to a constant torque of magnitude $M_T = bF$. The median line of the profile encloses a trapezoid with the area $A_m = \frac{1}{2}(2\,b + b)b = \frac{3}{2}b^2$. With $t_{\min} = t$ we obtain from (5.21) the maximum shear stresses

$$\underline{\underline{\tau_{\max}}} = \frac{M_T}{W_T} = \frac{b\,F}{3\,b^2\,t} = \underline{\underline{\frac{1}{3}\frac{F}{b\,t}}} \,.$$

They occur in the bottom chord and in the webs which have the same wall thickness t. It should be noted that normal stresses due to bending are also present in the beam. They can be calculated using (4.26).

The angle of twist ϑ_l at the free end follows from the torsion constant evaluated according to (5.25)

$$I_T = \frac{4\left(\frac{3}{2}\,b^2\right)^2}{\frac{2\,b}{2\,t} + \frac{b}{t} + 2\cdot\frac{1}{2}\sqrt{5}\,\frac{b}{t}} = \frac{9\,b^3\,t}{2+\sqrt{5}}$$

and substituted into (5.26) to give:

$$\underline{\underline{\vartheta_l}} = \frac{(b\,F)\,l}{G I_T} = \underline{\underline{\frac{2+\sqrt{5}}{9}\,\frac{Fl}{G\,b^2\,t}}}\,.$$

The deflection of the point of application of F consists of two parts. The deflection f_T as a consequence of torsion (small angle of twist) is

$$f_T = b\,\vartheta_l = \frac{F\,b^2\,l}{G I_T}\,.$$

The deflection f_B due to bending is listed in Table 4.3:

$$f_B = \frac{F\,l^3}{3\,EI}\,.$$

With the moment of inertia I, see Section 4.2, the final result is

$$\underline{\underline{f}} = f_T + f_B = \underline{\underline{\frac{F\,b^2\,l}{G I_T} + \frac{F\,l^3}{3\,EI}}}\,.$$

5.4

5.4 Thin-Walled Shafts with Open Cross Sections

As a further special case which can be treated with an elementary analysis, we consider thin-walled shafts with open cross sections. Here we restrict ourselves to profiles with sectionally constant wall thicknesses; this is typical for so-called T-, L-, U- or Z-profiles. All such profiles can be composed from narrow rectangles. Such a rectangle ($t \ll h$) is segmented in individual thin-walled hollow sections. A typical hollow section is depicted by the green area in Fig. 5.12a. We assume that the shear stresses (which are constant in each individual hollow section) vary linearly with y from zero at its center to a maximum value τ_0 at its outer surface (Fig. 5.12b):

$$\tau(y) = \tau_0 \, \frac{y}{t/2} \,. \tag{5.29}$$

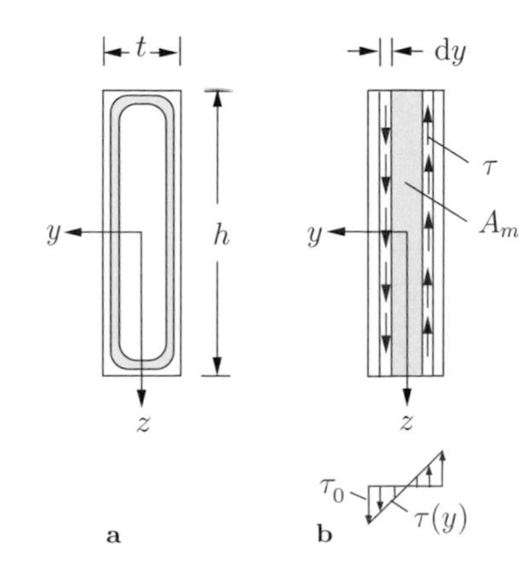

Fig. 5.12

Now we apply Bredt's first formula (5.20) to every hollow section of thickness dy. If we neglect the small deviations due to the "redirection" of the shear flow at the lower and upper ends of the rectangle, we can approximate $A_m(y)$ with $A_m = 2\,y\,h$, see Fig. 5.12b. The infinitesimal shear flow $\mathrm{d}T = \tau(y)\mathrm{d}y$ yields the infinitesimal torque

$$\mathrm{d}M_T = 2\,A_m\,\mathrm{d}T = 8\,\frac{\tau_0}{t}\,h\,y^2\,\mathrm{d}y$$

associated with a hollow section. An integration performed over the entire cross section leads to

$$M_T = \int\limits_{y=0}^{t/2} \mathrm{d}M_T = \frac{1}{3}\,\tau_0\,h\,t^2 \,. \tag{5.30}$$

According to (5.29) we obtain the maximum shear stress at the outer boundary: $\tau_{\max} = \tau(y = t/2) = \tau_0$. As in (5.9) we introduce a section modulus of torsion W_T. For a narrow rectangle (5.30) then gives

$$\tau_{\max} = \frac{M_T}{W_T} \qquad \text{with} \qquad W_T = \frac{1}{3} h\, t^2 \,. \tag{5.31}$$

Performing an analogous analysis we can determine the torsion constant using (5.25) to write

$$\mathrm{d}I_T = \frac{4(2\,y\,h)^2}{2\dfrac{h}{\mathrm{d}y}} = 8\,h\,y^2\,\mathrm{d}y\,,$$

which yields

$$I_T = \int\limits_0^{t/2} \mathrm{d}I_T \qquad \rightarrow \qquad I_T = \frac{1}{3} h\, t^3 \,. \tag{5.32}$$

We can generalize (5.32) to profiles which are composed of narrow rectangles and obtain

$$I_T \approx \frac{1}{3} \sum h_i\, t_i^3 \,. \tag{5.33}$$

Here we have to perform the summation over the individual rectangles with lengths h_i and thicknesses t_i. Omitting the derivation, we state that the corresponding section modulus of torsion is

$$W_T \approx \frac{1}{3} \frac{\sum h_i\, t_i^3}{t_{\max}} \,. \tag{5.34}$$

Thus, the maximum shear stress occurs in the part of the profile which has the *largest* wall thickness (compare Table 5.1). If a part of the profile has a curved median line (e.g. half circle profile), the corresponding length h_i can be approximated by the length of the curved median line.

Solid noncircular shafts cannot be analysed with the aforementioned relations. To determine the shear stress distributions in such cross sections we have to apply the theory of torsion named after de Saint Venant (1797–1886). This leads to the so-called potential theory, where the fundamental relations are described by a second-order partial differential equation. The underlying analysis requires a deeper mathematical background. Therefore, we omit the derivations in this text and refer to Volume 4, Section 2.6.

To conclude this chapter we summarize the most important formulas needed for the solution of torsional problems in Table 5.1. All formulas in the table are valid for constant I_T. They can also be applied to problems with slightly varying GI_T.

Table 5.1 Basic formulas for torsion

x M_T $\vec{\vartheta}$

$$\tau_{\max} = \frac{M_T}{W_T}, \qquad \frac{d\vartheta}{dx} = \frac{M_T}{GI_T}$$

Cross section	W_T	I_T	Remarks
solid circle (r, R)	$\frac{\pi R^3}{2}$	$\frac{\pi R^4}{2}$	$\tau(r) = \frac{M_T}{I_T} r$ Maximum shear stress at boundary $r = R$
solid ellipse (b, a)	$\frac{\pi ab^2}{2}$	$\frac{\pi a^3 b^3}{a^2 + b^2}$	Maximum shear stress at end points of minor semi axis
solid square (a, a)	$0,208\, a^3$	$0,141\, a^4$	Maximum shear stress in the middle of boundary edges
thick-walled circular tube (R_a, R_i) $\alpha = \frac{R_i}{R_a}$	$\frac{\pi R_a^3}{2}(1-\alpha^4)$	$\frac{\pi R_a^4}{2}(1-\alpha^4)$	Maximum shear stress at outer boundary R_a

Table 5.1 (Continuation)

Cross section	W_T	I_T	Remarks
thin-walled tube with closed cross section $t_{\min}$	$2\,A_m\,t_{\min}$	$\dfrac{(2\,A_m)^2}{\oint \dfrac{\mathrm{d}s}{t}}$	A_m is the area enclosed by profile median line. $\oint \mathrm{d}s/t$ is the line integral along profile median line. Shear flow $T = \dfrac{M_T}{2\,A_m} = \text{const}\,.$ Maximum shear stress at point with *smallest* wall thickness $t_{\min}$.
thin-walled circular tube $t = \text{const}$ t R_m	$2\,\pi\,R_m^2\,t$	$2\,\pi\,R_m^3\,t$	
narrow rectangle t h	$\dfrac{1}{3}\,h\,t^2$	$\dfrac{1}{3}\,h\,t^3$	
profile composed from narrow rectangles h_1 t_1 h_2 t_2	$\approx \dfrac{1}{3}\,\dfrac{\sum h_i\,t_i^3}{t_{\max}}$	$\approx \dfrac{1}{3}\sum h_i\,t_i^3$	Maximum shear stress at *largest* wall thickness $t_{\max}$.

Example 5.5 A torque acts a) on a closed thin-walled circular tube and b) on a thin-walled circular tube with a slit (Fig. 5.13). Both tubes are made of the same material and have the same length. **E5.5**

Determine the ratios of the maximum shear stresses and angles of twist at the free ends.

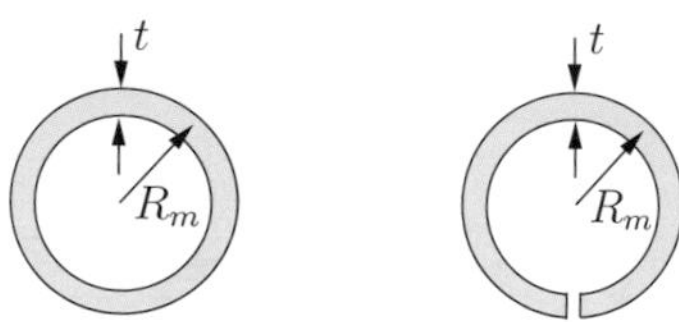

Fig. 5.13

Solution For a closed thin-walled circular tube – subscript c – we obtain from Table 5.1 (compare with (5.12)):

$$W_{T_c} = 2\,\pi\,R_m^2\,t, \qquad I_{T_c} = 2\,\pi\,R_m^3\,t\,.$$

For an open thin-walled circular tube – subscript o – we have to apply the formulas for open cross sections. With $h = 2\,\pi\,R_m$ we obtain from (5.31) and (5.32)

$$W_{T_o} = \frac{1}{3}\,t^2\,2\,\pi\,R_m \quad \text{and} \quad I_{T_o} = \frac{1}{3}\,t^3\,2\,\pi\,R_m\,.$$

A comparison of both cases yields the shear stress ratio

$$\underline{\underline{\frac{\tau_{\max_c}}{\tau_{\max_o}}}} = \frac{W_{T_o}}{W_{T_c}} = \frac{\frac{1}{3}\,t^2\,2\,\pi\,R_m}{2\,\pi\,R_m^2\,t} = \underline{\underline{\frac{1}{3}\,\frac{t}{R_m}}}$$

and the ratio of the angles of twist

$$\underline{\underline{\frac{\vartheta_c}{\vartheta_o}}} = \frac{I_{T_o}}{I_{T_c}} = \frac{\frac{1}{3}\,t^3\,2\,\pi\,R_m}{2\,\pi\,R_m^3\,t} = \underline{\underline{\frac{1}{3}\left(\frac{t}{R_m}\right)^2}}\,.$$

These results show that for the profile with the closed cross section the stresses are t/R_m-times smaller and the angles of twist are even $(t/R_m)^2$-times smaller than those for the profile with the open cross section. Therefore, it is useful to make use of closed profiles in problems with torsion.

E5.6

Example 5.6 A horizontal frame is clamped at A, simply supported at B and loaded by a torque M_D at D (Fig. 5.14a). The ratios of the bending stiffnesses and torsional rigidity are $EI_2 = 2\,EI_1$ and $GI_T = EI_1/2$. Assume $b = l/3$.

Determine the support reactions.

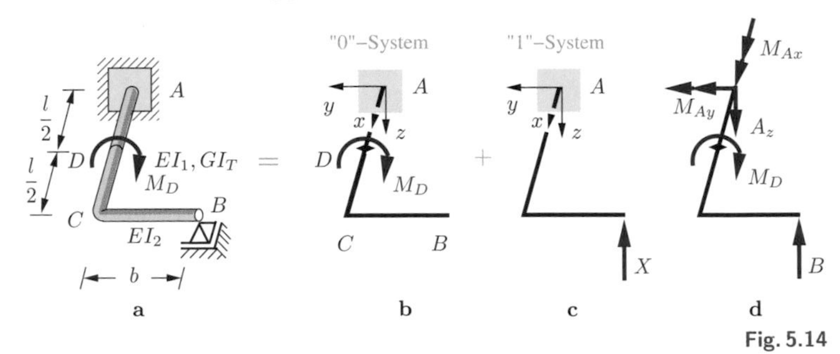

Fig. 5.14

Solution The frame has one degree of static indeterminacy. We first determine one support reaction with the method of superposition. If we remove the support at B, we obtain the "0"-system depicted in Fig. 5.14b. As a result of the loading M_D the cross section at C exhibits the same rotation as the cross section at D. With the help of (5.7) we obtain

$$\vartheta_C = \frac{M_D}{GI_T}\frac{l}{2}.$$

The attached beam $\overline{BC}$ rotates as a rigid body. Therefore point B exhibits the vertical deflection (small rotations)

$$w_B^{(0)} = b\,\vartheta_C = \frac{M_D\,l}{2\,GI_T}\,b\,.$$

The vertical displacement $w_B^{(1)}$ in the "1"-system (Fig. 5.14c) consists of three parts:

a) deflection $w_{B_1}^{(1)}$ of the beam $\overline{BC}$,
b) deflection $w_{B_2}^{(1)}$ (= deflection w_C of the beam $\overline{AC}$),
c) deflection $w_{B_3}^{(1)}$ as a result of the rotation of cross section C.

In order to calculate the deflections we use the formulas in Table 4.3 (deflection curves). With

$$w_{B_1}^{(1)} = -\frac{X\,b^3}{3\,EI_2}\,, \qquad w_{B_2}^{(1)} = -\frac{X\,l^3}{3\,EI_1}\,, \qquad w_{B_3}^{(1)} = -\frac{(X\,b)l}{GI_T}\,b$$

we obtain

$$w_B^{(1)} = -\left(\frac{b^3}{3\,EI_2} + \frac{l^3}{3\,EI_1} + \frac{b^2\,l}{GI_T}\right)X\,.$$

The vertical displacement of point B is zero in the original system. Therefore, we are able to determine the unknown support reaction $X = B$ by using the compatibility condition

$$w_B^{(0)} + w_B^{(1)} = 0\,.$$

Solving for X yields

$$\underline{\underline{X = B}} = \frac{\dfrac{M_D\,l}{2\,GI_T}\,b}{\dfrac{b^3}{3\,EI_2} + \dfrac{l^3}{3\,EI_1} + \dfrac{b^2\,l}{GI_T}} = \underline{\underline{\frac{54}{91}\frac{M_D}{l}}}\,.$$

Evaluating the force equilibrium condition in z-direction and the equilibrium of moments about the x- and the y-axis yield the support reactions at A (Fig. 5.14d):

$$A_z - B = 0 \quad \rightarrow \quad \underline{\underline{A_z = \frac{54}{91}\frac{M_D}{l}}}\,,$$

$$M_{Ax} - M_D + b\,B = 0 \quad \rightarrow \quad \underline{\underline{M_{Ax} = \frac{73}{91}M_D}}\,,$$

$$M_{Ay} + l\,B = 0 \quad \rightarrow \quad \underline{\underline{M_{Ay} = -\frac{54}{91}M_D}}\,.$$

5.5

5.5 Supplementary Examples

Detailed solutions to the following examples are given in (**A**) D. Gross et al. *Formeln und Aufgaben zur Technischen Mechanik 2*, Springer, Berlin 2010, or (**B**) W. Hauger et al. *Aufgaben zur Technischen Mechanik 1-3*, Springer, Berlin 2008.

E5.7

Example 5.7 The solid circular shaft in Fig. 5.15 consists of three segments. The radii of the segments ① and ③ are constant; segment ② has a linear taper. The shaft is subjected to a torque M_0.

Determine the angle of twist ϑ_E at the free end.

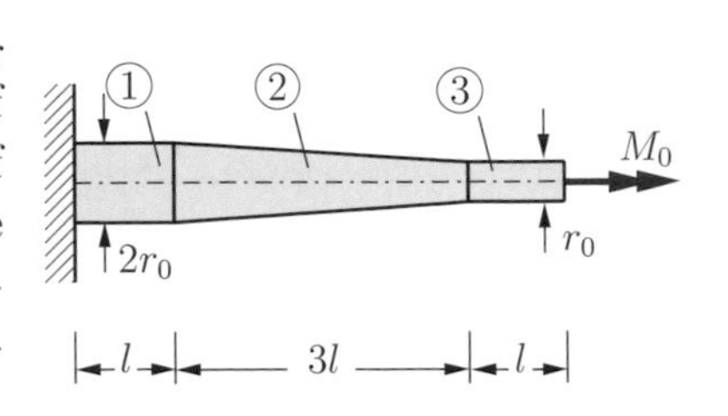

Fig. 5.15

Result: see (**B**) $\vartheta_E = \dfrac{62\,M_0\,l}{\pi\,G\,r_0^4}$.

E5.8

Example 5.8 A thin-walled tube (Fig. 5.16) is subjected to a torque M_T. Given: $a = 20$ cm, $t = 2$ mm, $\tau_{\text{allow}} = 40$ MPa, $l = 5$m, $G = 0.8 \cdot 10^5$ MPa.

Determine the allowable magnitude $M_{T\text{allow}}$ of the torque and the corresponding angle of twist ϑ for a) a closed cross section and b) an open cross section.

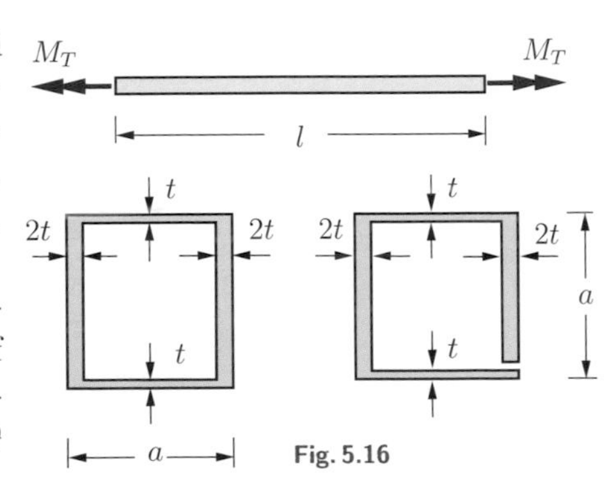

Fig. 5.16

Results: see (**A**) a) $M_{T\text{allow}} = 6400\,\text{Nm}, \quad \vartheta = 1.07°$,
b) $M_{T\text{allow}} = 96\,\text{Nm}, \quad \vartheta = 35.8°$.

E5.9

Example 5.9 The assembly shown in Fig. 5.17 consists of a thin-walled elastic tube (shear modulus G) and a rigid lever. The lever is subjected to a couple.

Calculate the displacement of point D.

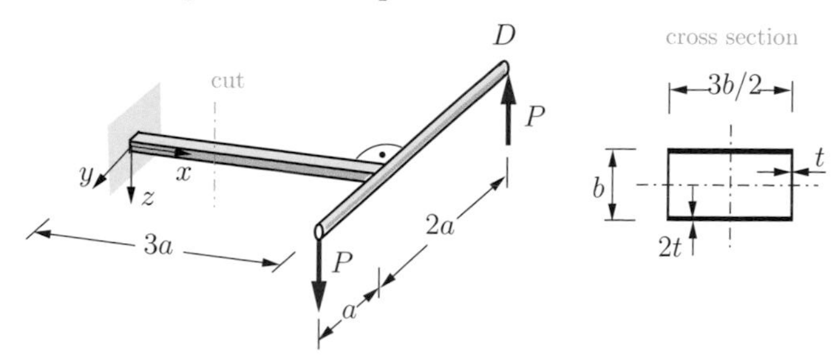

Fig. 5.17

Result: see (**B**) $w_D = 7\dfrac{P\,a^3}{G\,b^3\,t}$.

E5.10

Example 5.10 A rectangular lever has the thin-walled closed cross section (wall thickness $t = h/20$) shown in Fig. 5.18. It is loaded by a force F at the free end.

Determine the equivalent stress σ_e at point P. Use the maximum-shear-stress theory.

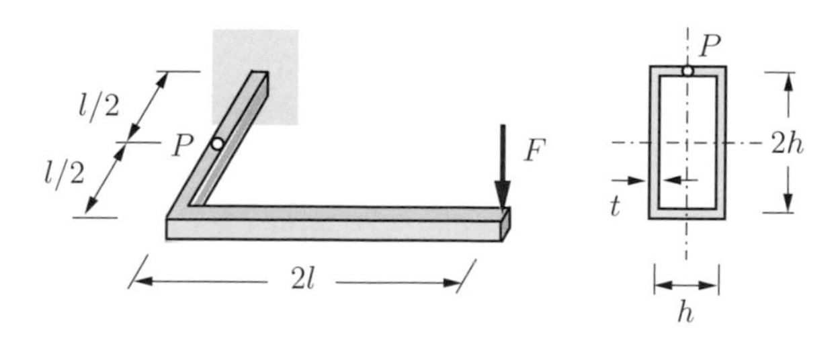

Fig. 5.18

Result: see (**B**) $\sigma_e = 20.2\dfrac{F\,l}{h^3}$.

E5.11

Example 5.11 A leaf spring ($t \ll b$) is subjected to an eccentrically acting force F as shown in Fig. 5.19.

Determine the displacement f of the point of application of F. Calculate the maximum normal stress and the maximum shear stress.

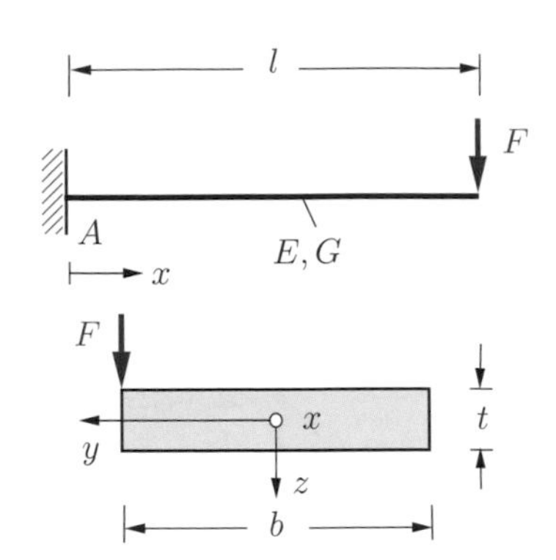

Fig. 5.19

Results: see (**A**) $f = \dfrac{4\,F\,l^3}{E\,b\,t^3}\left(1 + \dfrac{3\,E\,b^2}{16\,G\,l^2}\right)$,

$$\sigma_{\max} = \frac{3\,F\,l}{b\,t^2}\left\{1 + \sqrt{1 + \frac{b^2}{4\,l^2}}\right\}, \qquad \tau_{\max} = \frac{3\,F\,l}{b\,t^2}\sqrt{1 + \frac{b^2}{4\,l^2}}.$$

E5.12

Example 5.12 The thin-walled cantilever beam (length $l = 20b$, wall thickness t) shown in Fig. 5.20 is subjected to an eccentrically acting uniform line load q_0.

Determine the components of the stress tensor at point P. Calculate the principal stresses and the principal directions.

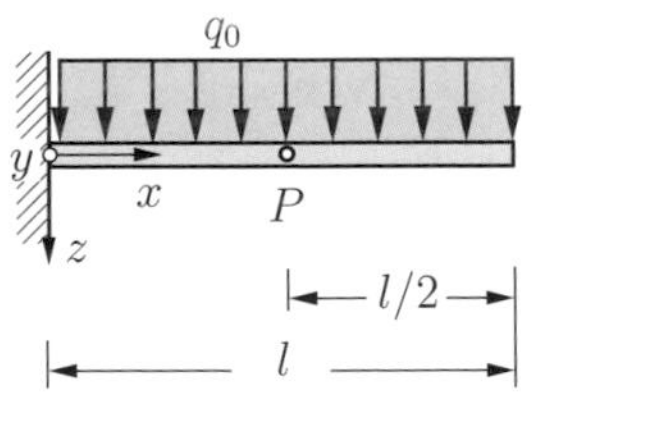

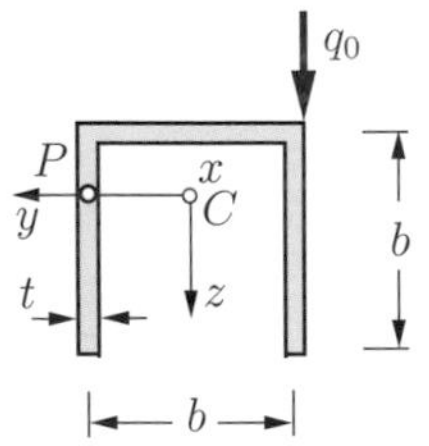

Fig. 5.20

Results: see (**B**) $\sigma_x = 0, \quad \sigma_z = 0, \quad \tau_{xz} = -\dfrac{q_0\,l}{4\,t^2}$,

$$\sigma_{1,2} = \pm\frac{q_0\,l}{4\,t^2}, \quad \varphi^* = 45^\circ.$$

E5.13

Example 5.13 A thin-walled tube with a closed quadratic cross section (Fig. 5.21) is subjected to a torque M_T.

Determine the warping of the cross section.

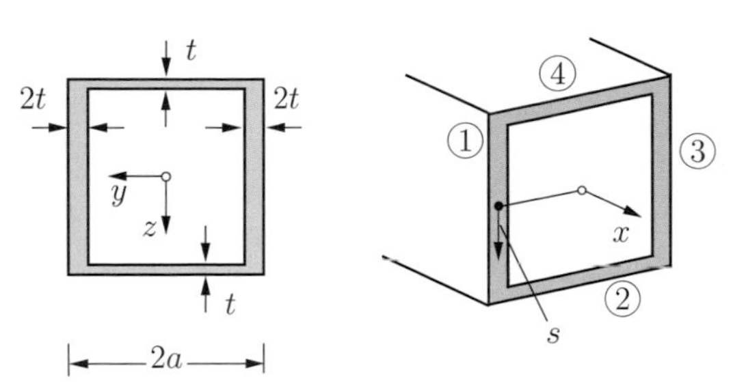

Fig. 5.21

Results: see (**A**) Selected values:
$$u_1(s) = -\frac{M_T}{32\,G\,a^2\,t}\,s,$$
$$u_2(s) = \frac{M_T}{32\,G\,a^2\,t}\,(s-2a).$$

E5.14

Example 5.14 The solid shaft in Fig. 5.22 is made from two segments with circular cross sections. It is fixed at its ends A and B and subjected to a torque M_0.

Calculate the support reactions M_A and M_B due to the applied torque and the angle of twist ϑ_C at point C.

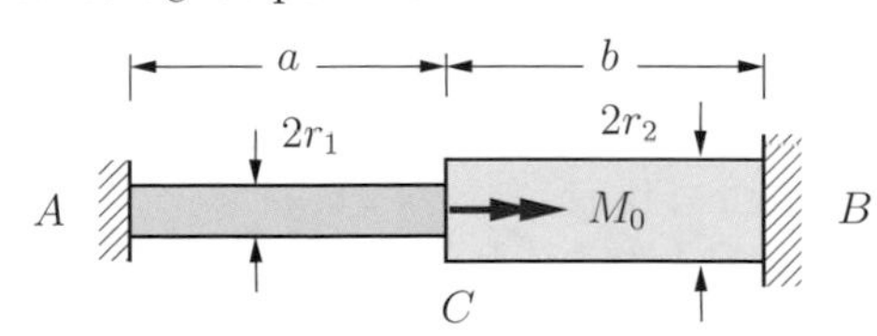

Fig. 5.22

Results: see (**A**)
$$M_A = M_0\,\frac{1}{1+\frac{r_2^4\,a}{r_1^4\,b}},\quad M_B = M_0\,\frac{1}{1+\frac{r_1^4\,b}{r_2^4\,a}},$$
$$\vartheta_C = \frac{2\,M_0\,a\,b}{\pi\,G\,(b\,r_1^4 + a\,r_2^4)}.$$

E5.15

Example 5.15 A steel frame ($E/G \approx 8/3$) consists of three solid circular beams (radius r), see Fig. 5.23.

Calculate the support reactions due to an applied force F.

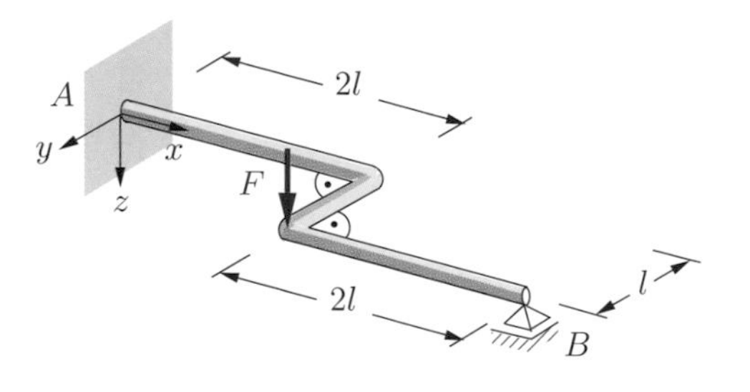

Fig. 5.23

Results: see (**B**) $B = 29\,F/89, \quad A = 60\,F/89,$

$$M_{Ax} = 60\,F\,l/89, \quad M_{Ay} = 62\,F\,l/89.$$

E5.16

Example 5.16 The system shown in Fig. 5.24 consists of a circular tube ① and a solid circular shaft ②. They are connected with a bolt at point A.

Determine the torque M_T acting in the system and the angle β of the bolt (see the figure) if the ends of the two parts make the angle α in the stress-free state (i.e., before the bolt is attached).

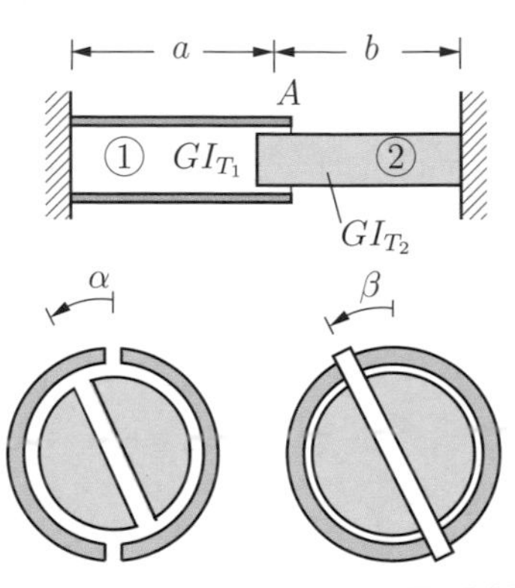

Fig. 5.24

Results: see (**A**) $M_T = GI_{T_1}\dfrac{\alpha}{a}\dfrac{1}{1+\frac{b}{a}\frac{I_{T_1}}{I_{T_2}}}, \quad \beta = \dfrac{\alpha}{1+\frac{b}{a}\frac{I_{T_1}}{I_{T_2}}}.$

E5.17

Example 5.17 Two rigid levers ① and ② are attached to the free end of a thin-walled rectangular tube (shear modulus G), see Fig. 5.25. The end points of the levers are at a distance a from the end points of two springs (spring constant c).

Determine the torque M_T in the tube and the maximum shear stress $\tau_{\max}$ if the springs are connected to the levers.

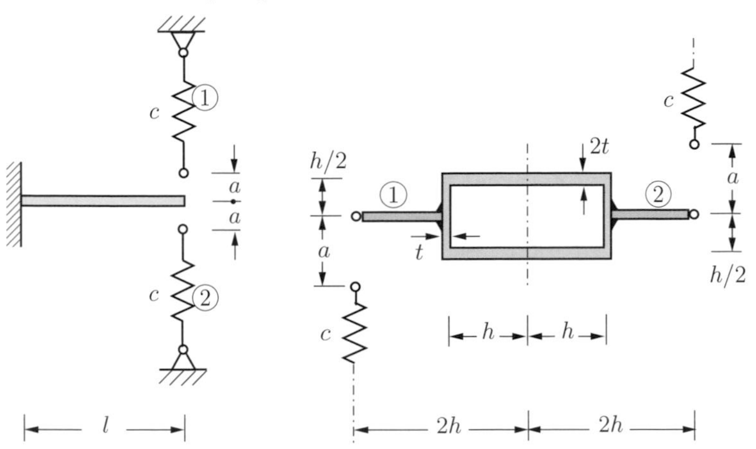

Fig. 5.25

Results: see (**B**) $M_T = \dfrac{4\,a\,c\,h}{1+\frac{2\,c\,l}{G\,h\,t}}$, $\tau_{\max} = \dfrac{a\,c\,G}{G\,h\,t + 2\,c\,l}$.

E5.18

Example 5.18 The shaft shown in Fig. 5.26 consists of a tube (shear modulus G_2) which is bonded to a core (shear modulus G_1). It is subjected to a torque M_T at its free end.

Determine the maximum shear stresses in the core ① and in the tube ② and the angle of twist.

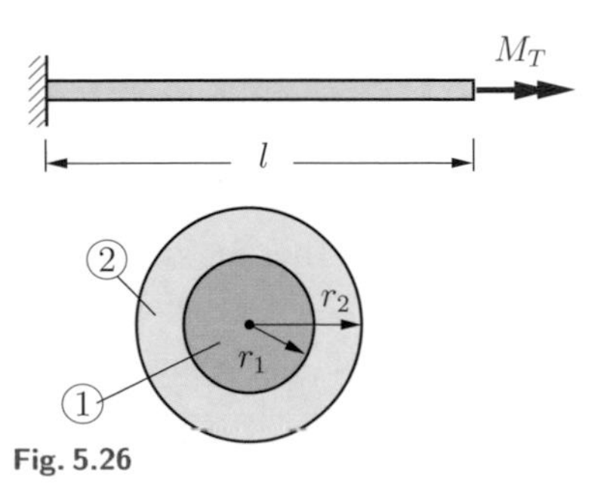

Fig. 5.26

Results: see (**A**) $\tau_{\max_1} = \dfrac{M_T G_1 r_1}{G_1 I_{p_1} + G_2 I_{p_2}}$, $\tau_{\max_2} = \dfrac{M_T G_2 r_2}{G_1 I_{p_1} + G_2 I_{p_2}}$,

$$\vartheta = \frac{M_T\,l}{G_1 I_{p_1} + G_2 I_{p_2}}.$$

E5.19

Example 5.19 The assembly shown in Fig. 5.27 consists of a thin-walled elastic tube (shear modulus G) and a rigid lever. The lever is subjected to a couple.

Given the allowable shear stress τ_{allow}, calculate the allowable forces and the corresponding angles of twist for the cross sections ① and ②.

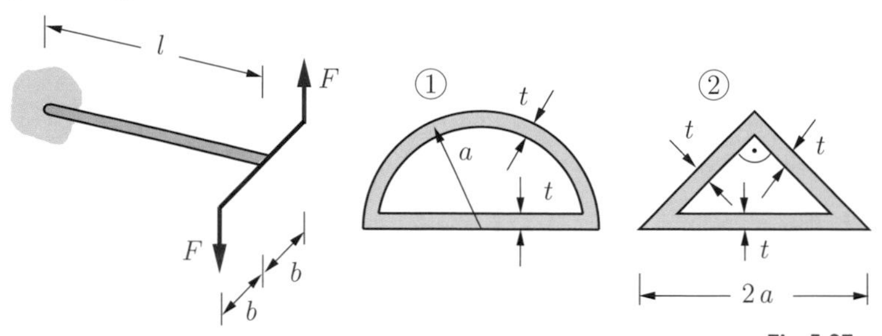

Fig. 5.27

Results: see (**A**) $F_{\text{allow}_1} = \dfrac{\pi}{2}\dfrac{a^2 t}{b}\tau_{\text{allow}}$, $F_{\text{allow}_2} = \dfrac{a^2 t}{b}\tau_{\text{allow}}$,

$$\Delta\vartheta_1 = \frac{2+\pi}{\pi}\,\frac{l\,\tau_{\text{allow}}}{a\,G}, \quad \Delta\vartheta_2 = (1+\sqrt{2})\frac{l\,\tau_{\text{allow}}}{a\,G}.$$

E5.20

Example 5.20 A solid circular shaft has a linear taper from $4a$ at the left end to $2a$ at the right end (Fig. 5.28). A torque M_T is applied at its free end.

Determine the angle of twist ϑ and the maximum shear stress $\tau_{\max}$ in the shaft at a location x along the shaft's axis.

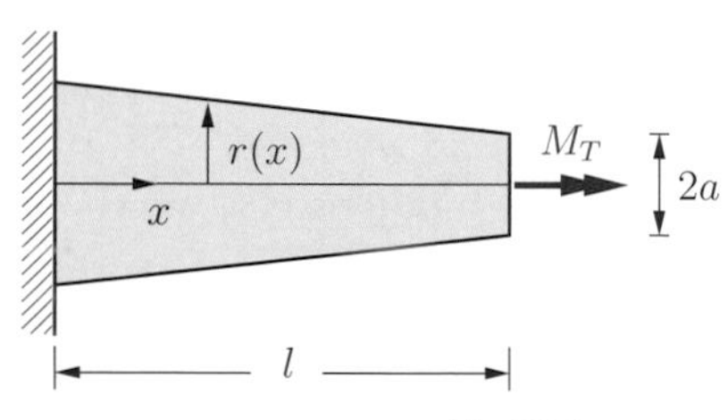

Fig. 5.28

Results: see (**A**) $\vartheta(x) = \dfrac{M_T\,l}{12\,\pi\,G\,a^4}\left\{\dfrac{1}{\left(1-\dfrac{x}{2l}\right)^3} - 1\right\}$,

$$\tau_{\max}(x) = \frac{M_T}{W_T} = \frac{2M_T}{\pi\,a^3\left(2-\dfrac{x}{l}\right)^3}.$$

Example 5.21 Determine the torsion constant I_T and the section modulus of torsion W_T for each of the depicted thin-walled cross sections ($t \ll b$). Calculate the ratio $\tau_{\max,o}/\tau_{\max,c}$ of the maximum shear stresses for the open cross sections (o) and the closed cross sections (c) assuming that the sections are subjected to a torque M_T.

E5.21

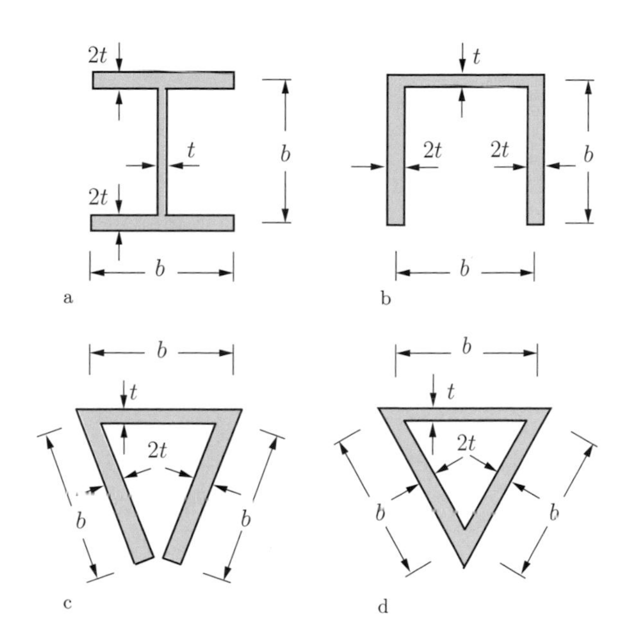

Fig. 5.29

Results: see (**B**) a,b,c) $I_T = \dfrac{17}{3} bt^3\,, \quad W_T = \dfrac{17}{6} bt^2\,,$

d) $I_T = \dfrac{3}{8} b^3 t\,, \quad W_T = \dfrac{\sqrt{3}}{2} b^2 t\,; \quad \dfrac{\tau_{\max,o}}{\tau_{\max,c}} = \dfrac{3\sqrt{3}b}{17\,t}\,.$

5.6

5.6 Summary

- Maximum shear stress:

$$\tau_{\max} = \frac{M_T}{W_T}\,,$$

M_T torque, W_T section modulus of torsion.

- The angle of twist ϑ is obtained through integration of

$$\vartheta' = \frac{M_T}{GI_T}\,,$$

GI_T torsional rigidity.
Special case $M_T = \text{const}$, $GI_T = \text{const}$: $\vartheta_l = \dfrac{M_T\,l}{GI_T}$.

- Solid circular shaft:

$$\tau = \frac{M_T}{I_T}\,r\,, \qquad I_T = \frac{\pi}{2}\,R^4\,, \qquad W_T = \frac{\pi}{2}\,R^3\,.$$

- Thin-walled closed cross section:

  ◇ Bredt's first formula:

$$\tau(s) = \frac{M_T}{2A_m t(s)}\,,$$

  A_m area enclosed by profile median line, s arc length.
  The maximum shear stress $\tau_{\max} = M_T/W_T$ occurs at the point with the *smallest* wall thickness $t_{\min}$.

  ◇ Bredt's second formula:

$$I_T = \frac{4\,A_m^2}{\displaystyle\oint \frac{\mathrm{d}s}{t}}\,.$$

  ◇ Noncircular shafts *warp*.

- Thin-walled open cross section:

$$I_T = \frac{1}{3}\sum h_i\,t_i^3\,, \qquad W_T = \frac{1}{3}\,\frac{\sum h_i\,t_i^3}{t_{\max}}\,.$$

The maximum shear stress $\tau_{\max} = M_T/W_T$ occurs at the point with the *largest* wall thickness $t_{\max}$.

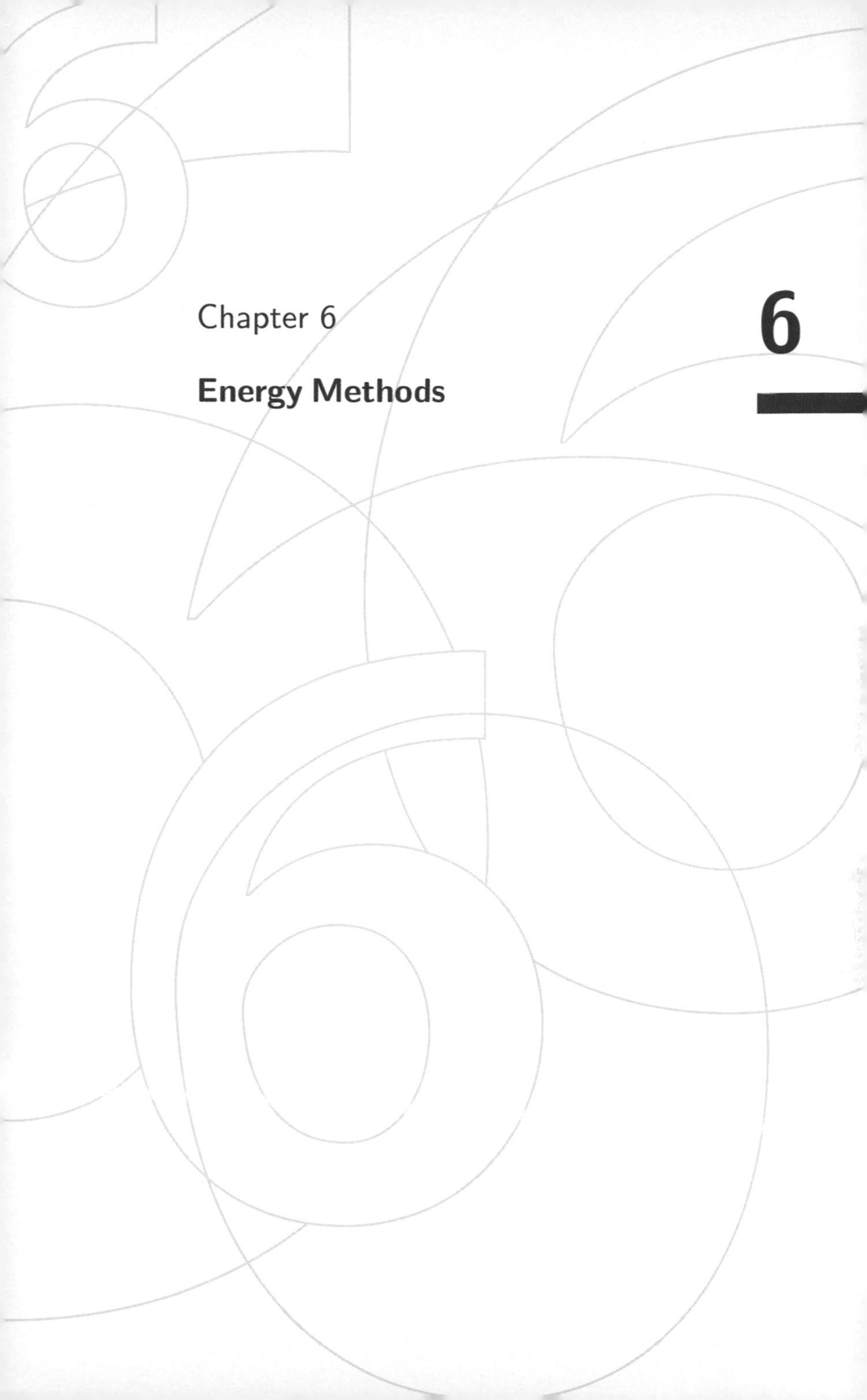

Chapter 6

Energy Methods

6

6 Energy Methods

6.1	Introduction	231
6.2	Strain Energy and Conservation of Energy	232
6.3	Principle of Virtual Forces and Unit Load Method	242
6.4	Influence Coefficients and Reciprocal Displacement Theorem	261
6.5	Statically Indeterminate Systems	265
6.6	Supplementary Examples	279
6.7	Summary	286

Objectives: It is often convenient to determine displacements/rotations or forces/moments with the aid of energy methods. The pertinent equations are presented in this chapter. For example, it will be shown how the displacement of an arbitrary point of a structure can be calculated using energy methods. These methods also enable us to calculate redundant support reactions of statically indeterminate systems in a simple way. The students will learn how to apply these methods to specific problems.

6.1 Introduction

In the preceding chapters we have always applied three different types of equations to determine forces/stresses or deformations:

a) The *equilibrium conditions* establish a relation between the applied loads and the internal forces (stress resultants).
b) *Kinematic equations* connect the displacements and the strains.
c) The *constitutive equations* (Hooke's law) connect the stresses and the strains.

Table 6.1 shows the corresponding equations for the three cases tension/compression, bending and torsion. In addition, the bottom row presents the differential equations for the displacements/rotations, assuming constant rigidities, that follow from the three types of equations given in the rows above.

Table 6.1. Basic Equations of Elastostatics

	Tension/ Compression	Bending	Torsion
Equilibrium	$N' = -n$	$M' - V = 0$ $V' = -q$	$M_T' = -m_T$
Kinematics	$\varepsilon = u'$	$\varkappa_B = -\psi'$ $\psi = -w'$	$\varkappa_T = \vartheta'$
Hooke's Law	$N = EA\,\varepsilon$	$M = -EI\,\varkappa_B$	$M_T = GI_T\,\varkappa_T$
	$EA\,u'' = -n$ cf. (1.20b)	$EI\,w^{IV} = q$ cf. (4.34b)	$GI_T\,\vartheta'' = -m_T$ cf. (5.14)

In Volume 1 it was shown how the equilibrium of a rigid body can be investigated with the aid of energy considerations: the principle of virtual work is equivalent to the equilibrium conditions (Volume 1, Section 8.2). Since there are no real displacements in the statics of rigid bodies, we introduced virtual displacements (i.e., imaginary displacements) in order to be able to apply the principle of virtual work. In contrast, the points of elastic structures undergo real displacements. To calculate these displacements it is often advantageous to use energy methods. These methods will be derived in the following sections.

6.2

6.2 Strain Energy and Conservation of Energy

Let us first consider a bar that is subjected to an external tensile force. We assume that the magnitude of this force "slowly" (quasistatically, no dynamic effects due to motion) increases from the initial value zero to its final value F. An arbitrary value between 0 and F is denoted by $\bar{F}$. After the loading process the point of application of the force is displaced by the amount u (Fig. 6.1a). Therefore, the work done by the force is given by

$$U_e = \int_0^u \bar{F}\,\mathrm{d}\bar{u} \tag{6.1}$$

(the subscript "e" refers to "external" force). If the functional relation between the force $\bar{F}$ and the corresponding displacement $\bar{u}$ is known, the integral in (6.1) can be evaluated. In this chapter it is always assumed that the material behaviour is linearly elastic. Then this relation is given by (1.18) for a bar with length l and constant axial rigidity EA:

$$\bar{u} = \frac{\bar{F}\,l}{EA} \qquad \rightarrow \qquad \bar{F} = \frac{EA}{l}\,\bar{u}. \tag{6.2}$$

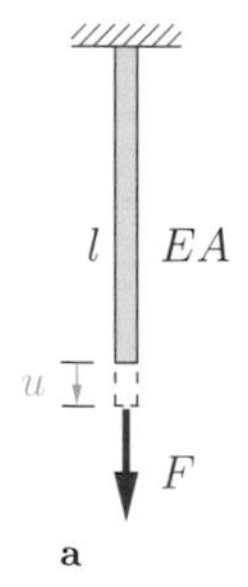

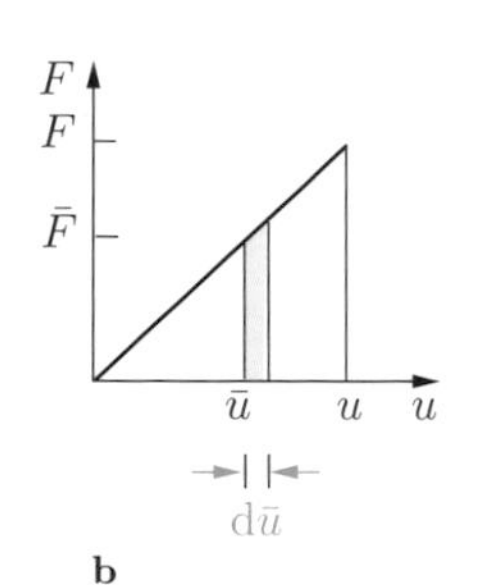

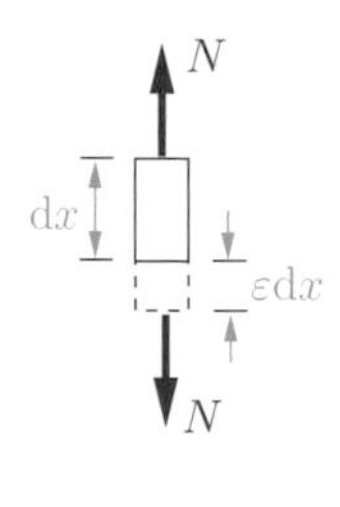

Fig. 6.1

Inserting (6.2) into (6.1) yields

$$U_e = \frac{EA}{l}\frac{u^2}{2} = \frac{1}{2}\frac{F^2\,l}{EA} = \frac{1}{2}\,F\,u. \tag{6.3}$$

The load-displacement diagram is shown in Fig. 6.1b. It illustrates the result (6.3): the integral over the elements of work $\mathrm{d}U_e = \bar{F}\mathrm{d}\bar{u}$ is equal to the area $\frac{1}{2}Fu$ of the triangle.

We will now determine the work U_i that is done by the *internal* forces in the bar. An element of length $\mathrm{d}x$ changes its length by the amount $\varepsilon\,\mathrm{d}x$ under the action of the normal force N (Fig. 6.1c). This force also slowly increases from the initial value of zero to its final value. Since the relation between force and elongation is linear, the work done by the internal forces is (compare to (6.3))

$$\mathrm{d}U_i = \frac{1}{2}\,N\,\varepsilon\,\mathrm{d}x. \tag{6.4}$$

The work done by internal forces (here: the normal force N) is called *internal work* or *strain energy*. The strain energy is stored in the structure just as the potential energy V is stored in a spring (see Volume 1, Equation (8.9)). Note that the strain energy is always positive (also in the case of compression). With the constitutive equation $\varepsilon = N/EA$ (see Table 6.1) we obtain

$$\mathrm{d}U_i = \frac{1}{2}\frac{N^2}{EA}\,\mathrm{d}x = {U_i}^*\,\mathrm{d}x$$

where

$${U_i}^* = \frac{1}{2}\frac{N^2}{EA} \tag{6.5}$$

is the strain energy per unit length. Integration of (6.5) over the length of the bar yields the total strain energy

$$U_i = \int\limits_0^l {U_i}^*\mathrm{d}x = \frac{1}{2}\int\limits_0^l \frac{N^2}{EA}\,\mathrm{d}x. \tag{6.6}$$

In the special case of a constant axial rigidity and a constant normal force $N = F$ we find

$$U_i = \frac{1}{2}\frac{F^2}{EA}\int\limits_0^l \mathrm{d}x = \frac{1}{2}\frac{F^2\,l}{EA}. \tag{6.7}$$

Comparison of (6.7) and (6.3) leads to

$$U_e = U_i\,. \tag{6.8}$$

This equation represents the *principle of conservation of energy.* It was derived for the particular case of a bar under tension or compression; however, it is valid for an arbitrary elastic system (note that only mechanical energy is considered here; energy developed by heat, chemical reactions etc, is disregarded). In words: the work U_e done by the external loads is stored as strain energy U_i in the elastic system. It is regained in the process of unloading: no energy is lost. The principle (6.8) which states that the work done by the external loads is equal to the strain energy is also referred to as *Clapeyron's theorem* (Benoit Paul Emile Clapeyron, 1799–1864).

In order to be able to apply (6.8) to an arbitrary elastic system, we need expressions for the work U_e of the external loads and the strain energy U_i. A force F that is applied to a structure does the work

$$U_e = \frac{1}{2}\,F\,f \tag{6.9a}$$

(compare to (6.3)), where f is the displacement component in the direction of the force at the point where the force is applied (Fig. 6.2a). In the case of an external couple moment M_0, the

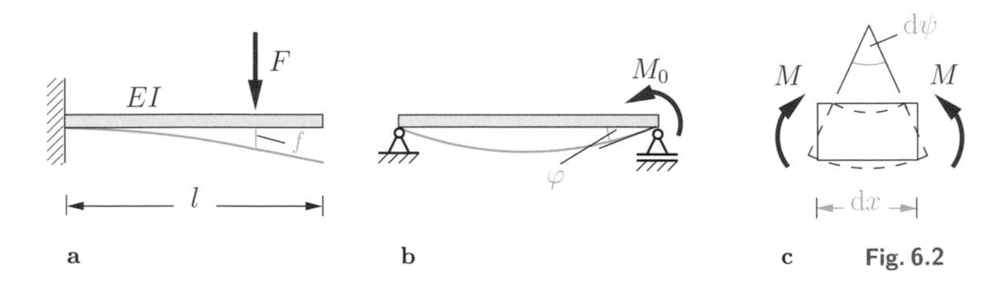

Fig. 6.2

work is given by

$$U_e = \frac{1}{2}\, M_0\, \varphi \tag{6.9b}$$

where φ is the angle of rotation in the direction of M_0 at the point of application of M_0 (Fig. 6.2b).

In contrast to the work of the external loads, the strain energy is obtained using different equations for different types of loading, i.e., for members subjected to a normal force, a bending moment or a torque. We will now derive the corresponding equation for the case of bending. Consider a beam element of length $\mathrm{d}x$. The two end cross sections undergo a relative rotation $\mathrm{d}\psi$ due to the bending moment M (Fig. 6.2c). The bending moment does the work

$$\mathrm{d}U_i = \frac{1}{2} M\, \mathrm{d}\psi = \frac{1}{2} M\, \psi'\, \mathrm{d}x \tag{6.10}$$

during the rotation. If we insert the constitutive equation $M = EI\psi'$ (see (4.24)) we obtain the strain energy per unit length

$$\mathrm{d}U_i = \frac{1}{2}\frac{M^2}{EI}\, \mathrm{d}x = U_i{}^*\, \mathrm{d}x \qquad \rightarrow \qquad U_i{}^* = \frac{1}{2}\frac{M^2}{EI}. \tag{6.11}$$

Integration over the length l of the beam yields the strain energy

$$U_i = \int\limits_0^l U_i{}^*\, \mathrm{d}x = \frac{1}{2}\int\limits_0^l \frac{M^2}{EI}\, \mathrm{d}x. \tag{6.12}$$

Analogous considerations can be applied to torsion and shear. With the torsional moment (torque) $M_T = GI_T\, \vartheta'$ (see (5.5)) and the shear force $V = GA_S\, \tilde{\gamma}$ (compare to (4.41)) we obtain the strain energies per unit length

$$U_i{}^* = \frac{1}{2}\frac{M_T^2}{GI_T} \qquad \text{and} \qquad U_i{}^* = \frac{1}{2}\frac{V^2}{GA_S}. \tag{6.13}$$

Various forms of the strain energies for different types of loading are presented in Table 6.2.

Note that it is assumed that only one load acts on the structure. This load may cause different stress resultants in a member, e.g.,

Table 6.2. Strain Energy $U_i{}^*$ per Unit Length

Tension	Bending	Shear	Torsion
$\frac{1}{2} N \varepsilon$	$\frac{1}{2} M \psi'$	$\frac{1}{2} V \tilde{\gamma}$	$\frac{1}{2} M_T \vartheta'$
$\frac{1}{2} EA \varepsilon^2$	$\frac{1}{2} EI \psi'^2$	$\frac{1}{2} GA_S \tilde{\gamma}^2$	$\frac{1}{2} GI_T \vartheta'^2$
$\frac{1}{2} \frac{N^2}{EA}$	$\frac{1}{2} \frac{M^2}{EI}$	$\frac{1}{2} \frac{V^2}{GA_S}$	$\frac{1}{2} \frac{M_T^2}{GI_T}$

a bending moment M and a shear force V (see Example 6.1) or a bending moment M and a torque M_T (see Example 6.2). In this case we can use the principle of superposition to obtain the strain energy. Hence, the total strain energy of a member which experiences bending, torsion and tension is given by

$$U_i = \frac{1}{2} \int \frac{M^2}{EI} \,\mathrm{d}x + \frac{1}{2} \int \frac{M_T^2}{GI_T} \,\mathrm{d}x + \frac{1}{2} \int \frac{N^2}{EA} \,\mathrm{d}x. \qquad (6.14)$$

If the structure is composed of different parts, the total strain energy is the sum of the strain energies of the various parts. Note, however, that the total strain energy due to more than one load is not the sum of the strain energies due to the individual loads acting separately (the strain energy is not a linear function of the loads).

The principle of conservation of energy in the form (6.8) can be applied to statically determinate systems that are subjected to only *one* force or *one* couple moment. It allows only the determination of the displacement at the point and *in* the direction of the external force or the angle of rotation at the point and *in* the direction of the external couple moment. Therefore, its importance in solving practical problems is rather limited. Frequently, structures are subjected to more than only one load and the displacements/rotations have to be determined at arbitrary points of a structure. For these more general cases we have to use an extension of the strain-energy method (6.8) which is also based on the principle of conservation of energy. This method will be derived in Section 6.3.

As an example for the application of (6.8) let us consider the cantilever beam in Fig. 6.2a. The deflection f at the point where the force F is applied follows with (6.9a) and (6.12):

$$U_e = U_i \qquad \rightarrow \qquad \frac{1}{2}\,F\,f = \frac{1}{2}\int_0^l \frac{M^2}{EI}\,\mathrm{d}x. \tag{6.15a}$$

Here, M is the bending moment due to the external force F. Similarly, (6.9b) and (6.12) yield the angle of slope at the point of application of the external couple moment M_0 (Fig. 6.2b):

$$U_e = U_i \qquad \rightarrow \qquad \frac{1}{2}\,M_0\,\varphi = \frac{1}{2}\int_0^l \frac{M^2}{EI}\,\mathrm{d}x \tag{6.15b}$$

where M is the bending moment caused by M_0.

In a truss, the only internal forces (stress resultants) are the normal forces $N_i = S_i$ in the individual members, which are constant. The strain energy in the i-th member is given by $\frac{1}{2}S_i^2\,l_i/E_iA_i$. Consider a truss which consists of n members and which is subjected to only *one* force F. Then the displacement at the point and in the direction of the force follows from

$$U_e = U_i \qquad \rightarrow \qquad \frac{1}{2}\,F\,f = \frac{1}{2}\sum_{i=1}^{n} \frac{S_i^2\,l_i}{EA_i}. \tag{6.16}$$

Here, the axial rigidity E_iA_i is denoted by EA_i.

As an example we can calculate the vertical displacement v of the point of application of the force F for the two-bar truss shown in Fig. 6.3a. According to (6.16) this displacement follows from

$$\frac{1}{2}\,F\,v = \frac{1}{2}\left(\frac{S_1^2\,l_1}{EA} + \frac{S_2^2\,l_2}{EA}\right).$$

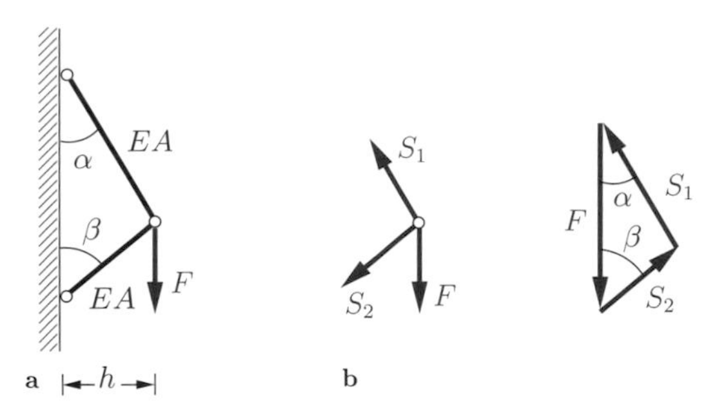

Fig. 6.3

The forces in the members can be taken from the force triangle in Fig. 6.3b. The law of sines yields

$$S_1 = F\frac{\sin\beta}{\sin(\alpha+\beta)}, \qquad S_2 = -\,F\frac{\sin\alpha}{\sin(\alpha+\beta)}.$$

If we also insert the lengths $l_1 = h/\sin\alpha$ and $l_2 = h/\sin\beta$ of the members we obtain

$$v = \frac{F}{EA}\,\frac{h}{\sin^2(\alpha+\beta)}\left(\frac{\sin^2\beta}{\sin\alpha}+\frac{\sin^2\alpha}{\sin\beta}\right).$$

A graphic-analytical method to determine the displacement of one of the pins of a truss was given in Section 1.5. It can be seen that the strain-energy method avoids the often quite cumbersome geometrical considerations.

We will now show how one can obtain an approximation for the shear correction factor $\varkappa$ of the cross section of a beam with the aid of energy considerations. The shear correction factor was introduced in the constitutive equation (4.25) for the shear force (see also Section 4.6):

$$V = G\,\varkappa\,A(w' + \psi) = GA_S\,(w' + \psi). \tag{6.17}$$

It was assumed that the shear force causes the average shear strain $\tilde{\gamma} = w' + \psi$ in the cross section. The shear area $A_S = \varkappa\,A$ is now obtained by equating the strain energy U^*_{iV} due to the shear force and the strain energy $U^*_{i\tau}$ due to the shear stress τ that acts in the cross section. According to Table 6.2 we have

$$U_{i}{}^{*}_{V} = \frac{1}{2} V \tilde{\gamma} = \frac{1}{2} \frac{V^2}{GA_S}. \tag{6.18}$$

Similarly, the force $\tau \,\mathrm{d}A$ that acts on an element $\mathrm{d}A$ of the cross section leads (with $\tau = G\gamma$) to

$$\mathrm{d}U_{i}{}^{*}_{\tau} = \frac{1}{2}(\tau \,\mathrm{d}A)\,\gamma = \frac{1}{2}\frac{\tau^2}{G}\,\mathrm{d}A.$$

Integration over the cross section yields the strain energy per unit length

$$U_{i}{}^{*}_{\tau} = \frac{1}{2}\int \frac{\tau^2}{G}\,\mathrm{d}A. \tag{6.19}$$

We now equate (6.18) and (6.19):

$$\frac{1}{2}\frac{V^2}{GA_S} = \frac{1}{2}\int \frac{\tau^2}{G}\,\mathrm{d}A. \tag{6.20}$$

If the distribution of the shear stress τ in the cross section is known, the integral in (6.20) can be evaluated and thus the shear area A_S and the shear correction factor $\varkappa$ can be calculated.

To illustrate the method we consider a rectangular cross section. According to (4.39) the distribution of the shear stress is given by

$$\tau = \frac{3}{2}\frac{V}{A}\left(1 - 4\,\frac{z^2}{h^2}\right)$$

(see Fig. 4.35b). Inserting τ into (6.20) and using $\mathrm{d}A = b\,\mathrm{d}z$ and $A = b\,h$ yields

$$\frac{1}{A_S} = \frac{9}{4}\frac{1}{A^2}\int_{-h/2}^{h/2}\left(1 - 4\,\frac{z^2}{h^2}\right)^2 b\,\mathrm{d}z = \frac{6}{5}\frac{1}{b\,h}.$$

Hence, we obtain

$$A_S = \frac{5}{6}\,b\,h, \qquad \varkappa = \frac{A_S}{A} = \frac{5}{6} \tag{6.21}$$

for a rectangular cross section. The average shear strain

$$\tilde{\gamma} = w' + \psi = \frac{V}{GA_S} = 1.2\,\frac{V}{GA}$$

is therefore 20% larger than the shear strain that would be caused by a uniform distribution $\tau = V/A$ of the shear stress.

Similar considerations lead to values between 0.8 and 0.9 of the shear correction factor $\varkappa$ for solid cross sections. In the case of an I-beam (see Fig. 4.42), the shear force is essentially supported by the web. Therefore we have to a good approximation

$$A_S \approx A_{\text{web}} = t\,h.$$

For a thin-walled circular cross section Equation (6.20) leads to

$$A_S = \frac{1}{2}A \qquad \text{with} \qquad A = 2\,\pi\,r\,t.$$

It should be noted that the values of the shear correction factor vary considerably depending on the type of the cross section, e.g., solid cross section, thin-walled open or thin-walled closed cross section.

E6.1

Example 6.1 A cantilever beam is subjected to a force F as shown in Fig. 6.4a.

Calculate the deflection f at the free end taking into account the shear deformation.

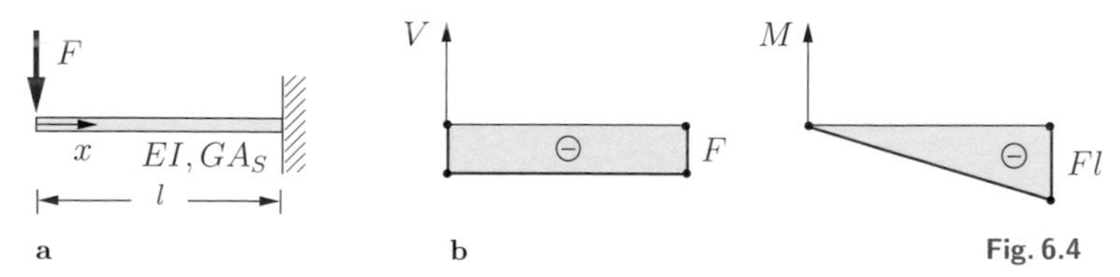

Fig. 6.4

Solution The principle of conservation of energy (6.8) reads

$$\frac{1}{2}\,F\,f = \frac{1}{2}\int \frac{M^2}{EI}\,\mathrm{d}x + \frac{1}{2}\int \frac{V^2}{GA_S}\,\mathrm{d}x$$

where the integrals are taken from Table 6.2. Using the coordinate x as shown in Fig. 6.4a, the stress resultants are given by (Fig. 6.4b)

$$V = -\,F, \qquad M = -\,F\,x.$$

Since the rigidities EI and GA_S are constant, we obtain

$$\begin{aligned}\frac{1}{2}\,F\,f &= \frac{1}{2}\int\limits_0^l \frac{F^2\,x^2}{EI}\,\mathrm{d}x + \frac{1}{2}\int\limits_0^l \frac{F^2}{GA_S}\,\mathrm{d}x \\ &= \frac{1}{2}\,F^2\frac{l^3}{3\,EI} + \frac{1}{2}\,F^2\frac{l}{GA_S} \qquad \rightarrow \qquad \underline{\underline{f = \frac{F\,l^3}{3\,EI} + \frac{F\,l}{GA_S}}}.\end{aligned}$$

This problem was already solved in Section 4.6.2 with the aid of the differential equations for the deflection of the Bernoulli beam and the deflection due to shear, respectively. Also, the influence of the shear rigidity was discussed there.

E6.2

Example 6.2 An angled member carries a load F at the free end (Fig. 6.5a).

Determine the deflection at the point of application of the force.

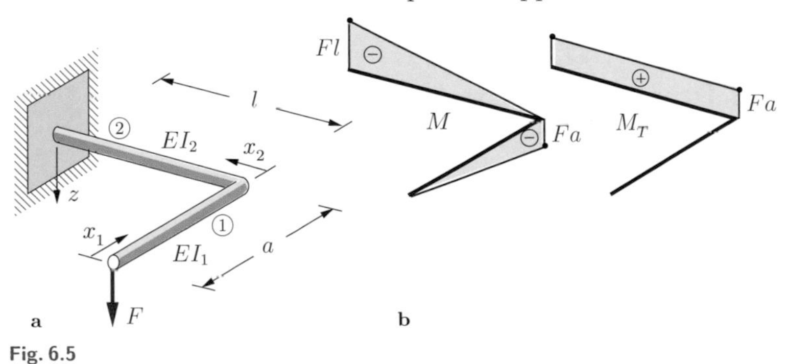

Fig. 6.5

Solution Part ① of the structure is subjected to bending; part ② is subjected to bending and torsion. Therefore, the principle of conservation of energy (6.8) is given by

$$\frac{1}{2} F f = \frac{1}{2} \int \frac{M^2}{EI} \, dx + \frac{1}{2} \int \frac{M_T^2}{GI_T} \, dx.$$

We use the coordinates x_1 and x_2 as shown in Fig. 6.5a. Then, the stress resultants are the bending moment $M_1 = -F x_1$ in beam ① and the bending moment $M_2 = -F x_2$ and the torque $M_{T2} = F a$ in beam ② (Fig. 6.5b). Thus,

$$\frac{1}{2} F f = \frac{1}{2} \int_0^a \frac{F^2 x_1^2}{EI_1} \, dx_1 + \frac{1}{2} \int_0^l \frac{F^2 x_2^2}{EI_2} \, dx_2 + \frac{1}{2} \int_0^l \frac{F^2 a^2}{GI_T} \, dx_2$$

$$= \frac{1}{2} \frac{F^2}{EI_1} \frac{a^3}{3} + \frac{1}{2} \frac{F^2}{EI_2} \frac{l^3}{3} + \frac{1}{2} \frac{F^2}{GI_T} a^2 l$$

and the deflection is obtained as

$$f = F \left\{ \frac{a^3}{3\,EI_1} + \frac{l^3}{3\,EI_2} + \frac{a^2 l}{GI_T} \right\}.$$

6.3 Principle of Virtual Forces and Unit Load Method

The principle of conservation of energy (6.8) enables us to calculate the displacement in the direction of an external force. For example, the *vertical* displacement v of the pin of the two-bar truss in Fig. 6.3 under the action of the *vertical* force F follows from (see (6.16))

$$U_e = U_i \quad \rightarrow \quad \frac{1}{2} F v = \frac{1}{2} \sum \frac{S_i^2 l_i}{EA_i}. \tag{6.22}$$

If a *horizontal* force Q is applied to the same truss instead of the vertical force F, the *horizontal* displacement u can be obtained from

$$\frac{1}{2}\,Q\,u=\frac{1}{2}\sum\frac{S_i^2\,l_i}{EA_i}\,, \tag{6.23}$$

where now S_i are the internal forces in the members due to the force Q. However, we also want to be able to determine the horizontal displacement caused by the vertical force F and the vertical displacement due to the horizontal force Q. In order to achieve this goal we introduce *virtual forces.* These are fictitious forces which are introduced only for the purpose of the calculation. Just as we may determine real forces with the aid of *virtual displacements* (see Volume 1, Section 8.2), we will be able to determine real displacements with the aid of virtual forces.

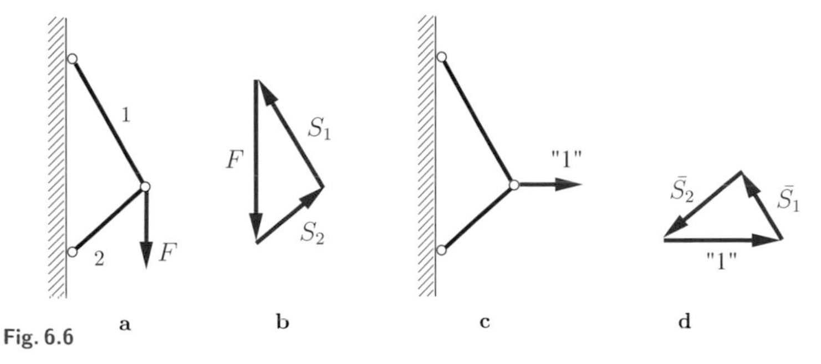

Fig. 6.6

In this section we restrict ourselves to statically determinate systems and illustrate the method with the example of the two-bar truss in Fig. 6.6a. The truss is subjected to the vertical force F as shown; the internal forces in the members are called S_i (Fig. 6.6b). If we want to determine the displacement of the pin in the *horizontal* direction, we first subject the truss only to the virtual force “1“ in the *horizontal* direction (Fig. 6.6c). This force is assumed to be gradually increasing to its final magnitude 1. A force triangle (Fig. 6.6d) yields the corresponding internal forces $\bar{S}_i$. Here and in what follows, forces or kinematical quantities due to a virtual force are always marked by a bar. The horizontal virtual force causes a displacement of the pin; its horizontal component is denoted by $\bar{u}$. During the displacement, the force “1“ does the work

$$U_{e,1} = \frac{1}{2} \cdot 1 \cdot \bar{u}. \tag{6.24}$$

Subsequently, in addition to the already acting virtual force "1", we apply the vertical force F. Then the corresponding displacement of the pin has the vertical component v and the force F does the work

$$U_{e,2} = \frac{1}{2} F\, v. \tag{6.25}$$

The horizontal component of the displacement due to F is u. Since the virtual force "1" has been applied before and therefore has the constant magnitude 1 (and since the force "1" and the displacement u have the same direction), it does the work

$$U_{e,3} = 1 \cdot u \tag{6.26}$$

during the displacement. The total work of the forces during the process described above is given by the sum of the three terms:

$$U_e = \frac{1}{2} \cdot 1 \cdot \bar{u} + \frac{1}{2} F\, v + 1 \cdot u. \tag{6.27}$$

According to the principle of superposition, the total internal forces in the bars are $\bar{S}_i + S_i$. Thus, the total strain energy in the truss is (compare (6.16))

$$\begin{aligned} U_i &= \frac{1}{2} \sum \frac{(\bar{S}_i + S_i)^2\, l_i}{EA_i} \\ &= \frac{1}{2} \sum \frac{\bar{S}_i^2\, l_i}{EA_i} + \frac{1}{2} \sum \frac{S_i^2\, l_i}{EA_i} + \sum \frac{S_i\, \bar{S}_i\, l_i}{EA_i} \end{aligned} \tag{6.28}$$

and the principle of conservation of energy (6.8) yields

$$\frac{1}{2} \cdot 1 \cdot \bar{u} + \frac{1}{2} F\, v + 1 \cdot u = \frac{1}{2} \sum \frac{\bar{S}_i^2\, l_i}{EA_i} + \frac{1}{2} \sum \frac{S_i^2\, l_i}{EA_i} + \sum \frac{S_i\, \bar{S}_i\, l_i}{EA_i}.$$

According to (6.22), the second term on the left-hand side is equal to the second term on the right-hand side. Similarly, if we use

(6.23) and set $Q = 1$, we see that the first terms are equal. This leaves the result

$$1 \cdot u = \sum \frac{S_i\,\bar{S}_i\,l_i}{EA_i}. \tag{6.29}$$

Thus, the virtual force "1" in the horizontal direction enables us to determine the real horizontal displacement u due to the vertical force F.

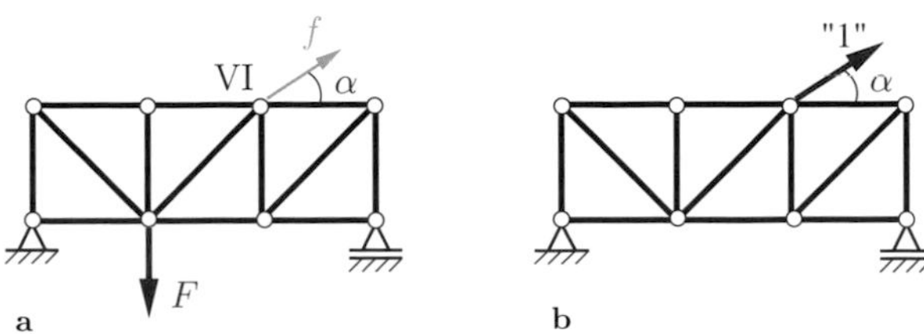

Fig. 6.7

Similar considerations lead to the displacement component in any given direction of an arbitrary pin of a general truss. Assume, for example, that the displacement component f of pin VI of the truss in Fig. 6.7a has to be determined (the direction of f is given by the angle α). In the first step, the forces S_i in the members due to the applied load F have to be calculated. In the second step, the truss is subjected only to the virtual force "1" at pin VI in the direction of f (Fig. 6.7b) and the corresponding forces $\bar{S}_i$ are calculated. Then, (6.29) yields

$$f = \sum \frac{S_i\,\bar{S}_i\,l_i}{EA_i}. \tag{6.30}$$

To obtain (6.30) we have divided (6.29) by the force 1. Thus, the forces $\bar{S}_i$ in (6.30) are due to a dimensionless force 1; hence, they are from now on also dimensionless quantities. Note that they must have the dimension of a force in (6.28) so that S_i and $\bar{S}_i$ can be added.

According to the principle of superposition, Equation (6.30) is valid for a truss subjected to arbitrarily many forces. In general, the quantities S_i are the forces in the members due to the total loading. Equation (6.30) is called the *principle of virtual forces* and its use for the evaluation of displacements is known as the *unit load method*. In summary: if we want to determine the component f of the displacement in a given direction at an arbitrary pin k of the truss, then we have to apply a virtual force "1" in this direction at pin k. The displacement f is obtained from (6.30), where S_i are the forces in the members due to the external forces, $\bar{S}_i$ are the forces in the members due to the virtual force "1" and l_i, EA_i are the lengths and the axial rigidities of the individual members.

In general, we do not know the direction of the displacement of a pin. Therefore we have to apply the method twice: a horizontal force "1" yields the horizontal component of the displacement; a vertical force "1" yields the vertical component. A vector addition of the components leads to the displacement of the pin.

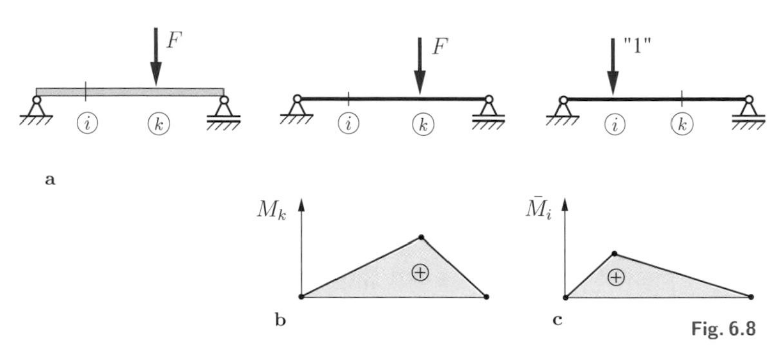

Fig. 6.8

The principle of virtual forces may also be applied to beams, frames, arches, etc. We will now derive the method for the bending of a beam with the aid of an example, namely, the simple beam in Fig. 6.8a subjected to a force at point k. We want to determine the deflection f at point i. For the sake of clarity we will use double subscripts: f_{ik} is the deflection at point i due to the force F at point k. In order to determine this deflection, we first apply a virtual force "1" at point i (Fig. 6.8c). Subsequently, the external force F is applied at point k. Using the same arguments as in the case of a truss, we obtain the total work of these two forces:

$$U_e = \frac{1}{2} \cdot 1 \cdot f_{ii} + \frac{1}{2} F\, f_{kk} + 1 \cdot f_{ik} . \tag{6.31a}$$

The force "1" causes the bending moment $\bar{M}_i$, the force F causes the bending moment M_k (Fig. 6.8b, c). Therefore, the total bending moment is given by $\bar{M}_i + M_k$ (here, $\bar{M}_i$ has the dimension of a moment). This yields the strain energy

$$U_i = \frac{1}{2}\int \frac{(\bar{M}_i + M_k)^2}{EI}\,dx$$

$$= \frac{1}{2}\int \frac{\bar{M}_i^2}{EI}\,dx + \frac{1}{2}\int \frac{M_k^2}{EI}\,dx + \int \frac{\bar{M}_i\, M_k}{EI}\,dx\,. \qquad (6.31b)$$

According to (6.15a), the first and the second terms, respectively, in U_e and U_i are equal. Therefore, the principle of conservation of energy (6.8) results in

$$f_{ik} = \int \frac{\bar{M}_i\, M_k}{EI}\,dx. \qquad (6.32)$$

Again, we have divided by the force 1. Thus, the moment $\bar{M}_i$ now has the dimension "length". Equation (6.32) represents the principle of virtual forces or unit load method for the bending of a beam: the deflection f_{ik} at point i due to a force F at point k is obtained by calculating the bending moment $\bar{M}_i$ due to the dimensionless force "1" at point i and the bending moment M_k due to the given force F at point k. Equation (6.32) then leads to f_{ik}.

The principle of virtual forces (6.32) is also valid for an arbitrary loading of the beam (several forces, couple moments, line loads). Then M_k is the bending moment due to all the given external loads. In this case, the subscripts i and k are omitted and (6.32) is written in the form

$$f = \int \frac{M\bar{M}}{EI}\,dx\,. \qquad (6.33)$$

Here, M is the bending moment due to the given loads and $\bar{M}$ (dimension: length) is the bending moment due to the dimensionless force “1“ which acts at the point where the deflection is to be determined.

If we want to calculate the angle φ of the slope of the deflection curve at a given point, we apply a dimensionless virtual couple moment “1“ at this point. The angle φ is then obtained from (6.33), where now $\bar{M}$ is the bending moment (dimensionless quantity) due to the virtual couple moment.

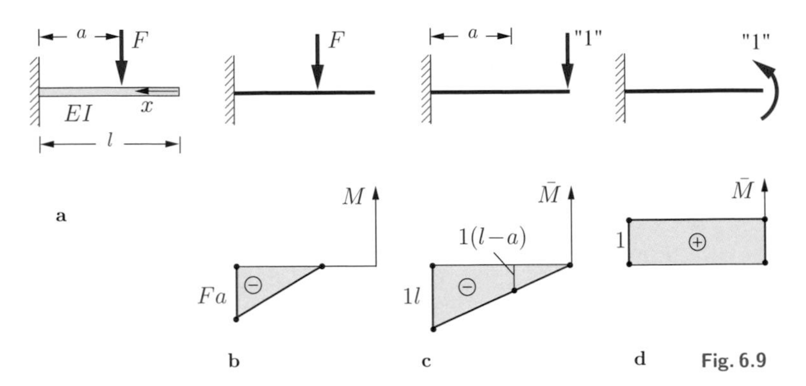

Fig. 6.9

To illustrate the method we consider the cantilever beam subjected to a force F as depicted in Fig. 6.9a. We will calculate the deflection and the angle of slope at the free end. If we use the coordinate x shown in Fig. 6.9a, the bending moment caused by the given force F is (Fig. 6.9b)

$$M = -\,F[x-(l-a)] \qquad \text{for} \qquad x \geq l-a.$$

In order to determine the deflection at the free end we apply the virtual force “1“ at this point. This yields the bending moment (Fig. 6.9c)

$$\bar{M} = -\,1 \cdot x \qquad \text{for} \qquad x \geq 0.$$

The bending moment $\bar{M}$ has the dimension “length”. The deflection follows from (6.33) (note that the bending moment M is zero in the region $0 \leq x \leq l - a$:

$$f = \int \frac{M\bar{M}}{EI}\,\mathrm{d}x = \frac{1}{EI}\int\limits_{l-a}^{l}(-F)[x-(l-a)](-x)\,\mathrm{d}x$$

$$= \frac{F}{EI}\left[\frac{x^3}{3}-(l-a)\frac{x^2}{2}\right]_{l-a}^{l} = \frac{F\,l^3}{6\,EI}\left[3\left(\frac{a}{l}\right)^2-\left(\frac{a}{l}\right)^3\right]. \quad (6.34)$$

To determine the angle of slope at the free end we apply the virtual couple moment "1" at this point. The corresponding bending moment is given by $\bar{M} = 1$ (Fig. 6.9d); it is dimensionless. Equation (6.33) yields

$$\varphi = \int \frac{M\bar{M}}{EI}\,\mathrm{d}x = \frac{1}{EI}\int\limits_{l-a}^{l}(-F)[x-(l-a)]\cdot 1\,\mathrm{d}x$$

$$= -\frac{F}{EI}\left[\frac{x^2}{2}-(l-a)x\right]_{l-a}^{l} = -\frac{F}{EI}\frac{a^2}{2}. \quad (6.35)$$

The negative sign indicates that the direction of the actual rotation is opposite to the direction that was chosen for the virtual moment.

In many problems the bending moments are linear, quadratic or cubic functions of x, respectively. Then the integrals in (6.33) can be computed in advance provided that the flexural rigidity is *constant*, and they can be listed in a table. Results of the integrations are presented in Table 6.3. Note that in order to evaluate the integrals, it is irrelevant which of the bending moments is due to the external loads and which one is caused by the virtual load. Therefore, the bar which is used to characterize the quantities due to virtual loads is omitted in Table 6.3. Now the subscripts i and k characterize two bending moments under the integral sign: $\int M_i\,M_k\,\mathrm{d}x$. For example, if M_i is represented by a quadratic parabola and M_k is linear, the notation used in Fig. 6.10 implies

$$M_i = 4\,i\left[\frac{x}{s}-\left(\frac{x}{s}\right)^2\right], \qquad M_k = k\,\frac{x}{s}.$$

Table 6.3. Integrals $\int M_i\, M_k\, \mathrm{d}x$

	M_i	M_k			
		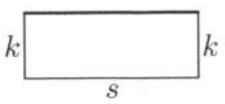 k, s, k	 s, k	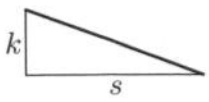 k, s	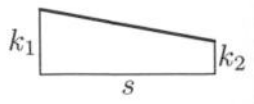 k_1, s, k_2
1	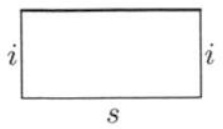 i, s, i	sik	$\frac{1}{2}sik$	$\frac{1}{2}sik$	$\frac{1}{2}si(k_1+k_2)$
2	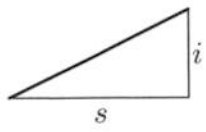 s, i	$\frac{1}{2}sik$	$\frac{1}{3}sik$	$\frac{1}{6}sik$	$\frac{1}{6}si(k_1+2\,k_2$
3	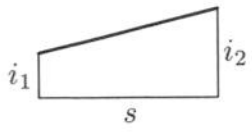 i_1, s, i_2	$\frac{1}{2}s(i_1+i_2)k$	$\frac{1}{6}s(i_1+2\,i_2)k$	$\frac{1}{6}s(2\,i_1+i_2)k$	$\frac{1}{6}s(2\,i_1\,k_1+2\,i_2\,k_2+i_1\,k_2$- $i_2\,k_1)$
4	quad. parabola 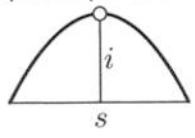 i, s	$\frac{2}{3}sik$	$\frac{1}{3}sik$	$\frac{1}{3}sik$	$\frac{1}{3}si(k_1+k_2)$
5	quad. parabola 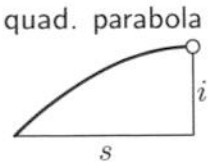 i, s	$\frac{2}{3}sik$	$\frac{5}{12}sik$	$\frac{1}{4}sik$	$\frac{1}{12}si(3\,k_1+5$
6	quad. parabola 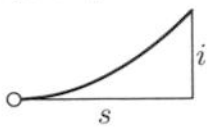 i, s	$\frac{1}{3}sik$	$\frac{1}{4}sik$	$\frac{1}{12}sik$	$\frac{1}{12}si(k_1+3\,k$
7	cub. parabola 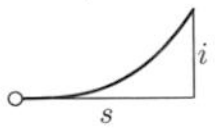 i, s	$\frac{1}{4}sik$	$\frac{1}{5}sik$	$\frac{1}{20}sik$	$\frac{1}{20}si(k_1+4\,k$
8	cub. parabola 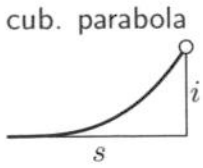 i, s	$\frac{3}{8}sik$	$\frac{11}{40}sik$	$\frac{1}{10}sik$	$\frac{1}{40}si(4\,k_1+11$
9	cub. parabola 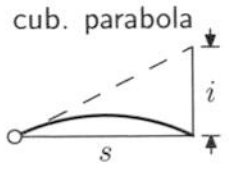 i, s	$\frac{1}{4}sik$	$\frac{2}{15}sik$	$\frac{7}{60}sik$	$\frac{1}{60}si(7\,k_1+8$

Quadratic parabola: ∘ indicates maximum
Cubic parabola: ∘ indicates zero value of the triangular load
Trapezium: i_1 and i_2 (k_1 and k_2) may have different algebraic signs

The integral is then obtained as

$$\int_0^s M_i\, M_k\, \mathrm{d}x = \int_0^s 4\, i \left[\frac{x}{s} - \left(\frac{x}{s}\right)^2\right] k\, \frac{x}{s}\, \mathrm{d}x$$

$$= 4\frac{i\, k}{s^2}\left(\frac{s^3}{3} - \frac{s^3}{4}\right) = \frac{1}{3} s\, i\, k.$$

This result can be taken directly from Table 6.3 without the need to integrate. It can be found in the fourth row/second column: $\frac{1}{3} s\, i\, k$.

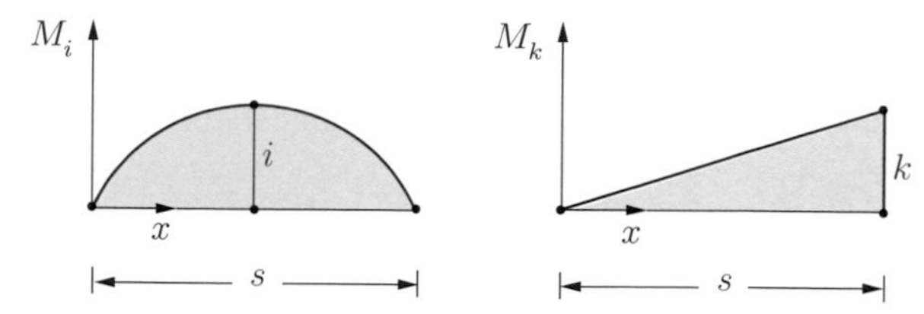

Fig. 6.10

The principle of virtual forces can also be applied to more general types of loading. In the case of a member being subjected to bending, torsion and tension, the deflection is obtained from

$$f = \int \frac{M \bar{M}}{EI}\, \mathrm{d}x + \int \frac{M_T\, \bar{M}_T}{GI_T}\, \mathrm{d}x + \int \frac{N \bar{N}}{EA}\, \mathrm{d}x \qquad (6.36)$$

where M, M_T and N are the bending moment, torque and normal force due to the given load. The virtual force "1" which is applied at the point where the displacement is to be determined leads to the stress resultants $\bar{M}, \bar{M}_T$ and $\bar{N}$. The integrals in (6.36) have to be evaluated for the entire system. If the rigidities GI_T and EA, respectively, are constant in a member, the corresponding integrals

$$\int M_T\, \bar{M}_T\, \mathrm{d}x, \qquad \int N\, \bar{N}\, \mathrm{d}x$$

can also be taken from Table 6.3.

Let us now consider a truss where the i-th member undergoes a change $\Delta\, T_i$ of its temperature. Then, in analogy with (1.17), the change $\alpha_{T_i}\,\Delta T_i\,l_i$ of the length caused by the change of the temperature has to be added to the elongation $\dfrac{S_i\,l_i}{EA_i}$ due to the internal force S_i in (6.30):

$$f = \sum \frac{S_i\,\bar{S}_i\,l_i}{EA_i} + \sum \bar{S}_i\,\alpha_{Ti}\,\Delta T_i\,l_i .$$

Similarly, a moment $M_{\Delta T}$ due to the temperature change (see (4.63)) has to be added to the bending moment M in (6.33) if a beam is subjected to a thermal load (see Section 4.9):

$$f = \int \frac{(M + M_{\Delta T})\,\bar{M}}{EI}\,\mathrm{d}x .$$

E6.3

Example 6.3 The truss in Fig. 6.11a consists of 17 members (axial rigidity EA).

Determine the vertical displacement f_V of pin V.

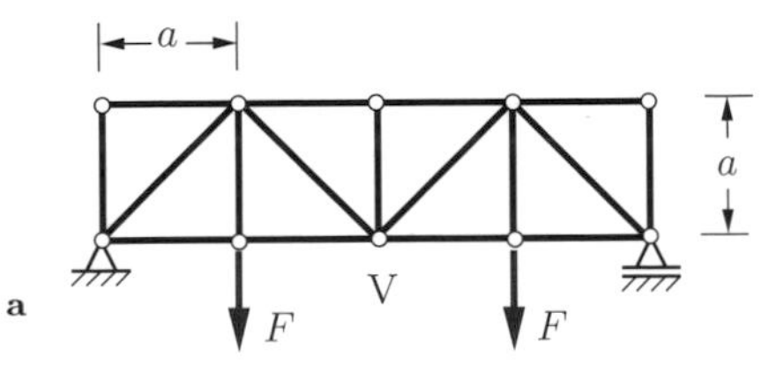

Truss subjected to given load

2 6 10 14
1 3 5 7 9 11 13 15 17
0 0 0 0
4 8 12 16
F F
A=F B=F
b

Truss subjected to virtual force

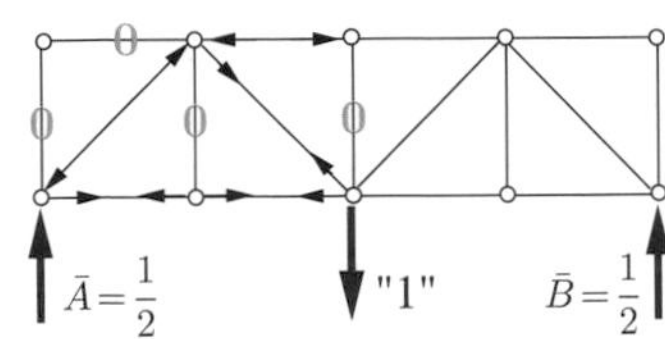

Fig. 6.11

Solution The truss has $j = 10$ joints, $m = 17$ members and $r = 3$ support reactions. Therefore, the necessary isostatic condition $2\,j = m + r$ is satisfied (see Volume 1, Section 6.1).

We determine the displacement with the aid of (6.30). In the first step, the internal forces S_i in the members due to the given loads (Fig. 6.11b) are calculated by the method of joints (Volume 1, Section 6.3.1). The results are presented in a table.

i	S_i	$\bar{S}_i$	l_i	$S_i\,\bar{S}_i\,l_i$
1	0	0	a	0
2	0	0	a	0
3	$-\sqrt{2}F$	$-\sqrt{2}/2$	$\sqrt{2}\,a$	$\sqrt{2}\,F\,a$
4	F	$1/2$	a	$Fa/2$
5	F	0	a	0
6	$-F$	-1	a	$F\,a$
7	0	$\sqrt{2}/2$	$\sqrt{2}\,a$	0
8	F	$1/2$	a	$Fa/2$
9	0	0	a	0
10	$-F$	-1	a	$F\,a$
11	0	$\sqrt{2}/2$	$\sqrt{2}\,a$	0
12	F	$1/2$	a	$F\,a/2$
13	F	0	a	0
14	0	0	a	0
15	$-\sqrt{2}F$	$-\sqrt{2}/2$	$\sqrt{2}\,a$	$\sqrt{2}\,F\,a$
16	F	$1/2$	a	$F\,a/2$
17	0	0	a	0

$\sum S_i\,\bar{S}_i\,l_i = (4 + 2\sqrt{2})\,F\,a$

Subsequently, the truss is subjected only to a vertical virtual force "1" at pin V (Fig. 6.11c). The resulting internal forces $\bar{S}_i$ are also recorded in the same table. The products $S_i\,\bar{S}_i\,l_i$ (l_i: length of the member i) are given in the last column of the table. With $EA_i = EA$, the displacement f_V follows from (6.30):

$$\underline{\underline{f_V}} = \sum \frac{S_i\,\bar{S}_i\,l_i}{EA} = \underline{\underline{(4 + 2\sqrt{2})\frac{F\,a}{EA}}}.$$

E6.4

Example 6.4 Determine the horizontal and the vertical components of the displacement of pin B of the truss in Fig. 6.12a. The axial rigidity of the members 1-3 is given by EA; member 4 has the rigidity $2\,EA$.

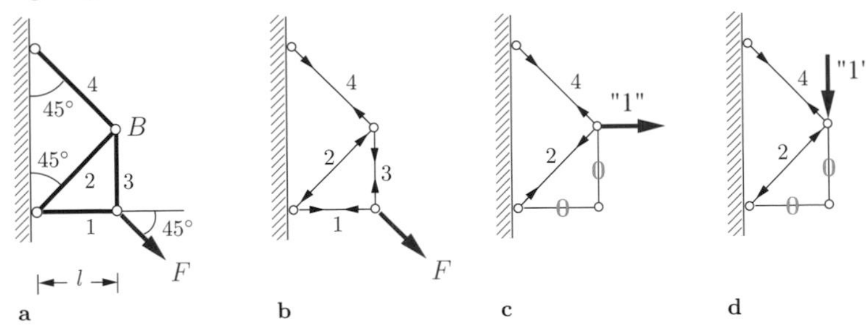

Fig. 6.12

Solution The truss is statically determinate. The internal forces S_i in the members due to the force F (Fig. 6.12b) can be calculated with the aid of the method of joints (Volume 1, Section 6.3.1).

To determine the horizontal displacement of B, we apply a horizontal force "1" at this point (Fig. 6.12c). The corresponding internal forces are denoted by $\bar{S}_{iH}$. Similarly, a vertical force "1" at B (Fig. 6.12d) produces the internal forces $\bar{S}_{iV}$ and leads to the vertical displacement. All the internal forces are given in a table.

i	l_i	S_i	$\bar{S}_{iH}$	$\bar{S}_{iV}$	$S_i\,\bar{S}_{iH}\,l_i$	$S_i\,\bar{S}_{iV}\,l_i$
1	l	$\dfrac{F}{\sqrt{2}}$	0	0	0	0
2	$\sqrt{2}\,l$	$-\dfrac{F}{2}$	$\dfrac{1}{2}\sqrt{2}$	$-\dfrac{1}{2}\sqrt{2}$	$-\dfrac{1}{2}F\,l$	$\dfrac{1}{2}F\,l$
3	l	$\dfrac{F}{\sqrt{2}}$	0	0	0	0
4	$\sqrt{2}\,l$	$\dfrac{F}{2}$	$\dfrac{1}{2}\sqrt{2}$	$\dfrac{1}{2}\sqrt{2}$	$\dfrac{1}{2}F\,l$	$\dfrac{1}{2}F\,l$

The components of the displacement at B follow from (6.30) (note the different axial rigidities of the members):

$$\underline{\underline{f_H}} = \sum \frac{S_i \bar{S}_{iH} l_i}{EA_i} = -\frac{1}{2}\frac{Fl}{EA} + \frac{1}{2}\frac{Fl}{2EA} = \underline{\underline{-\frac{1}{4}\frac{Fl}{EA}}},$$

$$\underline{\underline{f_V}} = \sum \frac{S_i \bar{S}_{iV} l_i}{EA_i} = \frac{1}{2}\frac{Fl}{EA} + \frac{1}{2}\frac{Fl}{2EA} = \underline{\underline{\frac{3}{4}\frac{Fl}{EA}}}.$$

The negative sign of f_H indicates that the horizontal displacement (to the left) is directed in the opposite direction to the force “1“. The vertical displacement is three times as large as the horizontal displacement.

E6.5

Example 6.5 The frame in Fig. 6.13a (flexural rigidity EI, axial rigidity $EA \to \infty$) is subjected to a constant line load q_0 and a force F.

Calculate the horizontal displacement u_B of the support B.

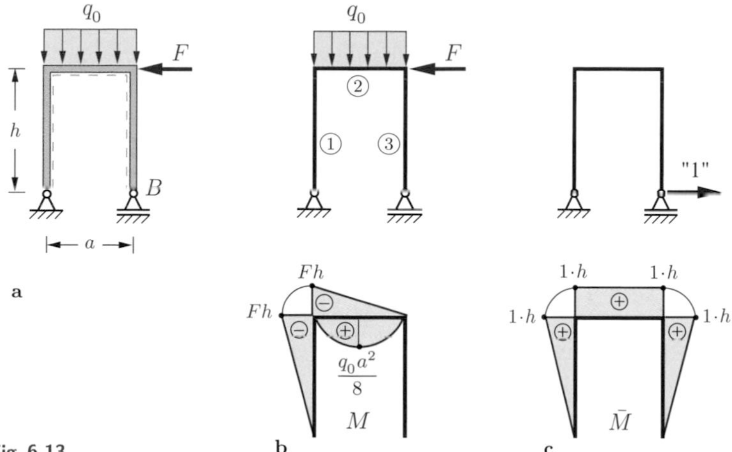

Fig. 6.13

Solution In the first step we determine the bending moment due to the external loads. It is advantageous for the integration that follows to present the moments due to q_0 and F in part ② of the frame in separate graphs (Fig. 6.13b).

Then we subject the frame only to the virtual force “1“ at B and determine the corresponding bending moment (Fig. 6.13c).

The displacement u_B is obtained according to (6.33) by multiplying the bending moments and integrating over the total frame:
Part ① : triangle and triangle

$$\int M\bar{M}\,\mathrm{d}x = \frac{1}{3}h(-F\,h)(1\cdot h) = -\,\frac{1}{3}F\,h^3$$

Part ② : rectangle and triangle

$$\int M\bar{M}\,\mathrm{d}x = \frac{1}{2}a(-F\,h)(1\cdot h) = -\,\frac{1}{2}F\,a\,h^2$$

rectangle and quadratic parabola

$$\int M\bar{M}\,\mathrm{d}x = \frac{2}{3}a\frac{q_0\,a^2}{8}(1\cdot h) = \frac{1}{12}q_0\,a^3\,h$$

Part ③ : $\int M\bar{M}\,\mathrm{d}x = 0$ since $M = 0$.

Equation (6.33) yields the displacement:

$$\underline{\underline{EI\,u_B = \frac{1}{12}q_0\,a^3\,h - \frac{1}{6}F\,h^2(2\,h+3\,a).}}$$

The algebraic signs indicate that the support is displaced to the right due to q_0 and to the left due to F.

E6.6

Example 6.6 The structure shown in Fig. 6.14a consists of the angled member BCD (flexural rigidity EI) and the two bars 1 and 2 (axial rigidity EA). A couple moment M_0 is applied at point C.

Determine the displacement v_B of the support B and the rotation φ_C at point C.

Solution First we calculate the bending moment M (Fig. 6.14b) and the forces $S_1 = M_0/2\,a$ and $S_2 = -M_0/2\,a$ in the bars due to the external load M_0. In order to find the displacement at B, we apply a virtual force “1“ in the direction of the displacement (Fig. 6.14c) which leads to $\bar{M} = 0$, $\bar{S}_1 = 0$ and $\bar{S}_2 = \sqrt{2}$. Equation (6.36) yields

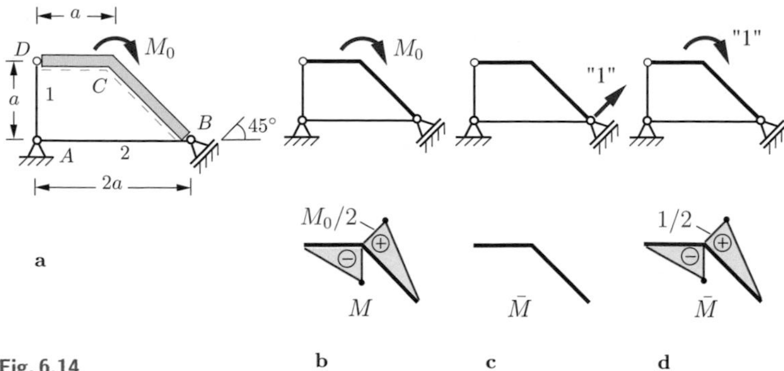

Fig. 6.14

$$\underline{\underline{v_B}} = \frac{S_2\,\bar{S}_2\,l_2}{EA} = -\,\frac{M_0}{2\,a}\sqrt{2}\,\frac{2\,a}{EA} = \underline{\underline{-\,\sqrt{2}\,\frac{M_0}{EA}}}.$$

To obtain the rotation at C we apply a virtual couple moment "1" at this point (Fig. 6.14d). This leads to $\bar{M}$ according to Fig. 6.14d and to $\bar{S}_1 = 1/2\,a$ and $\bar{S}_2 = -1/2\,a$. From

$$\varphi_C = \int \frac{M\bar{M}}{EI}\,\mathrm{d}x + \sum \frac{S_i\,\bar{S}_i\,l_i}{EA}$$

(see (6.30) and (6.36)), we obtain the rotation with the aid of Table 6.3:

$$\underline{\underline{\varphi_C}} = \frac{1}{EI}\left[\frac{1}{3}\left(-\frac{M_0}{2}\right)\left(-\frac{1}{2}\right)a + \frac{1}{3}\frac{M_0}{2}\frac{1}{2}\sqrt{2}\,a\right]$$

$$+\frac{1}{EA}\left[\frac{M_0}{2\,a}\frac{1}{2\,a}\,a + \left(-\frac{M_0}{2\,a}\right)\left(-\frac{1}{2\,a}\right)2\,a\right]$$

$$= \underline{\underline{\frac{M_0\,a}{12\,EI}\left[1+\sqrt{2}+9\frac{EI}{EA\,a^2}\right]}}.$$

E6.7

Example 6.7 Determine the displacement of point C for the frame in Fig. 6.15a.

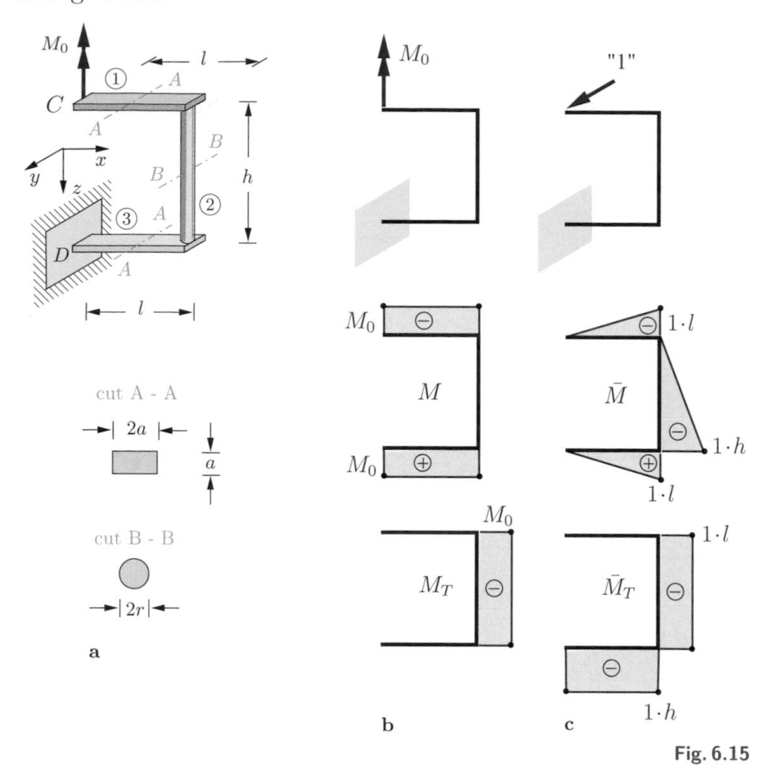

Fig. 6.15

Solution The parts ① and ③ are subjected to bending; part ② is subjected to torsion. One can see by inspection that point C undergoes a displacement v in the y-direction.

In the first step we determine the bending moment M and the torque M_T caused by the external couple moment M_0 (Fig. 6.15b). Note that the algebraic signs for the stress resultants in the individual parts of the system may be chosen arbitrarily. However, the same sign convention has to be applied to the system with the virtual force "1". The stress resultants $\bar{M}$ and $\bar{M}_T$ due to the virtual force in the y-direction are given in Fig. 6.15c. Integration (see (6.36)) leads to

$$v = \int \frac{M\bar{M}}{EI}\,\mathrm{d}x + \int \frac{M_T\bar{M}_T}{GI_T}\,\mathrm{d}x$$

$$= \frac{1}{2}\frac{M_0\,l}{EI_1}\,l + \frac{1}{2}\frac{M_0\,l}{EI_3}\,l + \frac{M_0\,l}{GI_{T2}}\,h.$$

Inserting the moments of inertia

$$I_1 = I_3 = \frac{a\,(2\,a)^3}{12} = \frac{2}{3}a^4, \qquad I_{T2} = \frac{\Pi}{2}r^4$$

yields

$$\underline{\underline{v = \frac{3}{2}\frac{M_0\,l^2}{E\,a^4} + \frac{2\,M_0\,l\,h}{G\,\Pi\,r^4}}}.$$

Example 6.8 Determine the displacement of point A of the lamp (weight W) in Fig. 6.16a. The weight of the arch (flexural rigidity EI) is negligible.

E6.8

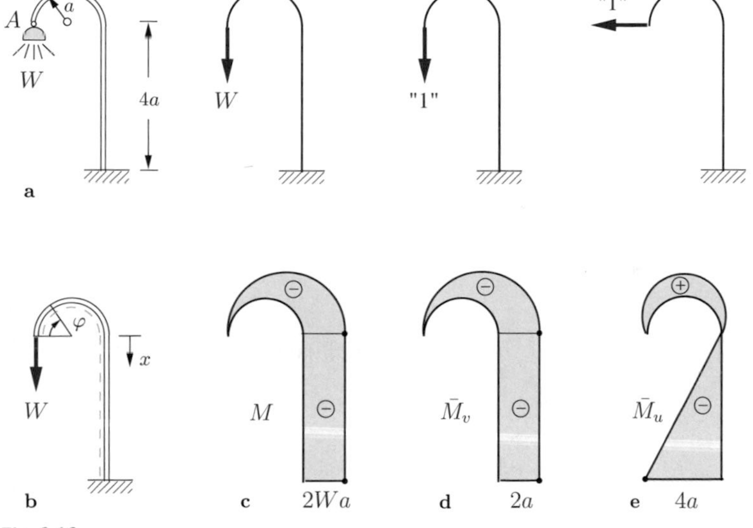

Fig. 6.16

Solution We introduce the dashed line (see Volume 1, Section 7.1) and the coordinates x and φ according to Fig. 6.16b. Then the bending moment M due to the load W is given by (see Fig. 6.16c)

$$M = \begin{cases} -Wa(1-\cos\varphi), & 0 \le \varphi \le \pi, \\ -W2a, & 0 \le x \le 4a. \end{cases}$$

In order to find the vertical displacement v of point A, we apply a vertical virtual force "1" at A (Fig. 6.16d). We obtain the bending moment $\bar{M}_v$ if we replace W with "1" in M:

$$\bar{M}_v = \begin{cases} -a(1-\cos\varphi), & 0 \le \varphi \le \pi, \\ -2a, & 0 \le x \le 4a. \end{cases}$$

Integration yields

$$v = \int \frac{M\bar{M}_v}{EI}\,\mathrm{d}s = \frac{1}{EI}\int_0^\pi -Wa(1-\cos\varphi)[-a(1-\cos\varphi)]a\,\mathrm{d}\varphi$$

$$+\frac{1}{EI}\int_0^{4a}(-2\,a\,W)(-2\,a)\mathrm{d}x$$

$$= \frac{Wa^3}{EI}\int_0^\pi (1-2\cos\varphi+\cos^2\varphi)\,\mathrm{d}\varphi + \frac{4\,Wa^2}{EI}4\,a$$

$$= \frac{Wa^3}{EI}\left(\frac{3\,\pi}{2}+16\right) \approx 20,7\frac{Wa^3}{EI}.$$

To determine the horizontal displacement u, we apply a horizontal force "1" at A (Fig. 6.16e). The corresponding bending moment $\bar{M}_u$ is given by

$$\bar{M}_u = \begin{cases} a\sin\varphi, & 0 \le \varphi \le \pi, \\ -x, & 0 \le x \le 4a. \end{cases}$$

Integration leads to

$$u = \int \frac{M\bar{M}_u}{EI}\,\mathrm{d}s = \frac{1}{EI}\int\limits_0^{\pi} -Wa(1-\cos\varphi)a\sin\varphi\, a\,\mathrm{d}\varphi$$

$$+\frac{1}{EI}\int\limits_0^{4a}(-2\,a\,W)(-x)\mathrm{d}x$$

$$= \frac{Wa^3}{EI}(-2+16) = 14\,\frac{Wa^3}{EI}.$$

The total displacement f_A is therefore found to be

$$\underline{\underline{f_A}} = \sqrt{u^2+v^2} \approx \frac{Wa^3}{EI}\sqrt{429+196} = \underline{\underline{25\,\frac{Wa^3}{EI}}}.$$

Note that the vertical load W causes a large horizontal displacement.

6.4

6.4 Influence Coefficients and Reciprocal Displacement Theorem

In Section 6.3 it was shown that the deflection f_{ik} of a beam at an arbitrary point i due to a force F_k at point k can be determined with the aid of the principle of virtual forces (see (6.32)). If the force F_k is the only external load, the deflection of the beam is proportional to this force. Therefore, we can write

$$f_{ik} = \alpha_{ik}\,F_k. \tag{6.37}$$

The proportionality factor α_{ik} is called the *influence coefficient*. It is equal to the deflection at point i due to the force “1“ at point k. As an illustrative example consider the cantilever beam in Fig. 6.9a which is subjected to a force F at point a. The influence coefficient for the deflection at the free end is obtained from (6.34):

$$\alpha_{la} = \frac{f}{F} = \frac{l^3}{6\,EI}\left[3\left(\frac{a}{l}\right)^2 - \left(\frac{a}{l}\right)^3\right].$$

Similarly, the influence coefficient for the deflection at point x for the beam subjected to the couple moment M_0 in Example 4.6 is found to be

$$\alpha_{xl} = \frac{w(x)}{M_0} = \frac{l^2}{6\,EI}\left[\left(\frac{x}{l}\right) - \left(\frac{x}{l}\right)^3\right].$$

Note that the two influence coefficients given here have different dimensions.

If a beam is subjected to n forces F_k, the deflection f at point i is obtained through superposition:

$$f = \sum_k f_{ik} = \alpha_{i1}\,F_1 + \alpha_{i2}\,F_2 + \alpha_{i3}\,F_3 + \ldots + \alpha_{in}\,F_n.$$

Let us now consider a beam that is subjected to two forces as shown in Fig. 6.17a: force F_i acts at point i, force F_k acts at point k. If we first apply F_k and subsequently apply F_i (see Fig. 6.17b), then the total work done by the external forces is given by

$$\begin{aligned} U &= \frac{1}{2}\,f_{kk}\,F_k + \frac{1}{2}\,f_{ii}\,F_i + F_k\,f_{ki} \\ &= \frac{1}{2}\,\alpha_{kk}\,F_k^2 + \frac{1}{2}\,\alpha_{ii}\,F_i^2 + F_k(\alpha_{ki}\,F_i). \end{aligned} \tag{6.38a}$$

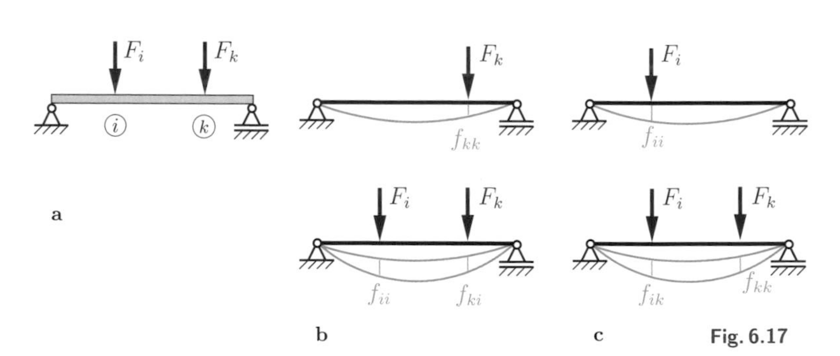

Fig. 6.17

If we first apply F_i and subsequently F_k (Fig. 6.17c), the total work is

$$\begin{aligned} U &= \frac{1}{2} f_{ii} F_i + \frac{1}{2} f_{kk} F_k + F_i f_{ik} \\ &= \frac{1}{2} \alpha_{ii} F_i^2 + \frac{1}{2} \alpha_{kk} F_k^2 + F_i(\alpha_{ik} F_k). \end{aligned} \tag{6.38b}$$

The total strain energy after the two forces have been applied is independent of the sequence of the application. According to the principle of conservation of energy (6.8), the total work of the external forces is also independent of this sequence. We therefore can equate the first lines of (6.38a) and (6.38b) to obtain

$$F_k f_{ki} = F_i f_{ik} . \tag{6.39}$$

This is referred to as the *reciprocal work theorem* or *Betti's theorem* (Enrico Betti, 1823–1892 and Lord Rayleigh, 1842–1919). It tells us that the work done by the force F_k during the displacement f_{ki} (which is caused by F_i) is equal to the work done by the force F_i during the displacement f_{ik} (which is caused by F_k). This statement can be generalized to arbitrary elastic systems.

If we equate the second lines of (6.38a) and (6.38b) we obtain the *reciprocal displacement theorem*, also called *Maxwell's reciprocal theorem* (James Clerk Maxwell, 1831–1879):

$$\alpha_{ik} = \alpha_{ki} . \tag{6.40}$$

It implies that the deflection α_{ik} at point i due to a force "1" at point k is equal to the deflection α_{ki} at point k due to a force "1" at point i.

Equation (6.40) can also be applied to a system that is subjected to a couple moment. Let us consider, for example, the beam in Fig. 6.18a which is subjected to a force F at point ① and to a

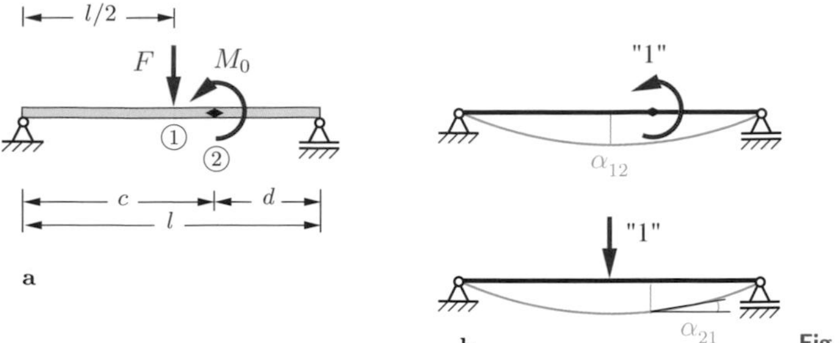

Fig. 6.18

moment M_0 at point ② . This moment causes the deflection

$$f_{12} = \alpha_{12}\, M_0 = -\,\frac{l^2}{6}\left\{\frac{1}{2}\left[3\left(\frac{d}{l}\right)^2 - 1\right] + \frac{1}{8}\right\}\frac{M_0}{EI} \qquad (6.41)$$

at point ① (see Table 4.3, Nr. 5, $\xi = 1/2$, $\beta = d/l$).

Now we determine the angle of slope of the deflection curve at point ② which is caused by F. From Table 4.3, Nr. 1, we find the slope at an arbitrary point ξ through differentiation:

$$EI\,w' = \frac{F\,l^2}{6}[\beta(1-\beta^2-3\,\xi^2)+3\langle\xi-\alpha\rangle^2].$$

In the present example we have to choose $\alpha = \beta = 1/2$ to obtain

$$EI\,w' = \frac{F\,l^2}{6}[\frac{1}{2}(1-\frac{1}{4}-3\,\xi^2)+3\langle\xi-\frac{1}{2}\rangle^2].$$

The angle has to be taken as positive if its sense of rotation coincides with the sense of rotation of the moment M_0. Thus, $\varphi_{21} = -w'$ at point ② . With $\xi = c/l$ we obtain

$$\varphi_{21} = \alpha_{21}\,F = -\,\frac{l^2}{6\,EI}\left[\frac{1}{2}\left(\frac{3}{4}-3\left(\frac{c}{l}\right)^2\right)+3\left(\left(\frac{c}{l}\right)-\frac{1}{2}\right)^2\right]F$$

$$= -\,\frac{l^2}{6\,EI}\left[\frac{3}{2}\left(\frac{c}{l}\right)^2-3\left(\frac{c}{l}\right)+\frac{9}{8}\right]F.$$

Using $c = l - d$, this can be written in the form

$$\varphi_{21} = \alpha_{21}\, F = -\, \frac{l^2}{6\,EI} \left\{ \frac{1}{2} \left[3 \left(\frac{d}{l} \right)^2 - 1 \right] + \frac{1}{8} \right\} F. \qquad (6.42)$$

Comparison of (6.41) and (6.42) yields

$$\alpha_{12} = \alpha_{21} .$$

That is, the displacement α_{12} at point ① due to the moment "1" at point ② is equal to the rotation α_{21} at point ② due to the force "1" at point ① (Fig. 6.18b). Note that α_{12} (displacement) and α_{21} (rotation) here have the same dimension.

Maxwell's reciprocal theorem has many useful applications. Note that knowing α_{12} (see (6.41)) in the preceding example, then α_{21} and therefore φ_{21} are also known according to (6.40). Thus, the rather cumbersome calculation to obtain φ_{21} in (6.42) was actually unnecessary.

6.5 Statically Indeterminate Systems

6.5

Statically indeterminate systems were investigated with the aid of the principle of superposition in the Sections 1.4, 1.6 and 4.5.4. In the case of a system which is externally statically indeterminate to the first degree, we removed one of the supports in order to obtain a statically determinate system. In the "0"-system we calculated the displacement $v^{(0)}$ due to the given load at the point where the support was removed. This displacement will now be denoted by α_{10}, i.e., $v^{(0)} = \alpha_{10}$. The new notation is similar to the notation used for the influence coefficients (see Section 6.4). Subsequently, the statically determinate structure was subjected only to the as yet unknown force "X" (the redundant) at the point of the removed support. This system was referred to as the "1"-system. The displacement caused by the force X is $v^{(1)} = X\,\alpha_{11}$, where α_{11} is the displacement caused by the force $X = 1$. The displacement in the given statically indeterminate structure has to be zero due to the actual support:

$$v = v^{(0)} + v^{(1)} = 0. \tag{6.43}$$

This compatibility condition yields the redundant force:

$$\alpha_{10} + X\,\alpha_{11} = 0 \qquad \rightarrow \qquad X = -\,\frac{\alpha_{10}}{\alpha_{11}}. \tag{6.44}$$

Analogous considerations are used if a statically indeterminate beam is made statically determinate by introducing a joint at a point G. In this case we have to apply a moment at G, and the displacement v in (6.43) has to be replaced with the angle φ_G (see Examples 4.11 and 6.12). Similar considerations are valid for a statically indeterminate truss with one redundant bar (internal statical indeterminacy). Then the redundant bar is removed and the displacements in the systems "0" and "1" are determined. Compatibility now requires that the change of the distance between the two joints from which the bar was removed is equal to the change of the length of this bar. The force in the statically redundant bar can also be calculated from (6.44), where the coefficients α_{ik} now are the corresponding influence coefficients of the truss (see (6.46)).

In this section we will also apply the principle of superposition, but in contrast to the calculations in the Sections 1.4, 1.6 and 4.5.4, we will now determine the displacements (rotations) with the aid of the principle of virtual forces.

As an illustrative example we consider the beam in Fig. 6.19a; it is statically indeterminate to the first degree. We first remove the support B and replace it with the as yet unknown support reaction X. This leads to the two systems "0" and "1" as shown in Fig. 6.19b. In the following derivations we will change the notation: bending moments in a "0"-system are called M_0 from now on and bending moments in a "1"-system are referred to as $\bar{M}_1$ (in Section 4.5.4 they were called $M^{(0)}$ and $M^{(1)} = X\bar{M}_1$). According

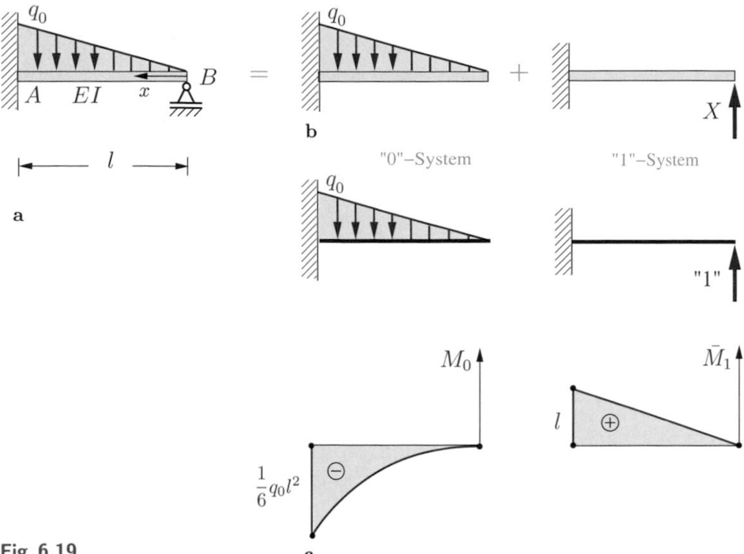

Fig. 6.19

to (6.44) the redundant support reaction X is obtained from

$$X = -\frac{\alpha_{10}}{\alpha_{11}} = -\frac{\displaystyle\int \frac{\bar{M}_1\, M_0}{EI}\,\mathrm{d}x}{\displaystyle\int \frac{\bar{M}_1^2}{EI}\,\mathrm{d}x}. \tag{6.45}$$

We use the coordinate x as shown in Fig. 6.19a. Then the bending moments in the systems "0" and "1" are given by

$$M_0 = -\frac{1}{2}x\left(q_0\frac{x}{l}\right)\frac{x}{3} = -\frac{q_0}{6\,l}\,x^3, \qquad \bar{M}_1 = x.$$

They lead to

$$\alpha_{10} = \int \frac{\bar{M}_1\, M_0}{EI}\,\mathrm{d}x = \frac{1}{EI}\int\limits_0^l x\left(-\frac{q_0}{6\,l}\,x^3\right)\mathrm{d}x = -\frac{q_0\, l^4}{30\,EI},$$

$$\alpha_{11} = \int \frac{\bar{M}_1^2}{EI}\,\mathrm{d}x = \frac{1}{EI}\int\limits_0^l x^2\,\mathrm{d}x = \frac{l^3}{3\,EI}.$$

Note that these results could also have been taken from Table 6.3 without performing the integrations. Equation (6.45) yields

$$X = B = -\frac{\alpha_{10}}{\alpha_{11}} = \frac{q_0\, l}{10}.$$

The bending moment in the beam is obtained by superposition:

$$M = M_0 + X\,\bar{M}_1 = -\frac{q_0}{6\,l}\,x^3 + \frac{q_0\, l}{10}\,x.$$

In particular, the moment at the clamped support ($x = l$) is found to be

$$M_A = M(l) = -\frac{q_0\, l^2}{15}.$$

We may also solve this problem using a different "0"-system: the clamped support is now replaced with a joint (Fig. 6.20). Then we have to apply a moment "1" at this point and the bending moments are given by

$$M_0 = \frac{q_0}{6}\,l\,x - \frac{q_0}{6\,l}\,x^3, \qquad \bar{M}_1 = \frac{x}{l}.$$

Compatibility analogous to (6.43) requires a vanishing slope w'_A at the left-hand side of the beam and leads to the moment at the

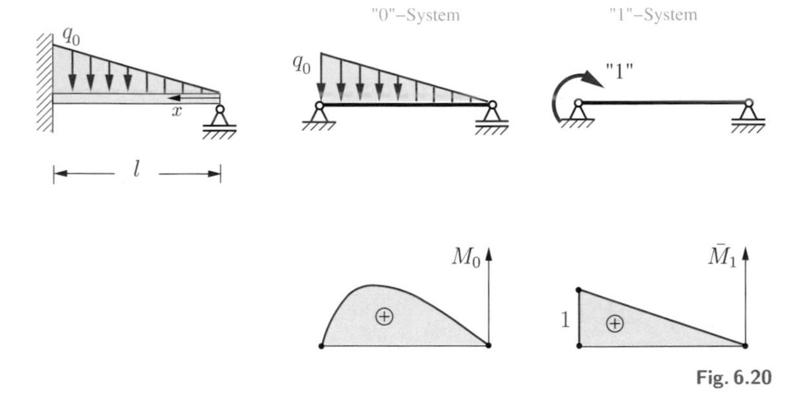

Fig. 6.20

clamping:

$$X = M_A = -\frac{\alpha_{10}}{\alpha_{11}} = -\frac{\int \frac{\bar{M}_1\, M_0}{EI}\,\mathrm{d}x}{\int \frac{\bar{M}_1^2}{EI}\,\mathrm{d}x} = \frac{\int\limits_0^l \frac{x}{l}\left(\frac{q_0}{6}\, l\, x - \frac{q_0}{6\, l}\, x^3\right)\mathrm{d}x}{\int\limits_0^l \left(\frac{x}{l}\right)^2 \mathrm{d}x}$$

$$= -\frac{q_0\, l^2}{15}$$

In the case of a truss which is statically indeterminate to the first degree, we may also use (6.44). The coefficients α_{10} and α_{11} follow according to (6.30):

$$X = -\frac{\sum \frac{\bar{S}_i\, S_i^{(0)}\, l_i}{EA_i}}{\sum \frac{\bar{S}_i^2\, l_i}{EA_i}}. \tag{6.46}$$

Here, $S_i^{(0)}$ are the forces in the members of the "0"-system and $\bar{S}_i$ are the forces in the members of the "1"-system.

If there are bending moments, torques and variable normal forces acting in a structure, the redundant follows from (see (6.14))

$$X = -\frac{\int \frac{\bar{M}_1\, M_0}{EI}\,\mathrm{d}x + \int \frac{\bar{M}_{T1}\, M_{T0}}{GI_T}\,\mathrm{d}x + \int \frac{\bar{N}_1\, N_0}{EA}\,\mathrm{d}x}{\int \frac{\bar{M}_1^2}{EI}\,\mathrm{d}x + \int \frac{\bar{M}_{T1}^2}{GI_T}\,\mathrm{d}x + \int \frac{\bar{N}_1^2}{EA}\,\mathrm{d}x}. \tag{6.47}$$

After having determined the unknown X, the other support reactions, the stress resultants and the displacements can be calculated.

Finally, we want to indicate the procedure in the case of a system with a statical indeterminacy of degree n. In this case we have to remove n constraints in order to obtain a statically determinate "0"-system. In addition, we have to consider n auxiliary systems to determine the n unknown redundants X_i which can be calculated from n compatibility conditions (see Example 6.12):

$$
\begin{aligned}
\alpha_{10} + X_1\,\alpha_{11} + \ldots + X_n\,\alpha_{1n} &= 0,\\
\alpha_{20} + X_1\,\alpha_{21} + \ldots + X_n\,\alpha_{2n} &= 0,\\
\ldots\ldots\ldots\ldots\ldots\ldots\ldots\ldots\ldots & \\
\alpha_{n0} + X_1\,\alpha_{n1} + \ldots + X_n\,\alpha_{nn} &= 0.
\end{aligned} \tag{6.48}
$$

The displacements in the "0"-system follow from

$$\alpha_{r0} = \int \frac{\bar{M}_r\, M_0}{EI}\,\mathrm{d}x$$

and the displacements in the auxiliary systems are given by

$$\alpha_{ri} = \int \frac{\bar{M}_r\, \bar{M}_i}{EI}\,\mathrm{d}x.$$

Here, M_0 is the bending moment in the "0"-system caused by the given load and $\bar{M}_i$ are the bending moments in the same system due to the virtual forces (moments) acting at the points i ($i = 1, 2, \ldots, n$).

E6.9

Example 6.9 Determine the forces in the members of the truss in Fig. 6.21a. The axial rigidity EA of the members is given.

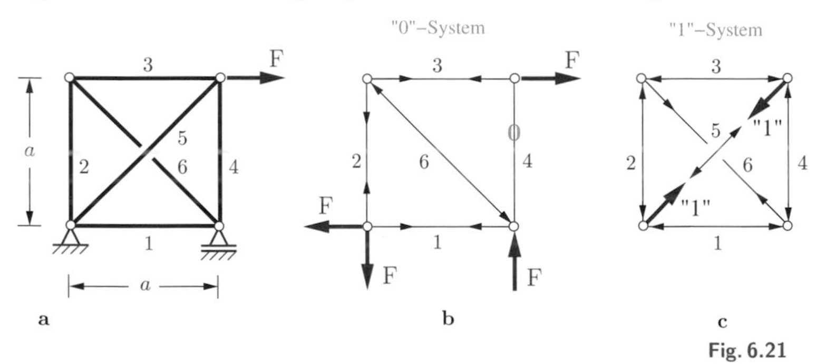

Fig. 6.21

Solution The truss consists of $m = 6$ members and $j = 4$ joints; it has $r = 3$ supports. According to Volume 1, Section 6.1, it is statically indeterminate to the first degree: $m + r - 2j = 1$. We obtain a statically determinate truss if we remove one of the bars.

We choose member 5 which leads to the "0"-system in Fig. 6.21b. Next we subject member 5 to the virtual force "1" ("1"-system, Fig. 6.21c). According to Newton's third law (action equals reaction), forces of the same magnitude act at the joints. The forces in the members of the "0"-system and of the "1"-system, respectively, are calculated and given in the following table.

i	$S_i^{(0)}$	$\bar{S}_i$	l_i	$\bar{S}_i\, S_i^{(0)}\, l_i$	$\bar{S}_i^2\, l_i$	S_i
1	F	$-1/\sqrt{2}$	a	$-F\,a/\sqrt{2}$	$\frac{1}{2}a$	$+0.40\,F$
2	F	$-1/\sqrt{2}$	a	$-F\,a/\sqrt{2}$	$\frac{1}{2}a$	$+0.40\,F$
3	F	$-1/\sqrt{2}$	a	$-F\,a/\sqrt{2}$	$\frac{1}{2}a$	$+0.40\,F$
4	0	$-1/\sqrt{2}$	a	0	$\frac{1}{2}a$	$-0.60\,F$
5	0	1	$\sqrt{2}\,a$	0	$\sqrt{2}\,a$	$+0.85\,F$
6	$-\sqrt{2}\,F$	1	$\sqrt{2}\,a$	$-2\,a\,F$	$\sqrt{2}\,a$	$-0.56\,F$

The unknown force in member 5 is obtained with $\sum \bar{S}_i\, S_i^{(0)}\, l_i = (-2-3/\sqrt{2})\,F\,a$ and $\sum \bar{S}_i^2\, l_i = 2(1+\sqrt{2})a$ from (6.46):

$$X = \underline{\underline{S_5}} = -\frac{\left(-2-\dfrac{3}{\sqrt{2}}\right)F\,a}{2(1+\sqrt{2})\,a} = \frac{3+2\sqrt{2}}{2(2+\sqrt{2})}\,F \approx \underline{\underline{0.85\,F}}.$$

The forces in the other members follow from

$$S_i = S_i^{(0)} + X\bar{S}_i.$$

They are given in the last column of the table.

Note that the support reactions can be calculated in advance in the case of a truss which is internally statically indeterminate.

E6.10

Example 6.10 The structure in Fig. 6.22a consists of an angled member (flexural rigidity EI) and two bars (axial rigidity EA). It is subjected to the force F.

Determine the bending moment in the angled member and the forces in the bars.

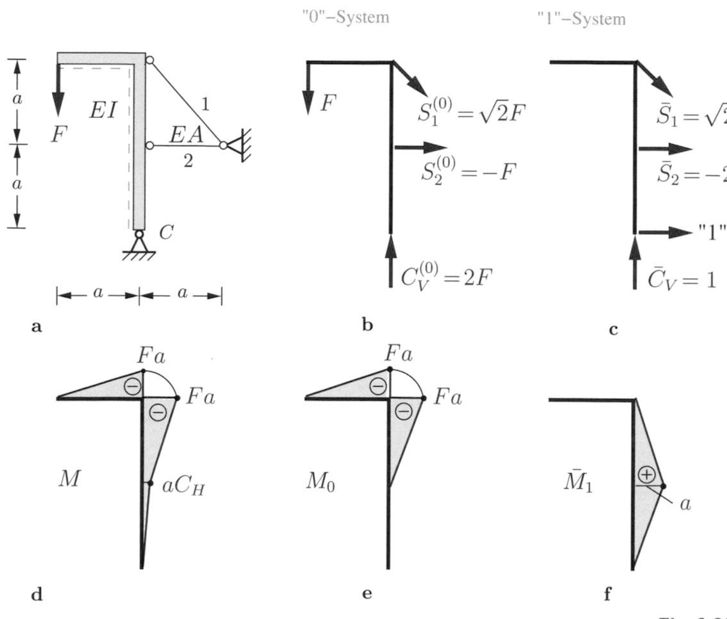

Fig. 6.22

Solution The structure is statically indeterminate to the first degree. To obtain a "0"-system we replace the hinged support at C with a roller support which can move in the horizontal direction. The equilibrium conditions in the "0"-system (Fig. 6.22b) yield

$$C_V^{(0)} = 2\,F, \qquad S_1^{(0)} = \sqrt{2}\,F, \qquad S_2^{(0)} = -\,F.$$

Now we apply a horizontal virtual force "1" at C ("1"-system, Fig. 6.22c) and obtain

$$\bar{C}_V = 1, \qquad \bar{S}_1 = \sqrt{2}, \qquad \bar{S}_2 = -\,2.$$

The resulting bending moments M_0 and $\bar{M}_1$ are shown in the Figs. 6.22e, f. Since the structure consists of beams and bars, the redundant force C_H follows from (6.47):

$$X = C_H = -\frac{\displaystyle\int \frac{\bar{M}_1 M_0}{EI}\,dx + \sum \bar{S}_i \frac{S_i^{(0)} l_i}{EA}}{\displaystyle\int \frac{\bar{M}_1^2}{EI}\,dx + \sum \bar{S}_i^2 \frac{l_i}{EA}}.$$

We use Table 6.3 (triangles with triangles) to calculate the integrals and introduce the parameter $\varkappa = EA\,a^2/EI$ to obtain

$$X = C_H = -\frac{\frac{1}{6}a(-F\,a)\frac{a}{EI} + \left(\sqrt{2}\sqrt{2}\,F\frac{a\sqrt{2}}{EA} + (-2)(-F)\frac{a}{EA}\right)}{2\cdot\frac{1}{3}\,a\frac{a^2}{EI} + \sqrt{2}\sqrt{2}\frac{\sqrt{2}\,a}{EA} + (-2)(-2)\frac{a}{EA}}$$

$$= \frac{\varkappa - 12(\sqrt{2}+1)}{4\,\varkappa + 12(\sqrt{2}+2)}\,F.$$

The forces in the bars follow from $S_i = S_i^{(0)} + X\,\bar{S}_i$:

$$\underline{\underline{S_1}} = \sqrt{2}\,F + \frac{\varkappa - 12(\sqrt{2}+1)}{4\,\varkappa + 12(\sqrt{2}+2)}\sqrt{2}\,F = \underline{\underline{\frac{5\sqrt{2}\,\varkappa + 12(\sqrt{2}+1)}{4\,\varkappa + 12(\sqrt{2}+2)}\,F}},$$

$$\underline{\underline{S_2}} = -\,F + \frac{\varkappa - 12(\sqrt{2}+1)}{4\,\varkappa + 12(\sqrt{2}+2)}(-2)\,F = \underline{\underline{\frac{-\,6\,\varkappa + 12\sqrt{2}}{4\,\varkappa + 12(\sqrt{2}+2)}\,F}}.$$

The bending moment is given by $M = M_0 + X\,\bar{M}_1$ and is displayed in Fig. 6.22d.

Frequently, the stiffness parameter $\varkappa$ is a large number. This is the case, for example, if the beams and the bars are made of the same material and their cross sectional areas are roughly the same. Then, $\varkappa \sim (a/r_g)^2$. Since the length a is much larger than the radius of gyration r_g, we have $\varkappa \gg 1$. In such a case we use $\varkappa \to \infty$ to obtain

$$C_H = \frac{F}{4}, \qquad C_V = \frac{9}{4}F, \qquad S_1 = \frac{5}{4}\sqrt{2}\,F, \qquad S_2 = -\frac{3}{2}F.$$

These are the support reactions and the forces in the bars if the bars are considered to be rigid ($EA \to \infty$).

E6.11

Example 6.11 The beam (flexural rigidity EI) in Fig. 6.23a is subjected to a moment M_D and a constant line load q_0.

Calculate the moment M_A at the clamped end A.

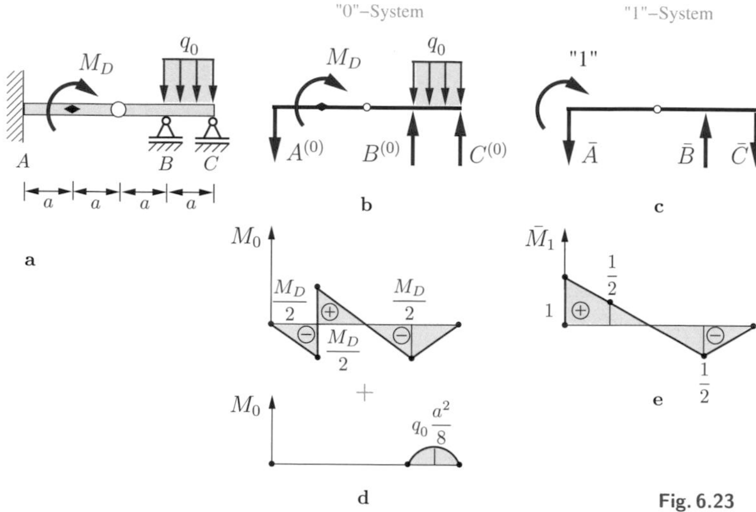

Fig. 6.23

Solution The beam is statically indeterminate to the first degree (see Volume 1, Section 5.3.3). To obtain a statically determinate "0"-system, we remove the clamping and replace it with a hinged support. The equilibrium conditions yield the support reactions (Fig. 6.23b)

$$A^{(0)} = \frac{M_D}{2\,a}, \qquad B^{(0)} = \frac{M_D}{a} + \frac{1}{2}q_0\,a, \qquad C^{(0)} = -\frac{M_D}{2\,a} + \frac{1}{2}q_0\,a.$$

The corresponding bending moment M_0 is displayed in Fig. 6.23d, where the moments caused by M_D and q_0 are given in separate graphs.

The moment “1“ in the “1“-system (Fig. 6.23c) causes the support reactions

$$\bar{A} = \frac{1}{2\,a}, \qquad \bar{B} = \frac{1}{a}, \qquad \bar{C} = \frac{1}{2\,a}.$$

The corresponding bending moment $\bar{M}_1$ is shown in Fig. 6.23e.

To determine the unknown moment M_A we apply (6.44): $X = -\alpha_{10}/\alpha_{11}$. The coefficients α_{ik} are calculated with the aid of Table 6.3. This gives

∘ due to q_0 (parabola and triangle):

$$EI\,\alpha_{10q} = \int \bar{M}_1\,M_{0q}\,\mathrm{d}x = \frac{1}{3}\,a\left(-\frac{1}{2}\right)\frac{q_0\,a^2}{8} = -\,\frac{1}{48}\,q_0\,a^3,$$

∘ due to M_D (triangle and trapezium, triangles and triangles):

$$EI\,\alpha_{10M} = \int \bar{M}_1\,M_{0M}\,\mathrm{d}x = \frac{1}{6}\,a\left(-\frac{M_D}{2}\right)\left(1+2\cdot\frac{1}{2}\right)$$

$$+\frac{1}{3}\,a\frac{M_D}{2}\frac{1}{2} + \frac{1}{3}\,a\left(-\frac{M_D}{2}\right)\left(-\frac{1}{2}\right)$$

$$+\frac{1}{3}\,a\left(-\frac{M_D}{2}\right)\left(-\frac{1}{2}\right) = \frac{1}{12}\,M_D\,a,$$

∘ due to “1“ (triangles and triangles):

$$EI\,\alpha_{11} = \int \bar{M}_1^2\,\mathrm{d}x = \frac{1}{3}\cdot 1\cdot 1\cdot 2\,a + \frac{1}{3}\cdot\frac{1}{2}\cdot\frac{1}{2}\cdot a$$

$$+\frac{1}{3}\cdot\frac{1}{2}\cdot\frac{1}{2}\cdot a = \frac{5}{6}\,a.$$

With $\alpha_{10} = \alpha_{10q} + \alpha_{10M}$ we obtain

$$X = \underline{\underline{M_A}} = -\,\frac{\alpha_{10}}{\alpha_{11}} = -\,\frac{\dfrac{-\,q_0\,a^3}{48} + \dfrac{1}{12}M_D\,a}{\dfrac{5}{6}\,a} = \underline{\underline{\frac{q_0\,a^2}{40} - \frac{M_D}{10}}}.$$

E6.12 **Example 6.12** Determine the support reactions for the frame (flexural rigidity EI) in Fig. 6.24a.

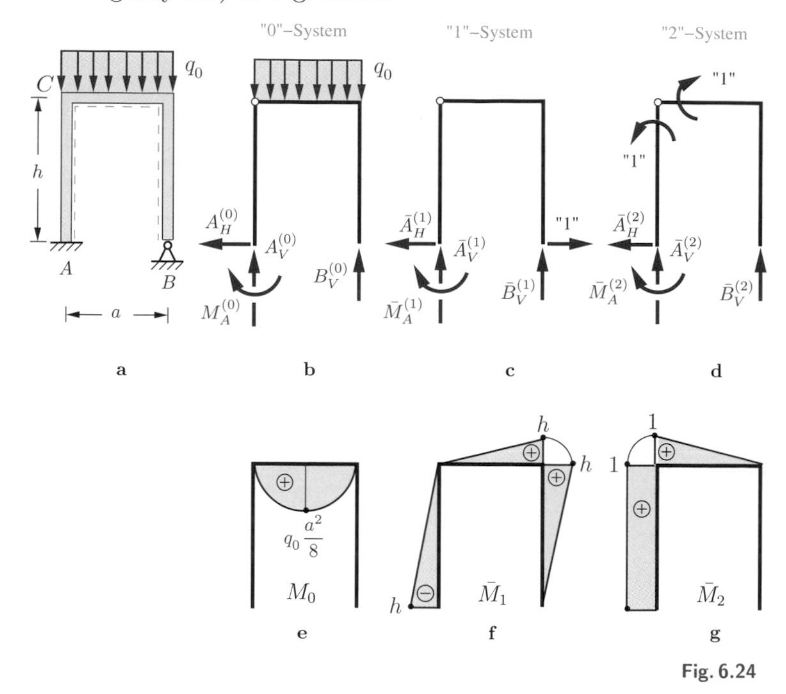

Fig. 6.24

Solution The frame has five support reactions (clamping and hinged support). Therefore it is statically indeterminate to the second degree. In order to obtain a statically determinate "0"-system, we replace the hinged support B with a roller support (which can move in the horizontal direction) and we introduce a hinge at the corner C.

The support reactions in the "0"-system (Fig. 6.24b) are obtained as

$$A_V^{(0)} = B_V^{(0)} = \frac{q_0\, a}{2}, \qquad M_A^{(0)} = 0, \qquad A_H^{(0)} = 0.$$

The corresponding bending moment M_0 is depicted in Fig. 6.24e.

Since we have two redundancies, we need two auxiliary systems. In the "1"-system (Fig. 6.24c) we apply the horizontal force "1" at point B. The support reactions

$$\bar{A}_V^{(1)} = -\,\bar{B}_V^{(1)} = \frac{h}{a}, \qquad \bar{M}_A^{(1)} = -\,h, \qquad \bar{A}_H^{(1)} = 1$$

lead to the bending moment $\bar{M}_1$ which is displayed in Fig. 6.24f. The virtual moment "1" (see Example 4.11) at the corner C in the "2"-system (Fig. 6.24d) causes the support reactions

$$\bar{A}_V^{(2)} = -\,\bar{B}_V^{(2)} = -\,\frac{1}{a}, \quad \bar{M}_A^{(2)} = 1, \quad \bar{A}_H^{(2)} = 0$$

and the bending moment $\bar{M}_2$ (Fig. 6.24g).

The two redundant reactions $X_1 = B_H$ and $X_2 = M_C$ can be determined from two compatibility conditions:

a) the horizontal displacement w_B at B has to be zero,
b) the right angle at C must remain unchanged ($\Delta\, w_C' = 0$).

These conditions are written as (see (6.48))

$$\begin{aligned} w_B &= \alpha_{10} + X_1\,\alpha_{11} + X_2\,\alpha_{12} = 0, \\ \Delta w_C' &= \alpha_{20} + X_1\,\alpha_{21} + X_2\,\alpha_{22} = 0. \end{aligned}$$

We calculate the coefficients α_{ik} with the aid of Table 6.3:

$$EI\,\alpha_{10} = \int \bar{M}_1\, M_0\, \mathrm{d}x = \frac{1}{3}a\,h\,\frac{q_0\,a^2}{8} = \frac{1}{24}q_0\,a^3\,h,$$

$$EI\,\alpha_{20} = \int \bar{M}_2\, M_0\, \mathrm{d}x = \frac{1}{3}a\,\frac{q_0\,a^2}{8} = \frac{1}{24}q_0\,a^3,$$

$$EI\,\alpha_{11} = \int \bar{M}_1^2\, \mathrm{d}x = \frac{1}{3}(h\cdot h^2 + a\,h^2 + h\cdot h^2) = \frac{h^2}{3}(2\,h + a),$$

$$EI\,\alpha_{22} = \int \bar{M}_2^2\, \mathrm{d}x = h + \frac{1}{3}a,$$

$$\begin{aligned} EI\,\alpha_{12} &= \int \bar{M}_1\, \bar{M}_2\, \mathrm{d}x = \frac{1}{2}(-h)h + \frac{1}{6}a\,h \\ &= \frac{1}{6}h(a - 3\,h) = EI\,\alpha_{21}. \end{aligned}$$

Thus, we obtain the system of equations

$$\frac{1}{24} q_0\, a^3\, h + X_1 \frac{h^2}{3}(2\,h + a) + X_2 \frac{1}{6} h(a - 3\,h) = 0,$$

$$\frac{1}{24} q_0\, a^3 + X_1 \frac{1}{6} h(a - 3\,h) + X_2 (h + \frac{1}{3} a) = 0$$

which has the solution

$$X_1 = B_H = -\frac{1}{4} q_0\, a^3 \frac{9\,h + a}{15\,h^3 + 26\,a\,h^2 + 3\,h\,a^2},$$

$$X_2 = M_C = -\frac{1}{4} q_0\, a^3 \frac{7\,h + a}{15\,h^2 + 26\,a\,h + 3\,a^2}.$$

The superposition of the three systems yields ($A_H^{(0)} = 0$, $M_A^{(0)} = 0$, $A_H^{(2)} = 0$)

$$\underline{\underline{A_V}} = A_V^{(0)} + X_1\,\bar{A}_V^{(1)} + X_2\,\bar{A}_V^{(2)} = \underline{\underline{\frac{15\,h^2 + 25\,a\,h + 3\,a^2}{15\,h^2 + 26\,a\,h + 3\,a^2} \frac{q_0\, a}{2}}},$$

$$\underline{\underline{A_H}} = X_1\,\bar{A}_H^{(1)} = \underline{\underline{-\frac{1}{4} \frac{9\,h + a}{15\,h^3 + 26\,a\,h^2 + 3\,h\,a^2}}} q_0\, a^3 = \underline{\underline{B_H}},$$

$$\underline{\underline{M_A}} = X_1\,\bar{M}_A^{(1)} + X_2\,\bar{M}_A^{(2)} = \underline{\underline{\frac{1}{4} \frac{2\,h}{15\,h^2 + 26\,a\,h + 3\,a^2} q_0\, a^3}},$$

$$\underline{\underline{B_V}} = B_V^{(0)} + X_1\,\bar{B}_V^{(1)} + X_2\,\bar{B}_V^{(2)} = \underline{\underline{\frac{15\,h^2 + 27\,a\,h + 3\,a^2}{15\,h^2 + 26\,a\,h + 3\,a^2} \frac{q_0\, a}{2}}}.$$

6.6 Supplementary Examples

Detailed solutions to the following examples are given in (**A**) D. Gross et al. *Formeln und Aufgaben zur Technischen Mechanik 2*, Springer, Berlin 2010, or (**B**) W. Hauger et al. *Aufgaben zur Technischen Mechanik 1-3*, Springer, Berlin 2008.

E6.13

Example 6.13 The truss in Fig. 6.25 consists of five bars (axial rigidity EA) of equal length l.

Determine the forces in the bars that are caused by the external load F.

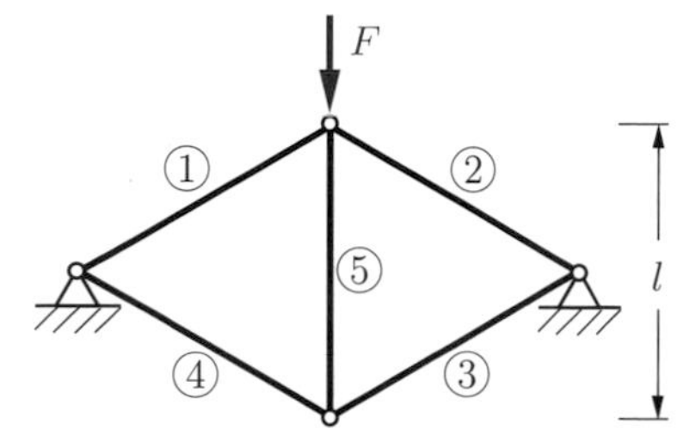

Fig. 6.25

Results: see (**B**) $S_1 = S_2 = -\frac{3F}{5}$, $S_3 = S_4 = \frac{2F}{5}$, $S_5 = -\frac{2F}{5}$.

E6.14

Example 6.14 The structure shown in Fig. 6.26 consists of six elastic bars (axial rigidity EA) of negligible weight and a rigid beam (weight W).

Calculate the support reaction at B due to the weight W.

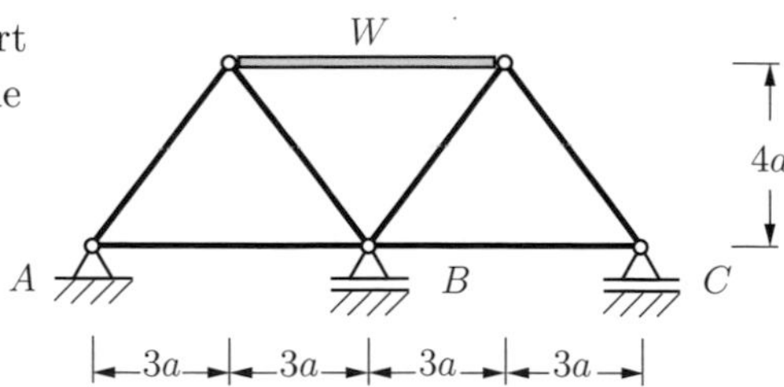

Fig. 6.26

Result: see (**B**) $B = \frac{179W}{304}$.

E6.15

Example 6.15 The structure in Fig. 6.27 consists of a beam (axial rigidity $EA \to \infty$, flexural rigidity EI) and two bars (axial rigidity EA). It is subjected to a force F.

Determine the displacement of the point of application of F.

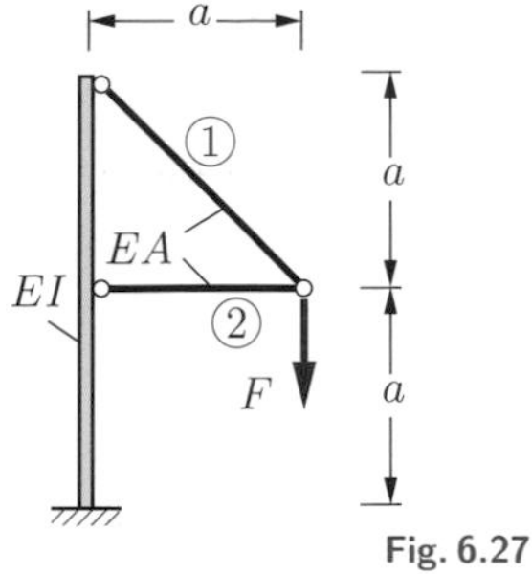

Fig. 6.27

Results: see (**A**) $f_H = \dfrac{Fa^3}{2EI} - \dfrac{Fa}{EA}, \; f_V = \dfrac{4}{3}\dfrac{Fa^3}{EI} + \dfrac{(1+2\sqrt{2})\,Fa}{EA}$.

E6.16

Example 6.16 The truss in Fig. 6.28 is subjected to a force F. The members of the truss have axial rigidity EA.

Determine the magnitude of F so that the vertical displacement of the point of application of F has the given value f_0.

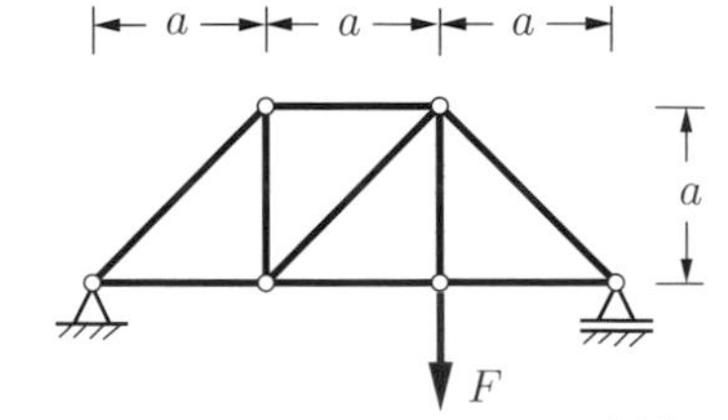

Fig. 6.28

Result: see (**A**) $F = \dfrac{9EAf_0}{4\,(5+3\sqrt{2})\,a}$.

E6.17

Example 6.17 Determine the support reaction B of the beam (flexural rigidity EI) and the angle of slope φ_B due to the applied moment M_A (Fig. 6.29).

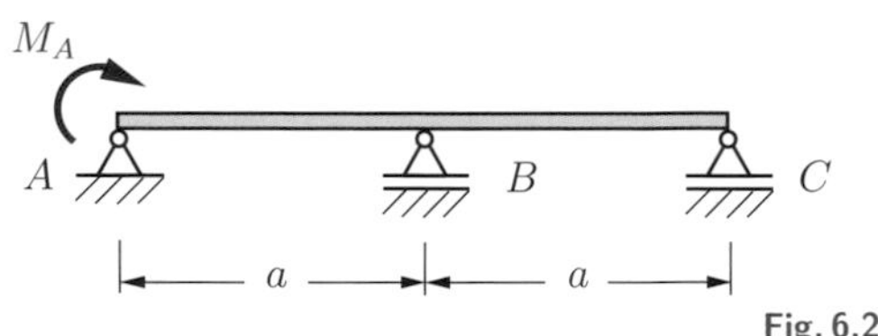

Fig. 6.29

Results: see (**B**) $B = \dfrac{3M_A}{2a}, \quad \varphi_B = -\dfrac{M_A a}{12EI}$.

E6.18

Example 6.18 A circular arch is subjected to a force F as shown in Fig. 6.30.

Determine the displacement of the point of application of the force due to bending.

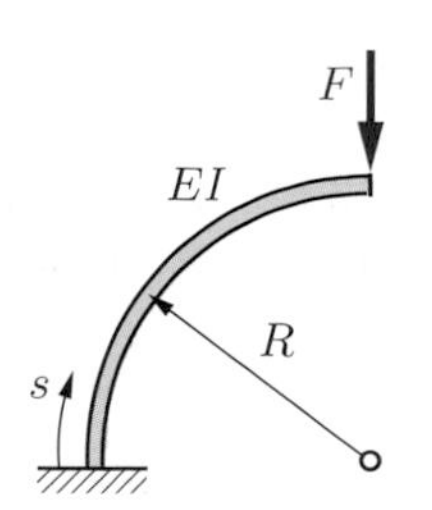

Fig. 6.30

Results: see (**A**) $f_H = \dfrac{FR^3}{4EI}, \quad f_V = \dfrac{\pi FR^3}{4EI}$.

E6.19

Example 6.19 The two beams (modulus of elasticity E) of the frame shown in Fig. 6.31 have rectangular cross sections with constant width b. The depth h is constant ($h = h_0$) in region AB, whereas in region BC it has a linear taper ($h = h(x)$). A constant line load q_0 acts in region BC.

Calculate the vertical displacement w_C of point C. Neglect axial deformations.

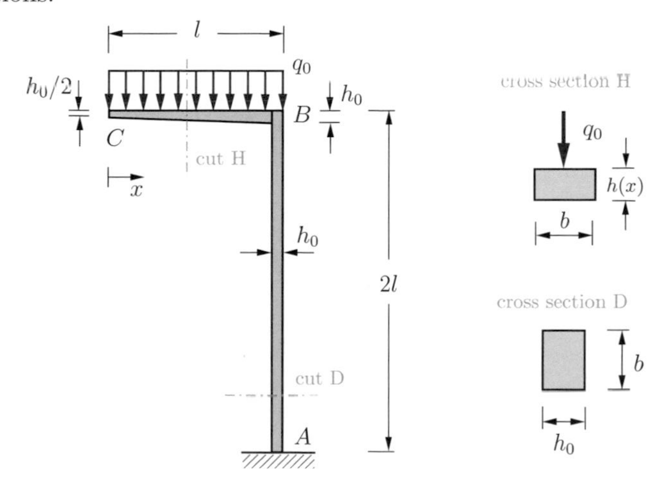

Fig. 6.31

Result: see (**B**) $w_C = 1.2\dfrac{q_0 l^4}{EI_0}$.

E6.20

Example 6.20 A rectangular frame (flexural rigidity EI, axial rigidity $\to \infty$) is subjected to a uniform line load q_0 (Fig. 6.32).

Determine the bending moment in the frame.

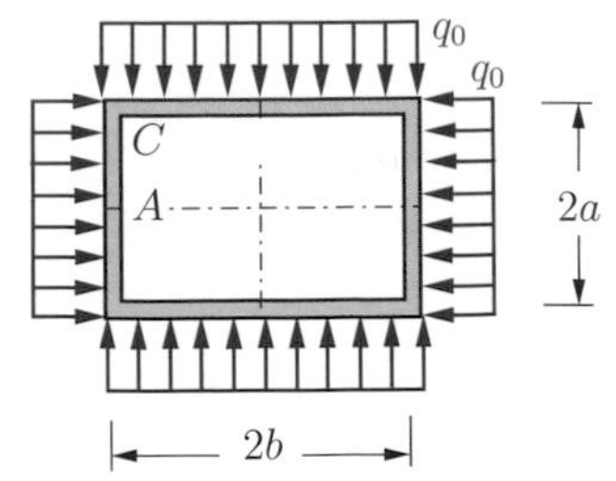

Fig. 6.32

Results: see (**B**) Selected values:

$$M_A = \left(a^2 + 2ab - 2b^2\right) q_0/6, \; M_C = \left(a^2 - ab + b^2\right) q_0/3 .$$

E6.21

Example 6.21 The assembly shown in Fig. 6.33 consists of a frame (flexural rigidity EI, torsional rigidity $GI_T = 3EI/4$, negligible weight) and a wheel (weight $W_1 = W$, radius $r = a/4$). The wheel is attached in a fixed manner to the frame at point C. A rope which is wrapped around the wheel carries a barrel (weight $W_2 = 8W$).

Determine the support reactions and the vertical displacement of point C due to the load.

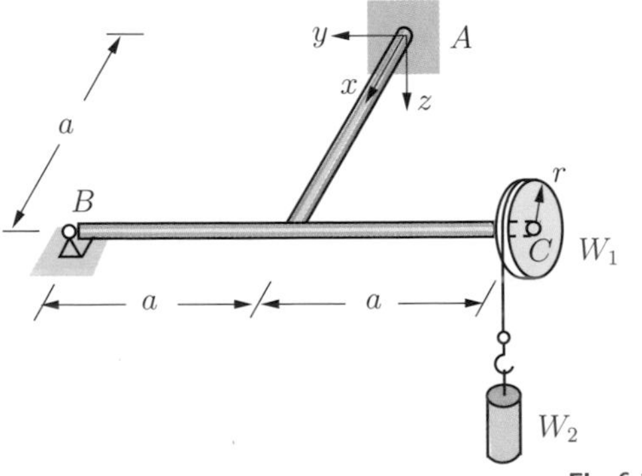

Fig. 6.33

Results: see (**B**)

$$A_z = -13W, \quad B = 4W, \quad M_{Ax} = 5Wa, \quad M_{Ay} = 15Wa,$$

$$w_C = 15\frac{Wa^3}{EI} .$$

E6.22

Example 6.22 The structure shown in Fig. 6.34 consists of a beam (flexural rigidity EI) and three bars (axial rigidity EA). It is subjected to a force F.

Find the vertical displacement w of the point of application of F.

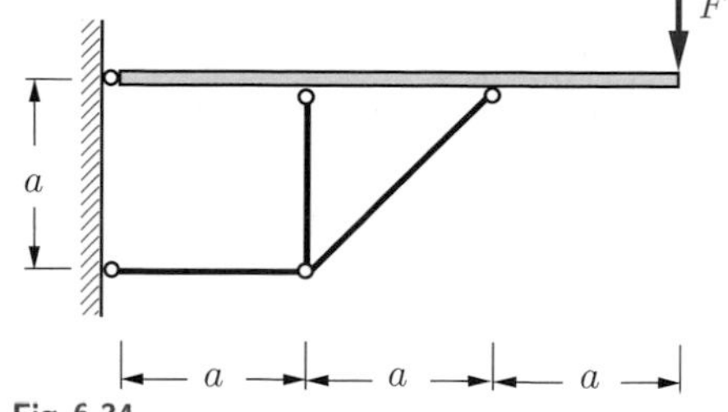

Fig. 6.34

Result: see (**B**) $w = 18\left(1+\sqrt{2}\right)\dfrac{Fa}{EA} + \dfrac{Fa^3}{EI}\,.$

E6.23

Example 6.23 The structure in Fig. 6.35 consists of a frame (axial rigidity $EA \to \infty$, flexural rigidity EI) and a bar (axial rigidity EA). It is subjected to a uniform line load q_0.

Determine the force S in the bar.

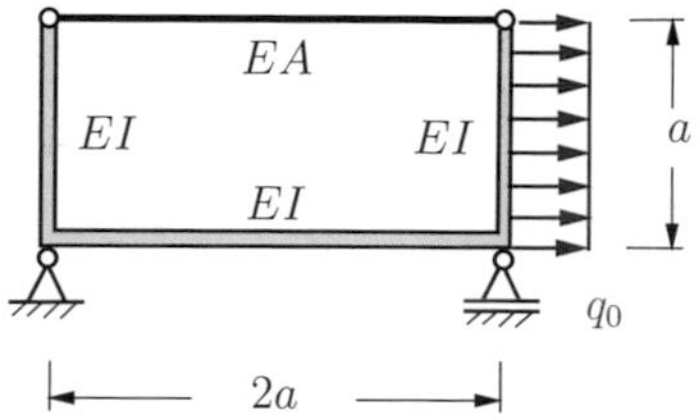

Fig. 6.35

Result: see (**A**) $S = \dfrac{15}{64}\left(1 + \dfrac{3EI}{4EAa^2}\right)^{-1} q_0 a\,.$

E6.24

Example 6.24 A rope S (axial rigidity EA_2) is attached to a cantilever beam (flexural rigidity EI_1) as shown in Fig. 6.36.

Calculate the force S in the rope due to an applied force F at the free end of the cantilever. Disregard axial deformations of the beam.

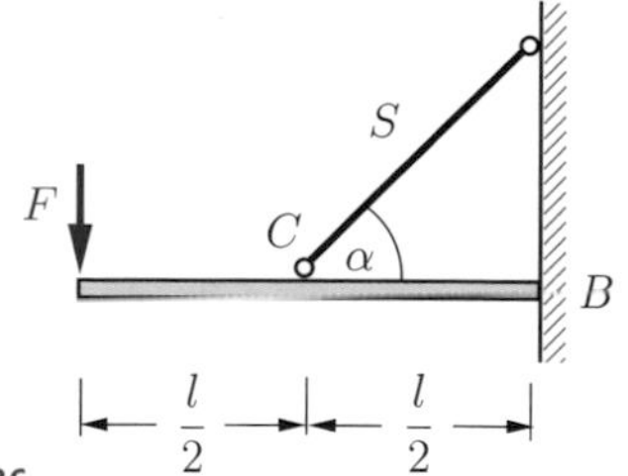

Fig. 6.36

Result: see (**B**) $S = \dfrac{5F\sin\alpha\cos\alpha}{2\sin^2\alpha\cos\alpha + 24\dfrac{(EI)_1}{l^2(EA)_2}}\,.$

E6.25

Example 6.25 The continuous beam in Fig. 6.37 is subjected to a force F at point G.

a) Determine the deflections f_D and f_G at points D and G.
b) Calculate the total deflection f at point G if, in addition to force F, a force $2F$ acts at point D.

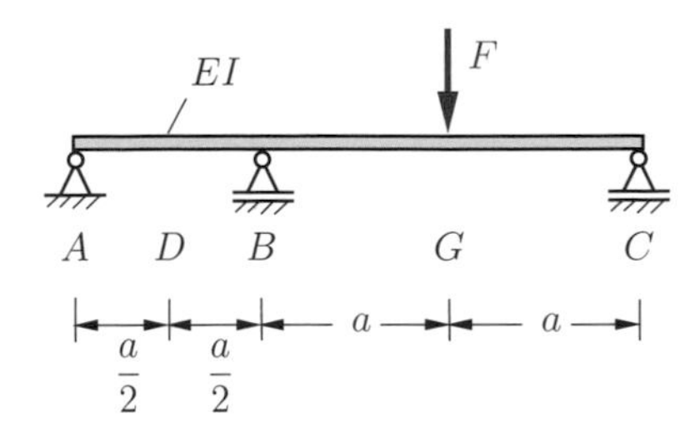

Fig. 6.37

Results: see (**A**) a) $f_D = -\dfrac{1}{64}\dfrac{Fa^3}{EI}$, $f_G = \dfrac{5}{48}\dfrac{Fa^3}{EI}$,

b) $f = \dfrac{7}{96}\dfrac{Fa^3}{EI}$.

E6.26

Example 6.26 Determine the forces S_i in the members (axial rigidity EA) of the truss shown in Fig. 6.38. Calculate the vertical displacement f_F of the point of application of the force F.

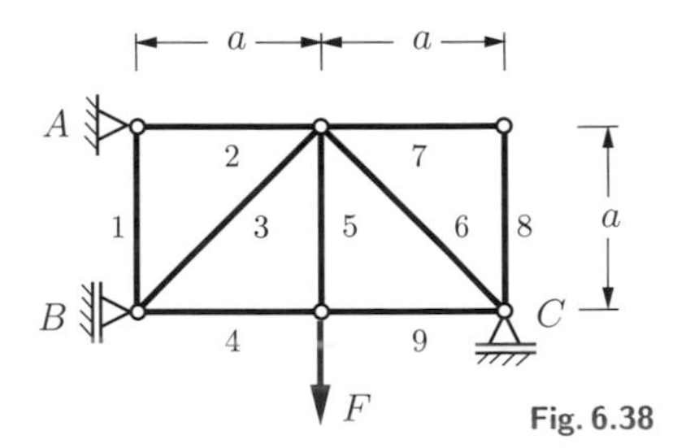

Fig. 6.38

Results: see (**A**) Selected values:

$$S_1 = \frac{4+2\sqrt{2}}{7+4\sqrt{2}}\,F,\; S_2 = \frac{1}{7+4\sqrt{2}}\,F,\; S_3 = -\frac{4+4\sqrt{2}}{7+4\sqrt{2}}\,F,$$

$$f_F = \frac{20+14\sqrt{2}}{7+4\sqrt{2}}\frac{Fa}{EA}.$$

E6.27

Example 6.27 The beam in Fig. 6.39 (flexural rigidity EI) is pin-supported at its left end and suspended by a rope (axial rigidity EA). It is subjected to a force F.

Determine the vertical displacement f of the point of application of F. Disregard axial deformations of the beam.

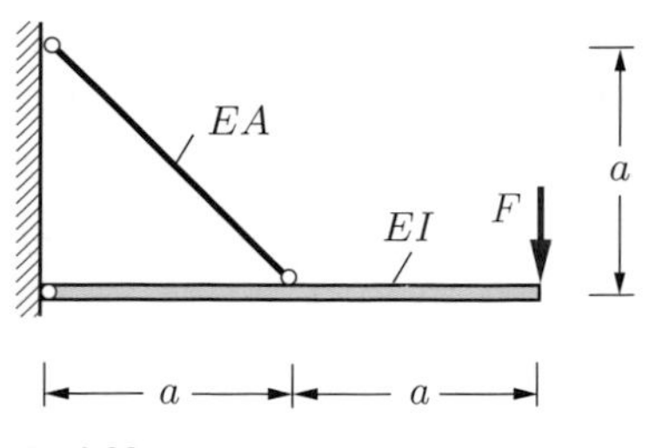

Fig. 6.39

Results: see (**A**) $f = \dfrac{2}{3}\dfrac{Fa^3}{EI} + 8\sqrt{2}\dfrac{Fa}{EA}$.

6.7

6.7 Summary

- Principle of conservation of energy

$$U_e = U_i\,,$$

$U_e = \dfrac{1}{2}\,Ff$ work of a force F acting on a linear-elastic bar/beam (analogous relations in the case of an applied moment),

$U_i = \dfrac{1}{2}\displaystyle\int \frac{M^2}{EI}\,\mathrm{d}x$ strain energy in bending (analogous relations for torsion, tension/compression).

- Principle of virtual forces (Unit load method)

  ◇ Statically determinate beam under bending (analogous relations for torsion, tension/compression):

$$f = \int \frac{M\bar{M}}{EI}\,\mathrm{d}x\,,$$

M bending moment due to the applied load,
$\bar{M}$ bending moment due to a virtual force (moment) “1“.

Special case truss: $f = \sum \dfrac{S_i \bar{S}_i l_i}{EA_i}\,.$

  ◇ Determination of the redundant X of a beam being statically indeterminate to the first degree:

$$X = -\frac{\alpha_{10}}{\alpha_{11}}\,, \qquad \alpha_{10} = \int \frac{\bar{M}_1 M_0}{EI}\,\mathrm{d}x\,, \quad \alpha_{11} = \int \frac{\bar{M}_1^2}{EI}\,\mathrm{d}x\,.$$

M_0 bending moment in the “0“-system,
$\bar{M}_1$ bending moment in the “1“-system.
If a system is subjected to bending, torsion and tension/compres the appropriate terms have to be considered.

- Influence coefficients

  ◇ α_{ik} displacement at x due to a load “1“ at k.

  ◇ Reciprocal displacement theorem

$$\alpha_{ik} = \alpha_{ki}\,.$$

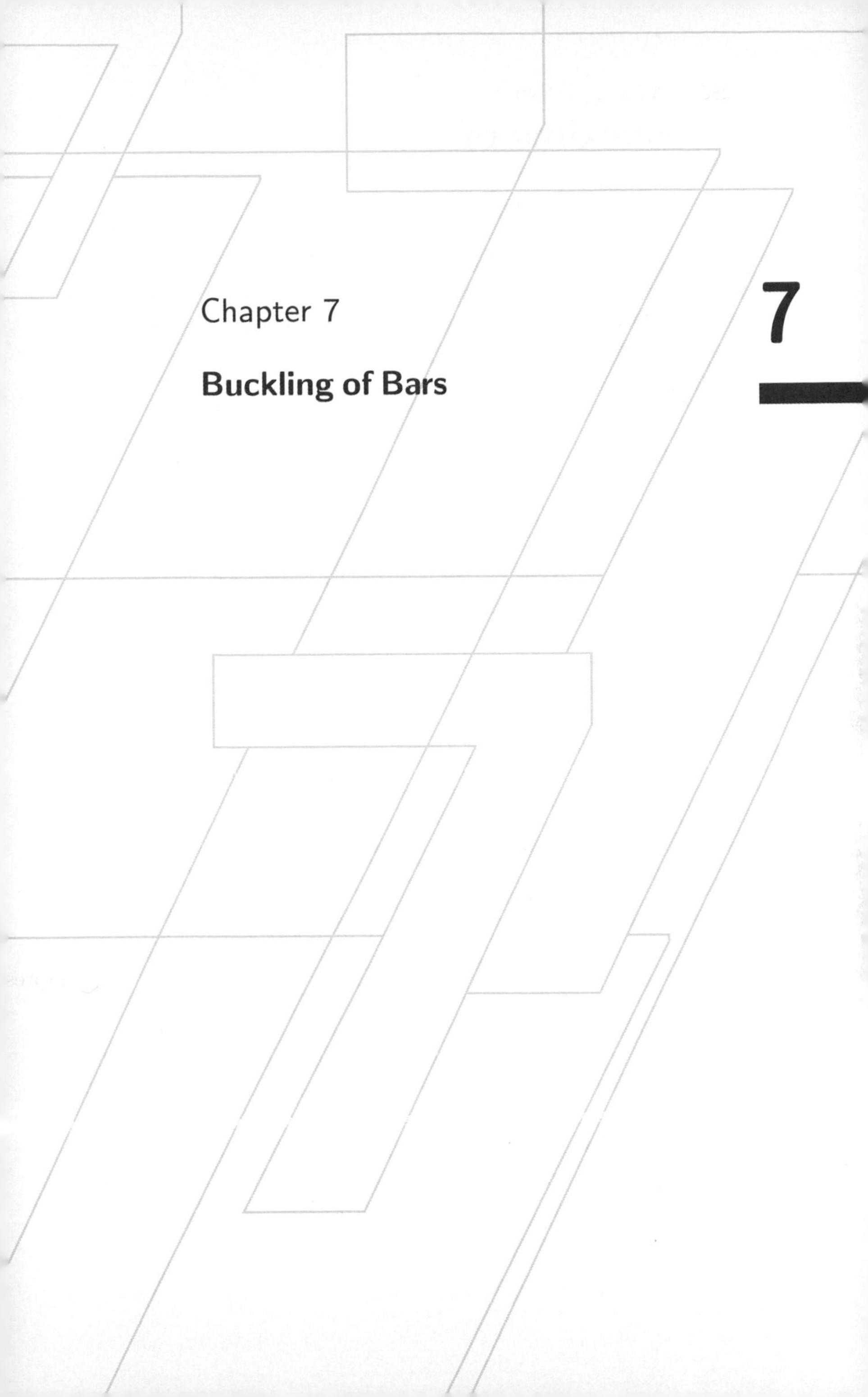

Chapter 7

Buckling of Bars

7 Buckling of Bars

7.1	Bifurcation of an Equilibrium State	**289**
7.2	Critical Loads of Bars, Euler's Column	**292**
7.3	Supplementary Examples	**302**
7.4	Summary	**305**

Objectives: In this chapter we analyse the stability of equilibrium positions of bars under compressive loads. Methods are presented which will enable us to determine the so-called *critical load* under which a bar deflects due to buckling. The aim is to enable students to apply the appropriate methods in order to determine critical loads.

7.1 Bifurcation of an Equilibrium State

If a bar is subjected to a tensile force, the relationship between the external load and the elongation of the bar is unique (cf. (1.18)). Thereby, the deformations are so small that the equilibrium conditions can be formulated with respect to the *undeformed* system. In contrast, a bar under a compressive force need not lead to a unique relationship between the load and the deformation. At a certain value of the compressive force, further equilibrium states emerge which are associated with lateral deflections. This phenomenon, which is especially observed in slender bars, is called *buckling*. In the following sections, we will determine the associated critical loads. In such problems, the equilibrium conditions have to be formulated with respect to the *deformed* configuration.

As an introduction to buckling, we will first consider a *rigid* rod on an elastic support. In Volume 1, Example 8.8, we already investigated the behavior of such a rod, held on each side by a spring, and we found that it may possess several equilibrium positions for the same compressive force. Now we consider a pin supported rigid column, held by an elastic torsion spring (stiffness k_T), which is subjected to a compressive force F (Fig. 7.1a). Since we assume that the force remains vertical during a lateral deflection, the force is said to be *conservative*, see Fig. 7.1b.

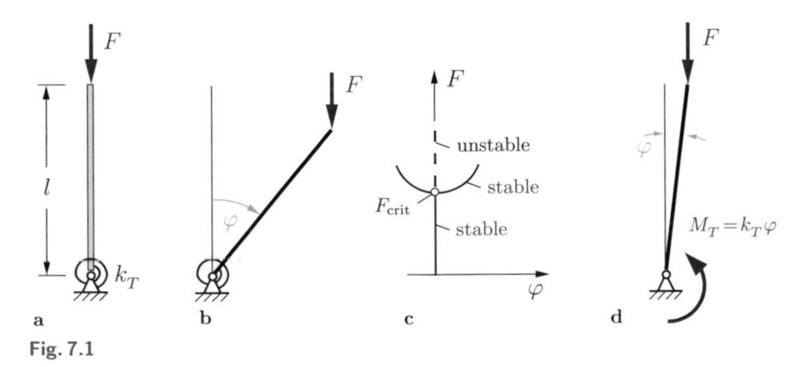

Fig. 7.1

In order to determine the equilibrium positions and to analyse their stability we use the total potential of the system. Choosing the zero-level of the potential of F at the height of the pin support, the potential of the system, rotated about φ (Fig. 7.1b), is given by

$$V = F\,l\,\cos\varphi + \frac{1}{2}k_T\,\varphi^2.$$

The equilibrium positions can be determined from (see Volume 1, Equation (8.13))

$$V' = \frac{\mathrm{d}V}{\mathrm{d}\varphi} = 0 \qquad \rightarrow \qquad -F\,l\,\sin\varphi + k_T\,\varphi = 0. \tag{7.1}$$

This equation is always fulfilled for $\varphi = 0$, i.e. regardless of the values of the parameters F, l and k_T we obtain as the first equilibrium state the vertical position which is characterized by no lateral displacement, that is

$$\varphi_1 = 0. \tag{7.2}$$

According to (7.1), a second equilibrium position follows from

$$\frac{\varphi_2}{\sin\varphi_2} = \frac{F\,l}{k_T}. \tag{7.3}$$

If $\varphi_2 \neq 0$ we have $\varphi_2/\sin\varphi_2 > 1$. Thus, a deformed position φ_2 can only occur for $F\,l/k_T > 1$. For $F\,l/k_T = 1$ we get $\sin\varphi_2 = \varphi_2 = 0$, i.e. both equilibrium states coincide.

In order to analyse the stability of the equilibrium positions we determine the second derivative of the total potential:

$$V'' = \frac{\mathrm{d}^2V}{\mathrm{d}\varphi^2} = -F\,l\,\cos\varphi + k_T. \tag{7.4}$$

First, we introduce the solution $\varphi_1 = 0$ of the first equilibrium position into (7.4):

$$V''(\varphi_1) = -F\,l + k_T = k_T\left(1 - \frac{F\,l}{k_T}\right).$$

The algebraic sign of V'' and therefore the stability of this equilibrium state depends on the algebraic sign of the bracketed term.

Therefore,

$$V''(\varphi_1) > 0 \qquad \text{for} \qquad \frac{F\,l}{k_T} < 1 \qquad \rightarrow \qquad \text{stable position,}$$

$$V''(\varphi_1) < 0 \qquad \text{for} \qquad \frac{F\,l}{k_T} > 1 \qquad \rightarrow \qquad \text{unstable position.}$$

Inserting the second equilibrium position φ_2 from (7.3) into (7.4), we obtain

$$V''(\varphi_2) = -\,F\,l\,\cos\varphi_2 + k_T = k_T\left(1 - \frac{\varphi_2}{\tan\varphi_2}\right).$$

Since $\varphi_2/\tan\varphi_2 < 1$, the condition $V''(\varphi_2) > 0$ always holds: the second equilibrium position is always stable.

The special case $F\,l/k_T = 1$ (corresponding to the angle $\varphi_2 = \varphi_1 = 0$) characterizes the *critical load*:

$$F_{\text{crit}} = \frac{k_T}{l}. \tag{7.5}$$

Let us summarise the results: when the rod is loaded by a sufficiently small force F, it remains in its original vertical position $\varphi_1 = 0$ (Fig. 7.1c). If we increase the load to the value F_{crit}, see (7.5), a bifurcation into a second equilibrium position φ_2 takes place. A further increase of the load leads to increasing deflections φ_2 and we obtain three possible positions for $F > F_{\text{crit}}$: an unstable position $\varphi_1 = 0$ and two stable positions $\pm\varphi_2$ (since $\varphi_2/\sin\varphi_2$ is an even function, (7.3) has the second solution $-\varphi_2$ in addition to φ_2). Usually, in engineering applications only the critical load is of interest since the large deflections that occur after F_{crit} is exceeded need to be avoided in most cases.

The critical load can also be obtained directly from the equilibrium conditions (without using the potential). If the rod is subjected to F_{crit}, it may undergo a slight lateral deflection. Therefore, we consider the rod in a position which is adjacent to the original vertical position. This adjacent position $\varphi \neq 0$ is also an equilibrium state. Thus, from the moment equilibrium condition with respect to the pin support (Fig. 7.1d) we again obtain (7.5) for small values of φ (adjacent equilibrium position!):

$$F\,l\,\varphi = k_T\,\varphi \qquad \rightarrow \qquad F = F_{\text{crit}} = \frac{k_T}{l}\,.$$

This procedure can be generalized. If we want to determine the critical load of an arbitrary structure, we have to consider the originally stable equilibrium position and an adjacent configuration. If the adjacent configuration is an equilibrium position, the associated load is the critical load.

7.2

7.2 Critical Loads of Bars, Euler's Column

In the previous section we considered a rigid rod. We now want to analyse an *elastic* rod which can deform due to its elasticity. Our first example will be a bar with a roller support on the left and a pin support on the right end, subjected to an applied compressive load F, see Fig. 7.2a. We assume that the unloaded bar is perfect-

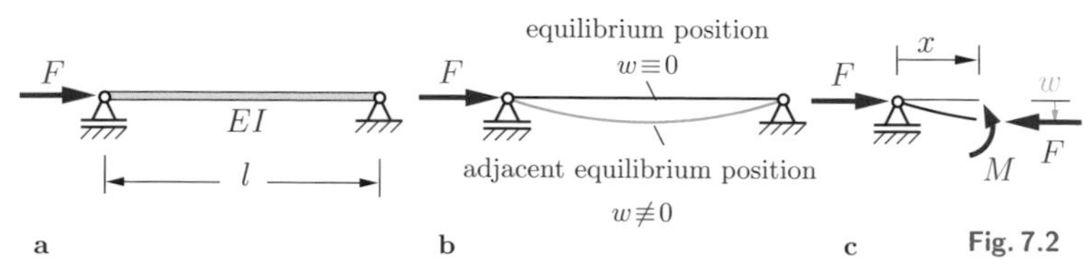

Fig. 7.2

ly straight and that the line of action of the external force passes through the centroid of the cross section. If the bar is subjected to the critical load F_{crit}, there exists (in addition to the undeformed configuration $w \equiv 0$) an adjacent equilibrium position with a lateral deflection $w \not\equiv 0$ (Fig. 7.2b). In order to determine F_{crit} we must formulate the equilibrium conditions for the deflected position, i.e. based upon the geometry of the deformed system. Thereby, we can neglect the axial deformation of the bar.

Let us now pass an imaginary section perpendicular to the axis of the deformed bar at a point x (Fig. 7.2c). Evaluating the moment equilibrium condition at the deflected position we obtain

$$M = F\,w. \tag{7.6}$$

Note that there exists no vertical force in the roller support under the horizontal force F.

Inserting (7.6) into the differential equation $EI\,w'' = -M$ of the deflection curve for the Euler-Bernoulli beam yields

$$EI\,w'' = -\,F\,w \qquad \text{or} \qquad EI\,w'' + F\,w = 0\,. \tag{7.7a}$$

With the abbreviation

$$\lambda^2 = F/EI$$

we obtain the *differential equation for the critical load*

$$w'' + \lambda^2\,w = 0. \tag{7.7b}$$

This is a homogeneous, second-order, linear differential equation with constant coefficients. Its general solution is

$$w = A\cos\,\lambda\,x + B\,\sin\,\lambda\,x. \tag{7.8}$$

The two constants of integration A and B can be determined from the boundary conditions. The deflections are zero at the supports ($x = 0,\, l$):

$$\begin{aligned} w(0) &= 0 \qquad \rightarrow \qquad A = 0, \\ w(l) &= 0 \qquad \rightarrow \qquad B\sin\,\lambda\,l = 0. \end{aligned}$$

In addition to the trivial solution $B = 0$ (no deflection), the second equation has the solution

$$\sin\,\lambda\,l = 0 \quad \rightarrow \quad \lambda_n\,l = n\,\pi \quad \text{with} \quad n = 1, 2, 3, \ldots\,. \tag{7.9}$$

Thus, we get an infinite number of values λ_n and the corresponding forces F_n for which non-trivial (i.e., deflected) equilibrium positions exist. The quantities λ_n are called the *eigenvalues* of the problem. Note that for $n = 0$ we have $\lambda = 0$ and hence $F = 0$. This solution is of no interest.

The only eigenvalue of technical interest is the smallest one, namely λ_1. The corresponding force F_1 causes the bar to buckle.

We therefore find the *critical load* F_{crit} from $\lambda_1 l = \pi$ as

$$F_{\text{crit}} = \lambda_1^2 \, EI = \pi^2 \frac{EI}{l^2}. \tag{7.10}$$

According to (7.8) the associated *buckled shape*, also called the *mode shape*, is

$$w_1 = B \sin \lambda_1 x = B \sin \pi \frac{x}{l}$$

since $A = 0$. When the bar buckles it takes the shape of a sinus half-wave, where the amplitude B remains indeterminate. This solution is also denoted as *eigenmode* or *eigenfunction*.

When the critical force is exceeded, the theory of small deflections is no longer sufficient to calculate the deflection of the bar. We must then apply a *higher order theory* (cf. Volume 4, Chapter 5.4.1). Within this basic course, however, we will not go into further detail.

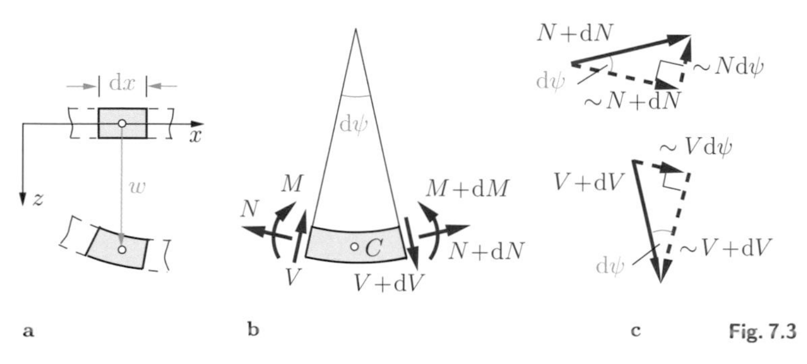

Fig. 7.3

We can describe the buckling of a bar with pinned ends by means of the differential equation (7.7a) and its solution (7.8). In order to determine the critical loads of bars with arbitrary supports we have to derive a more general differential equation. Here we have to take into account that shear forces may occur. Fig. 7.3b shows the free-body diagram of a beam element of infinitesimal length $\mathrm{d}x$ (Fig. 7.3a) in a buckled configuration $w \neq 0$. The equilibrium conditions must be formulated with respect to the deformed configuration. For this purpose we assume that the deformations are small, in particular the slope $w' = -\psi$ of the elastic curve is small, and the length of the deformed element is approximately the same as its original length. The directions of the normal forces N and $N + \mathrm{d}N$ (and of the shear forces V and $V + \mathrm{d}V$), respectively, do not coincide at the positive and the negative face of the infinitesimal element (see Fig. 7.3b). Therefore, we have to take into account the components $N\mathrm{d}\psi$ and $V\,\mathrm{d}\psi$ (cf. Fig. 7.3c) when writing down the equilibrium conditions:

$$\begin{aligned}
\rightarrow: &\quad \mathrm{d}N + V\,\mathrm{d}\psi = 0,\\
\downarrow: &\quad \mathrm{d}V - N\,\mathrm{d}\psi = 0,\\
\overset{\frown}{C}: &\quad \mathrm{d}M - V\,\mathrm{d}x = 0.
\end{aligned}$$

After inserting the third equation into the first equation and using the differential equation (4.24) for the deflection curve of a beam, we obtain

$$\frac{\mathrm{d}N}{\mathrm{d}x} = -\,V\,\frac{\mathrm{d}\psi}{\mathrm{d}x} = -\,\frac{\mathrm{d}M}{\mathrm{d}x}\frac{\mathrm{d}\psi}{\mathrm{d}x} = -\,\frac{\mathrm{d}}{\mathrm{d}x}\left(EI\frac{\mathrm{d}\psi}{\mathrm{d}x}\right)\frac{\mathrm{d}\psi}{\mathrm{d}x}.$$

On the right-hand side we have a product of infinitesimal kinematic quantities which is "small of higher order" compared with the left-hand side. Thus, we can neglect this "higher order" term and find that the derivative of the normal force is zero: $\mathrm{d}N/\mathrm{d}x = 0$. If we take into account that a compressive force F is applied to the bar, we get

$$N = \mathrm{const} = -\,F. \tag{7.11}$$

Inserting this result into the second equilibrium condition and applying the relations $V = \mathrm{d}M/\mathrm{d}x$, $M = EI\,\mathrm{d}\psi/\mathrm{d}x$ in combination with the kinematic relation $\psi = -w'$ (cf. (4.30)) leads to the homogeneous differential equation

$$(EI\,w'')'' + F\,w'' = 0\,. \tag{7.12}$$

Note that (7.12) may also be found by taking the second derivative of (7.7a). In the case of a constant bending stiffness EI we obtain, with the abbreviation $\lambda^2 = F/EI$, the differential equation

$$w^{IV} + \lambda^2 w'' = 0 . \tag{7.13}$$

This equation is a fourth-order, linear differential equation, just as the differential equation (4.34b) for the deflection curve of a beam. The general solution of (7.13) is

$$w = A \cos \lambda x + B \sin \lambda x + C \lambda x + D. \tag{7.14}$$

Note that the factor λ in the third term is introduced for convenience. Since λx is a dimensionless quantity, the constants A to D have the same dimension $[l]$.

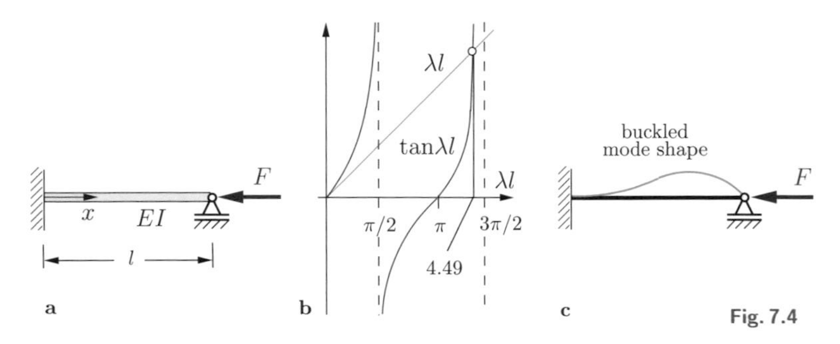

Fig. 7.4

The four constants of integration can be determined from the boundary conditions: two conditions at each boundary. In order to illustrate the procedure, let us consider a statically indeterminate bar as depicted in Fig. 7.4a. From (7.14) we obtain

$$w' = -A \lambda \sin \lambda x + B \lambda \cos \lambda x + C \lambda,$$

$$w'' = -A \lambda^2 \cos \lambda x - B \lambda^2 \sin \lambda x.$$

Applying the relationship $EI\, w'' = -M$, the boundary conditions are given by

$$\begin{aligned} w(0) = 0 \quad &\rightarrow \quad A + D = 0, \\ w'(0) = 0 \quad &\rightarrow \quad B + C = 0, \\ w(l) = 0 \quad &\rightarrow \quad A\cos\lambda l + B\sin\lambda l + C\lambda l + D = 0, \\ M(l) = 0 \quad &\rightarrow \quad A\cos\lambda l + B\sin\lambda l = 0. \end{aligned} \tag{7.15}$$

Using the first two equations, the constants C and D can be eliminated from the third equation. Thus, we obtain the system of equations

$$\begin{aligned} (\cos\lambda l - 1)A + (\sin\lambda l - \lambda l)B = 0, \\ \cos\lambda l A + \sin\lambda l \; B = 0 \end{aligned} \tag{7.16}$$

for the two unknowns A and B. This homogeneous system of equations has a non-trivial solution if the determinant Δ of the coefficient matrix vanishes:

$$\Delta = (\cos\lambda l - 1)\sin\lambda l - \cos\lambda l\,(\sin\lambda l - \lambda l) = 0,$$

that is

$$\lambda l\cos\lambda l - \sin\lambda l = 0 \qquad \rightarrow \qquad \tan\lambda l = \lambda l. \tag{7.17}$$

A graphical solution of this transcendental equation, called the *buckling equation*, is depicted in Fig. 7.4b; the smallest eigenvalue is $\lambda_1 l \approx 4.49$. Thus the critical load is given by

$$F_{\text{crit}} = \lambda_1^2 EI = (4.49)^2 \frac{EI}{l^2}. \tag{7.18}$$

Introducing (7.17) into (7.15) leads to $B = -A/\lambda l, C = -B = A/\lambda l$ and $D = -A$. Substituting these relations into (7.14) yields the corresponding buckled shape

$$w = A\left(\cos\lambda x - \frac{\sin\lambda x}{\lambda l} + \frac{x}{l} - 1\right),$$

which is depicted in Fig. 7.4c for $\lambda = \lambda_1$.

The Swiss mathematician Leonhard Euler (1707-1783) was the first scientist to analyse the buckling of columns. Therefore, the critical loads (7.10) and (7.18), see the cases II and III in Fig. 7.5, are called *Euler loads*. We also refer to the critical loads of the cases I and IV as Euler loads. In addition to the critical loads, Fig. 7.5 illustrates the mode shapes of the four cases. One can show that the cases I, II and IV are interrelated. For example, the buckled shape of case I (one fourth of a sine wave) is contained twice in the buckled shape of case II (half of a sine wave). If we therefore replace the length l with $2\,l$ in the formula for the critical load of case II, we obtain the critical load of a column of length l fixed at its base and free at the top (case I). If we introduce the so-called column's *effective length* l_e, we are able to write down the critical loads in analogy to the second case as

$$F_{\text{crit}} = \pi^2 \frac{EI}{l_e^2}. \tag{7.19}$$

The effective lengths of the four columns (Euler cases) are also given in Fig. 7.5.

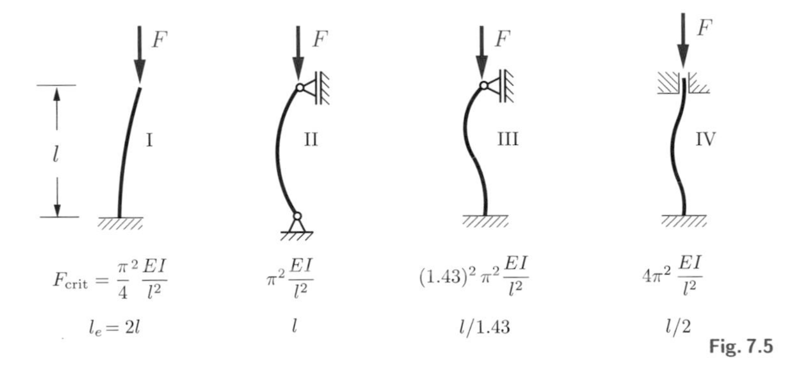

Fig. 7.5

In the preceding discussion we assumed that the material behaves linearly elastic before buckling, i.e. the stresses remain below the proportional limit. In particular, for columns with a small effective *slenderness ratio*, defined by l_e/r_g (effective length/smallest radius of gyration), the critical stress in the column may exceed the yield point of the material before buckling begins. Within this basic course we cannot go into details regarding the inelastic buckling of columns. Furthermore, in this introductory course we cannot analyse buckling due to torsion (torsional buckling) or the lateral buckling of a compression flange when a beam is loaded in flexure (lateral-torsional buckling). We also do not discuss the energy method which can be used for the determination of critical loads by analysing the change of the total potential of the system (potential of the external loads and internal elastic energy). This method is analogous to the procedure presented in Section 7.1.

It should be noted that the factors of safety specified by the codes have to be taken into account in the stability analysis and design of structures. Moreover, columns may undergo inadmissibly large deformations under loads below the critical value F_{crit} due to imperfections (e.g. unavoidable eccentricities in the application of the load and initial curvature of the column).

E7.1

Example 7.1 An elastic bar (flexural rigidity EI) is pin-supported at the left end and elastically clamped (spring constant k_T) at the right end, see Fig. 7.6.

Determine the buckling equation and calculate the critical load for $k_T\, l/EI = 10$.

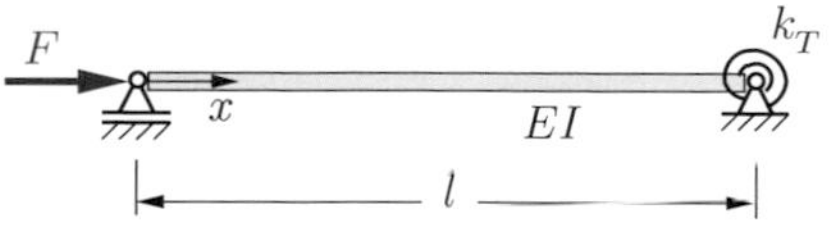

F x EI k_T l

Fig. 7.6

Solution We introduce the coordinate x as shown in Fig. 7.6. The general solution of the differential equation for the critical load, cf. (7.14), is

$$w = A\cos\lambda x + B\sin\lambda x + C\lambda x + D.$$

The four constants of integration are determined from the four boundary conditions

$$\begin{aligned}
&w(0) = 0 &&\rightarrow\quad A + D = 0 \\
&M(0) = 0 &&\rightarrow\quad \lambda^2 A = 0
\end{aligned}\Bigg\}\quad\rightarrow\quad A = D = 0,$$

$$\begin{aligned}
w(l) &= 0 &&\rightarrow\quad B\sin\lambda l + C\lambda l = 0, \\
M(l) &= k_T\, w'(l) &&\rightarrow\quad EI\,\lambda^2 B\sin\lambda l = k_T\,\lambda(B\cos\lambda l + C).
\end{aligned}$$

The elimination of C yields the buckling equation

$$\left(EI\,\lambda^2 + \frac{k_T}{l}\right)\sin\lambda l - k_T\lambda\cos\lambda l = 0 \quad\rightarrow\quad \underline{\underline{\tan\lambda l = \frac{\dfrac{k_T\,l}{EI}(\lambda l)}{(\lambda l)^2 + \dfrac{k_T\,l}{EI}}}}\,. \quad \text{(a)}$$

For the given stiffness ratio $k_T\,l/EI = 10$ we obtain the numerical approximation of the smallest eigenvalue as $\lambda_1\,l = 4.132$. Therefore the critical load is

$$\underline{\underline{F_{\text{crit}}}} = \lambda_1^2\,EI = 17.07\frac{EI}{l^2} = \underline{\underline{(1.31)^2\pi^2\frac{EI}{l^2}}}\,.$$

The buckling condition (a) includes the special cases
a) $k_T = 0$ (equivalent to a pin support)

$$\tan\lambda l = 0 \quad\rightarrow\quad F_{\text{crit}} = \pi^2\frac{EI}{l^2} \quad \text{(second Euler case)},$$

b) $k_T \rightarrow \infty$ (equivalent to a clamped support)

$$\tan\lambda l = \lambda l \quad\rightarrow\quad F_{\text{crit}} = (1.43)^2\pi^2\frac{EI}{l^2} \quad \text{(third Euler case)}.$$

E7.2

Example 7.2 A stress-free bar, supported as depicted in Fig. 7.7, is uniformly heated.

Determine the increase ΔT of the temperature at which the bar buckles.

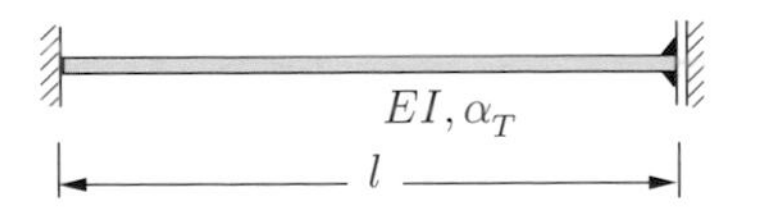

Fig. 7.7

Solution If we heat an unrestrained bar, a thermal expansion ε_T occurs, cf. (1.10). In the present example, the bar cannot expand due to the supports at both ends. The thermal expansion ε_T must therefore be compensated by a compression caused by a stress σ_T. Since $\varepsilon = 0$, the thermal stress follows from (1.12) as

$$\sigma_T = -\,E\,\alpha_T\,\Delta T.$$

Thus we obtain the force

$$F = \sigma_T\, A = EA\, \alpha_T\, \Delta T.$$

The buckling analysis of the bar subjected to this compressive force yields the critical increase of the temperature.

The general solution of the differential equation for the critical load, cf. (7.14), is

$$w = A^* \cos \lambda x + B \sin \lambda x + C \lambda x + D.$$

(An asterisk was added to the first constant of integration in order to avoid confusion with the cross sectional area A.) We introduce a coordinate system, where x starts at the left end of the bar. The evaluation of the boundary conditions yields with $\lambda^2 = F/EI$

$$\begin{aligned}
w(0) = 0 \quad &\rightarrow \quad A^* + D = 0,\\
w'(0) = 0 \quad &\rightarrow \quad B + C = 0,\\
w'(l) = 0 \quad &\rightarrow \quad -A^* \sin \lambda l + B \cos \lambda l + C = 0,\\
Q(l) = 0 \quad &\rightarrow \quad -A^* \sin \lambda l + B \cos \lambda l = 0.
\end{aligned}$$

After inserting $C = -B$, we obtain the two equations

$$\begin{aligned}
\sin \lambda l\, A^* &- (\cos \lambda l - 1)B = 0,\\
\sin \lambda l\, A^* &- \cos \lambda l\, B = 0
\end{aligned}$$

for the two unknowns A^* and B. This homogeneous system of equations has a non-trivial solution if the determinant of the coefficient matrix vanishes: $\sin \lambda l = 0$. The smallest eigenvalue $\lambda_1 = \pi/l$ yields the critical load

$$F_{\text{crit}} = \pi^2 \frac{EI}{l^2}.$$

With the radius of gyration $r_g^2 = I/A$, we obtain the critical increase of the temperature:

$$\underline{\underline{\Delta T_{\text{crit}}}} = \frac{F_{\text{crit}}}{EA\, \alpha_T} = \underline{\underline{\pi^2 \left(\frac{r_g}{l}\right)^2 \frac{1}{\alpha_T}}}.$$

Note that the increase of temperature is independent of Young's modulus. In order to get an idea of the magnitude of the temperature which causes the buckling of the bar, we consider a steel bar ($\alpha_T = 1.2 \cdot 10^{-5}/°$ C) with a slenderness ratio $l/r_g = 100$. It buckles at an increase of temperature of $\Delta T_{\text{crit}} \approx 80°$ C.

7.3

7.3 Supplementary Examples

Detailed solutions to the following examples are given in (**A**) D. Gross et al. *Formeln und Aufgaben zur Technischen Mechanik 2*, Springer, Berlin 2010, or (**B**) W. Hauger et al. *Aufgaben zur Technischen Mechanik 1-3*, Springer, Berlin 2008.

E7.3

Example 7.3 The structure in Fig. 7.8 consists of an arch AB and an elastic column BC ($h \ll a$, Young's modulus E).

Calculate the weight W_{crit} that causes buckling of the column.

Result: see (**B**) $W_{\text{crit}} = \dfrac{7\,\pi^2 h^4 E}{750\,a^2}$.

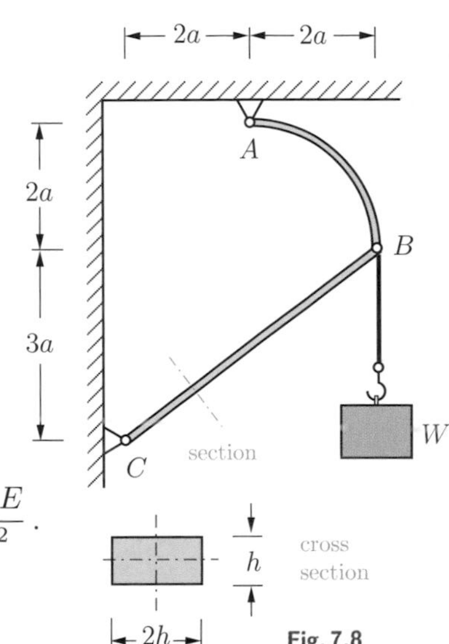

Fig. 7.8

E7.4

Example 7.4 The two systems shown in the Figs. 7.9a,b consist of rigid bars and elastic springs (spring constant c).

Determine the critical loads.

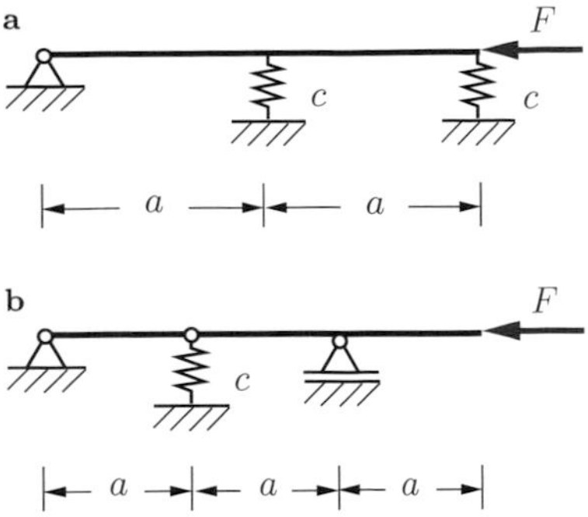

Fig. 7.9

Results: see (**A**)

a) $F_{\text{crit}} = \dfrac{5\,c\,a}{2}$, b) $F_{\text{crit}} = \dfrac{c\,a}{3}$.

E7.5

Example 7.5 The bar shown in Fig. 7.10 is composed of a rigid part and an elastic part (flexural rigidity EI). It is subjected to a force F.

Find the buckling equation and the critical load.

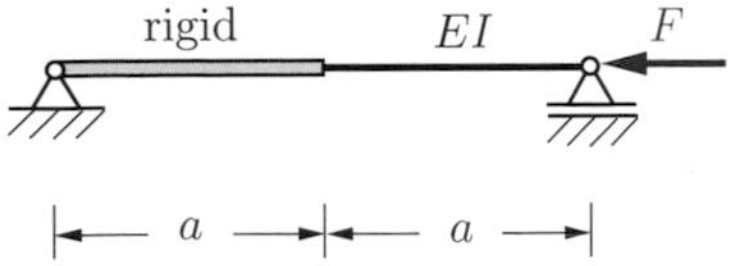

Fig. 7.10

Results: see (**A**) $\lambda a + \tan\lambda a = 0, \quad F_{\text{crit}} = 4.12\,\dfrac{EI}{a^2}$.

E7.6

Example 7.6 The bars depicted in Fig. 7.11 have different flexural rigidities. The structure is subjected to a force F.

Assume $EI_2 = 2\,EI_1$. Which bar buckles first if the load is increased?

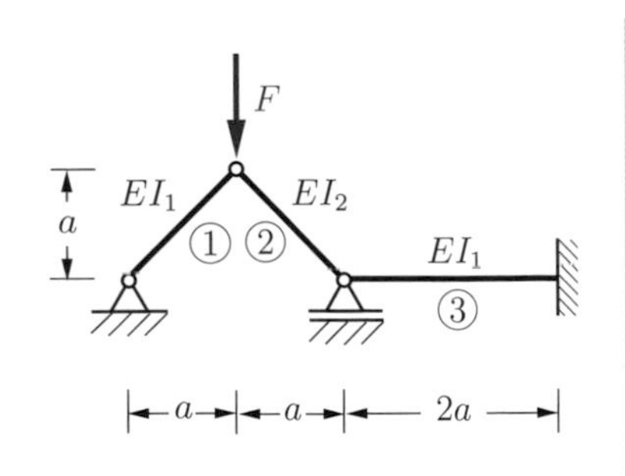

Fig. 7.11

lt: see (**A**) bar 1.

E7.7

Example 7.7 The truss in Fig. 7.12 consists of two equal elastic bars ($h \ll l$, Young's modulus E). A vertical force applied at pin K causes the bars to snap through from the upper position to the lower position (indicated by the broken lines in the figure).

Determine the maximum angle α for which the bars do not buckle.

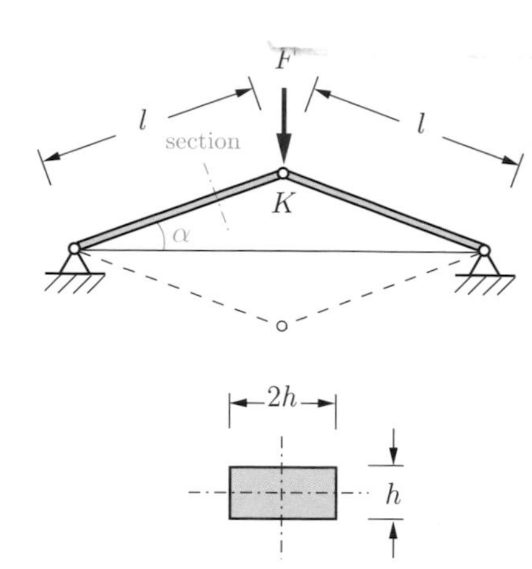

Fig. 7.12

Result: see (**B**) $\alpha < \dfrac{\pi h}{\sqrt{6}\, l}$.

E7.8

Example 7.8 An elastic column (flexural rigidity EI) is clamped at the base and pin-supported at the top by a spring (spring constant c), see Fig. 7.13.

Derive the buckling equation. Determine the critical loads F_{crit} for a) $c = 0$, b) $c = EI/l^3$, c) $c \to \infty$.

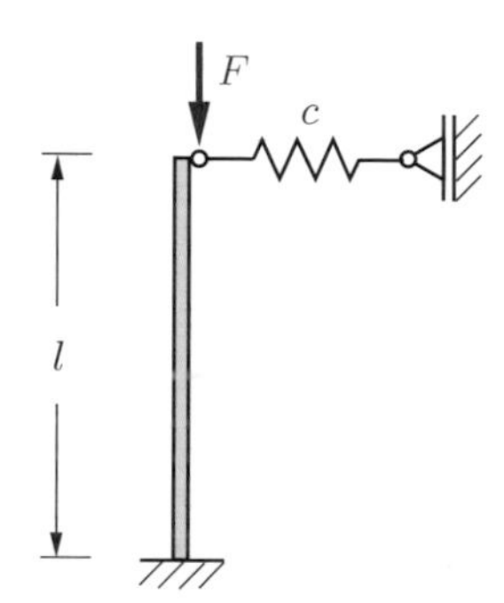

Fig. 7.13

Results: see (**A** and **B**) $\lambda l - \tan\lambda l - (\lambda l)^3 \dfrac{EI}{cl^3} = 0\,,$

a) $F_{\text{crit}} = 2.47\,\dfrac{EI}{l^2}$ (Euler case I), b) $F_{\text{crit}} = 3.27\,\dfrac{EI}{l^2}$,

c) $F_{\text{crit}} = 20.2\,\dfrac{EI}{l^2}$ (Euler case III).